D0455029

# THE KINDNESS OF STRANGERS

# THE KINDNESS OF STRANGERS

## HOW A SELFISH APE INVENTED A NEW MORAL CODE

MICHAEL E. McCULLOUGH

BASIC BOOKS
New York

Basic Books
Hachette Book Group
1290 Avenue of the Americas, New York, NY 10104
www.basicbooks.com

Printed in the United States of America

First Edition: May 2020

Published by Basic Books, an imprint of Perseus Books, LLC, a subsidiary of Hachette Book Group, Inc. The Basic Books name and logo is a trademark of the Hachette Book Group.

The Hachette Speakers Bureau provides a wide range of authors for speaking events. To find out more, go to www.hachettespeakersbureau.com or call (866) 376-6591.

The publisher is not responsible for websites (or their content) that are not owned by the publisher.

Print book interior design by Amy Quinn

Library of Congress Cataloging-in-Publication Data
Names: McCullough, Michael E., author.
Title: The kindness of strangers : how a selfish ape invented a new moral code / Michael McCullough.
Description: First edition. | New York : Basic Books, [2020] | Includes bibliographical references and index.
Identifiers: LCCN 2019051149 | ISBN 9780465064748 (hardcover) | ISBN 9781541617520 (ebook)
Subjects: LCSH: Kindness. | Social psychology. | Evolution (Biology)
Classification: LCC BJ1533.K5 M38 2020 | DDC 177/.7—c23
LC record available at https://lccn.loc.gov/2019051149ISBNs: 978-0-465-06474-8 (hardcover), 978-1-5416-1752-0 (ebook)

LSC-C

10 9 8 7 6 5 4 3 2 1

*To Joel William Michael McCullough*
*and*
*To Madeleine Elisabeth McCullough*
*with love.*
*Where did all the years go?*

# CONTENTS

CHAPTER 1

# A GOLDEN AGE OF COMPASSION

This book is about one of the great zoological wonders of the world. I'm not talking about the tears of the elephant, the smile of the dolphin, the politics of the chimpanzee, the consciousness of the octopus, the peacock's tail, the kingdom of the ants, or the wisdom of the birds or the bees or the dogs. I'm talking about a scrawny, brainy ape with the habit of helping strangers—often risking time and treasure and occasionally even life and limb to do so. It's about you and me, and how we treat everybody else. It's about the kindness of strangers.

When it comes to compassion for strangers, the human species is in a class of its own. Chimpanzees, like humans, regularly help kith and kin, but the number of chimpanzees who dive into swollen rivers to save drowning strangers, or send food to families of needy chimps in Tanzania, or perform weekend volunteer work at chimpanzee retirement homes, is zero. Year after year after year after year (do this 8 million times), no chimpanzee lifts a finger to help a stranger. No less a naturalist than Charles Darwin saw the gulf between humans' and chimpanzees' capacity for caring as one of the most blindingly obvious behavioral differences between the two species:

> There can be no doubt that the difference between the mind of the lowest man and that of the highest animal is immense. Some apes . . . might insist that they were ready to aid their fellow-apes of the same

troop in many ways, to risk their lives for them, and to take charge of their orphans; but they would be forced to acknowledge that disinterested love for all living creatures, the most noble attribute of man, was quite beyond their comprehension.[1]

Let's try to comprehend what the chimpanzees cannot. In contrast to our closest primate cousins, more than 150 people in the United States and nearly 100 in Great Britain donate a kidney to a complete stranger each year.[2] The World Holocaust Remembrance Center in Jerusalem honors more than 27,000 non-Jews who risked their lives and their liberty to rescue Jewish people during the Holocaust.[3] The Carnegie Corporation has recognized more than 10,000 ordinary Americans who knowingly put themselves in grave danger in order to rescue someone from death. One out of every five of those Carnegie medals was awarded posthumously because the honoree had died while trying to help.[4] Most heroes, of course, don't get a medal at all.

Humans also help strangers in a variety of less heroic ways. In the month after the World Trade Center attacks of September 11, 2001, 40,000 New Yorkers lined up to donate blood.[5] Each month, nearly 4 billion adults around the world help a stranger in need, 2.3 billion donate money to a charitable organization, and more than 1.6 billion perform volunteer work.[6] Americans alone commit $600 billion worth of cash and volunteer labor annually to organizations that promote health, education, and human welfare.[7] Two-thirds of British adults engage in a charitable activity at least once per month.[8]

Humans' generous spirit is also revealed by the activities of their governments on behalf of their most vulnerable citizens. On average, the rich nations of the developed world commit 21 percent of their gross domestic incomes (GDIs) to domestic social spending, which includes money for retirement pensions, health insurance, unemployment insurance, family benefits, disability benefits, food subsidies, and housing support, plus an additional 5 percent of their GDIs to education.[9] You might not think of domestic social spending as "generosity toward strangers": after all, we don't pay our taxes gladly. All the same, until 150 years ago, the notion that the state was responsible for meeting such a broad array of human needs didn't exist anywhere. Then it existed everywhere.

And let's not forget the $150 billion worth of official development assistance and humanitarian aid that the world's governments and nongovernmental agencies share with the world's neediest countries each year. Sure, these contributions amount to just a fraction of a percent of most donor nations' GDIs. Even so, $100 billion here, $100 billion there, and pretty soon you're talking about real money.[10]

Part of what makes modern generosity toward strangers so remarkable is the seemingly long odds against it. By most scientific accounts, the humans from whom we are descended were fanatically xenophobic, ready to greet needy strangers not with clean water, a hot meal, or a place to lay their heads for the night, but with spears and arrows instead. Just a few days before I sat down to draft these words, in fact, a Christian missionary from the United States was pinioned with arrows by members of a still-uncontacted tribe of hunter-gatherers on Sentinel Island in the Bay of Bengal. Later, the Sentinelese dragged the missionary's body out to the beach and buried it there. What makes our attitudes toward strangers so different from theirs? Modern humans' concern for the welfare of perfect strangers has no analog in the rest of the animal kingdom or even in most of our own history as a species. It's a true one-off. As such, it calls out for a special explanation.

Most modern historians have tried to help us understand the history of human generosity by serving it up one small bite at a time. One historian writes about philanthropy in ancient Athens, another about almsgiving in medieval Europe, another about the Elizabethan Poor Laws of sixteenth-century England, another about the nineteenth-century stirrings of the modern welfare state, and still another about twentieth-century efforts to eliminate poverty from the face of the Earth.

Other scholars, admittedly fewer in number today than a century ago, have sought to explain the rise of generosity as the result of historians' own version of Adam Smith's invisible hand: civilization. The Irish historian William Hartpole Lecky was one of the major exponents of the civilizationist theory of moral progress. In Lecky's view, humanity's regard for the welfare of strangers resulted from a centuries-long civilizing process in which superstition, xenophobia, and a stultifying satisfaction with the status quo were replaced by reason, cosmopolitanism, and a spirit of experimentation and self-improvement. A widening of the breadth of human charity came along for the ride. As Lecky wrote in his two-volume *History of European Morals* in 1869, "history tells us that, as civilisation advances, the charity of men becomes at once warmer and more expansive, their habitual conduct both more gentle and more temperate, and their love of truth more sincere."[11] Many other Victorian writers were civilizationists (also known as progressives) as well. Darwin himself was a civilizationist of sorts, as were many of the reformers responsible for the social innovations that we still rely on today to assist strangers.

Despite historians' many invaluable contributions to our understanding of the kindness of strangers, the one-bite-at-a-timers and the civilizationists alike have committed one important oversight: in their efforts to explain altruism and compassion using the standard tools of the historian, they have failed to grapple with the natural human faculties—our characteristic beliefs, desires, motivations, emotions, and cognitive powers—that were activated on a mass scale by the twists and turns of history to produce the penchant for helping strangers that we indulge today. To quote the cognitive scientist Pascal Boyer, minds make societies.[12] Thus, to fully explain how *Homo sapiens* came to greet needy strangers with compassion rather than cruelty, we have to understand why people believe what they believe, ponder what they ponder, and want what they want—and how they then figure out how to get it. We have to explain how a human mind that is built for a stone-aged world in which strangers were feared and killed can fashion a world for itself in which strangers are respected and assisted.

Complementing the historians' approach to explaining human generosity is an approach that comes from the evolutionary sciences. For several decades, evolutionary biologists and evolutionary social scientists have been obsessed with humans' generosity toward strangers, precisely because of its seeming implausibility. How could a penchant for wasting valuable resources on complete strangers evolve? After all, natural selection runs on reproductive fitness, and reproductive fitness runs on resources, so the more resources you keep for yourself, the better. Wouldn't evolution punish people who got into the habit of giving stuff away to nobody in particular?

Among modern Darwinians, explanations for generosity toward strangers come in two forms. First, there are those who argue that strangers were a prominent feature of the ancestral human environment, and that our ancestors in fact were able to obtain better Darwinian fitness by helping them. According to proponents of this "stranger-adaptation" hypothesis, we help strangers in the modern world because evolution designed us specifically to do so, circuitous though that design process must have been.

Second, there are those who argue that generosity toward strangers is merely a by-product of evolved instincts for taking care of our friends and relatives. When we help strangers in the modern world, these scholars argue, we are following ancient rules of thumb that worked well enough in a world in which meeting someone for the first time was a reasonably good indicator that you'd meet them again in the future (at which point they would have the opportunity to return your kindness). They argue further that in our modern world, those ancient rules of thumb cause us to mistakenly help strangers whom we will never meet again, which in the eyes of natural selection really does amount to throwing your help away. The biologist Richard Dawkins refers to these modern-day errors of altruism as blessed mistakes:

> An intelligent couple can read their Darwin and know that the ultimate reason for their [sexual] urges is procreation. They know that the woman cannot conceive because she is on the pill. Yet they find that their sexual desire is in no way diminished by that knowledge. Sexual desire is sexual desire and its force, in an individual's psychology, is independent of the ultimate Darwinian pressure that drove it. It is a strong urge which exists independently of its ultimate rationale.
>
> I am suggesting that the same is true of the urge to kindness—to altruism, to generosity, to empathy, to pity. In ancestral times, we had the opportunity to be altruistic only towards close kin and potential reciprocators. Nowadays that restriction is no longer there, but the rule of thumb persists. Why would it not? It is just like sexual desire. We can no more help ourselves feeling pity when we see a weeping unfortunate (who is unrelated and unable to reciprocate) than we can help ourselves feeling lust for a member of the opposite sex (who may be infertile or otherwise unable to reproduce). Both are misfirings, Darwinian mistakes: blessed, precious mistakes.[13]

Despite the differences between the stranger-adaptationists and the blessed-mistakers, members of both evolutionary camps end up predicting that humans will display an abiding interest in the welfare of strangers—either because we were in a profound sense designed to care about strangers or because the ubiquity of strangers in our daily lives causes us to mistake them (unconsciously) for friends and loved ones. We can no more stop ourselves from wanting to help strangers in need, if the evolutionists are right, than we can stop our stomachs from growling when we're hungry.

These evolutionary explanations have their own blind spots, albeit different ones from those of the historian. The weakness of the stranger-adaptationists' approach is its inability to account for the reams of empirical evidence that our minds are actually quite poorly designed for motivating us to look after the welfare of strangers. The research indicates instead that our intuitive interest in the welfare of strangers—particularly when weighed against the strength of our intuitive interest in our own welfare, as well as the welfare of our friends and loved ones—is fickle, reluctant, and easily distracted. The evidence that evolution has tuned our minds for active concern for the welfare of strangers, as we will soon see, is thin indeed.

Additionally, the blessed-mistakers' argument that compassion for strangers is the result of the cognitive systems that motivate us to care about family and friends must contend with an important principle of natural selection: those cognitive systems are likely to contain sophisticated fail-safes and identity-verification procedures that are designed to prevent us from helping strangers "by accident." Mistakes are costly—even the blessed ones—so natural selection designed us to avoid those mistakes whenever possible. And as we will see, we do try to avoid them whenever possible. Yes, we evolved to help our kin, but we also evolved to be able to distinguish kin from non-kin. And yes, we evolved to help people who would likely help us in return in the future, but as an appurtenance to that faculty, we also evolved to be able to distinguish people who are likely to reciprocate from those who aren't. Our evolved social instincts and sympathies are highly relevant to understanding the kindness of strangers, but stranger-adaptation and blessed-mistake theories are too simplistic.

Many evolutionary perspectives on generosity toward strangers also fail to take recent human history seriously enough, no doubt because evolutionists are primarily interested in natural selection, which needs eons to create complex functional design. As a result, they don't spend enough time considering the causal pathways by which our generosity toward strangers has effloresced over the past ten thousand years.

Finally, evolutionists also tend to overlook two important mental faculties. The first is our ability to follow our incentives. Like other animals, we can track the paths of action that will lead us to the things we care about—food, shelter, clothing, fame, a city free of disease and crime, a prosperous national economy, fidelity to our ethical convictions, a meaningful life. We can then construct courses of action that will lead us closer to those things we desire. Second, and relatedly, modern evolutionists often overlook our capacity for reason. Humans evolved both to produce reasons—that is, to offer justifications for their beliefs and convictions—and to process reasons—that is, to evaluate the justifications that others offer for their beliefs and convictions. For too long, evolutionists have been allergic to psychological explanations for behavior that rely on seemingly general-purpose cognitive abilities such as "tracking incentives" and "reasoning." However, as we will see, these faculties are indispensable for a complete account of our concern for strangers today.[14] You just can't explain it without them.

Fittingly enough, it was Darwin himself who fashioned our social instincts and our intellective powers into a scientific explanation for the vast expansion of human concern over the past ten millennia. "Any animal whatsoever, endowed with well-marked social instincts, the parental and filial affections being here included," Darwin began,

> would inevitably acquire a moral sense or conscience, as soon as its intellectual powers had become as well, or nearly as well developed, as in man. For, *firstly*, the social instincts lead an animal to take pleasure in the society of its fellows, to feel a certain amount of sympathy for them, and to perform various services for them. . . . But these feelings are by no means extended to all the individuals of the same species, only to those of the same association.[15]

Darwin surmised further that those "well-marked social instincts" that motivate our concern for kith, kin, and compatriots, in spite of their parochiality, were recruited into service more recently in human history to promote our regard for the welfare of all of humanity. And it was our capacity for reasoning, he averred, that did the recruiting:

> As man advances in civilization, and small tribes are united into larger communities, the simplest reason would tell each individual that he ought to extend his social instincts and sympathies to all the members of the same nation, though personally unknown to him. This point being once reached, there is only an artificial barrier to prevent his sympathies extending to men of all nations and races.[16]

Later in the same chapter from his 1871 book *The Descent of Man*, Darwin drove his argument home again, almost apologetically, as if he were worried that he had begun to beat a dead horse:

> The moral sense perhaps affords the best and highest distinction between man and the lower animals; but I need say nothing on this head, as I have so lately endeavoured to shew that the social instincts—the prime example of man's moral constitution—with the aid of active intellectual powers and the effects of habit, naturally lead to the golden rule, "As ye would that men should do to you, do ye to them likewise"; and this lies at the foundation of morality.[17]

The argument I tender in this book is similar to Darwin's own, and it is a straightforward one. I argue that the kindness of strangers is built upon a surprisingly small number—four, in fact—of our evolved human instincts. These include two of the instincts that Darwin called our "social instincts"—our instinct for helping others in hopes of receiving help in return, and our instinct for helping others in pursuit of glory—as well as the instincts Darwin called our "active intellectual powers," especially our ability to track incentives and our capacity for reason.

I argue further that the kindness of strangers emerged over the past ten millennia through seven different confrontations with mass suffering, and the solutions that our social instincts and our active intellective powers commended as solutions to those confrontations. To be sure, those seven historical encounters with want and woe engaged our basic social instincts—we helped in search of return favors and in search of glory—but they also engaged our ability to figure out what is important to us and our ability to reason our way to plans for how to obtain what is important to us. In short, our ancestors' encounters with mass suffering created threats and opportunities to which they applied their powers of reasoning to figure out how best to respond. And those responses turned out to be compassionate ones.

Although our social instincts and our capacity for reason furnished us with the *desire* to care about strangers, I argue that it was progress in three human endeavors—technology, science, and trade—that furnished us with the *ability* to care. We have Carnegie Heroes, Holocaust Rescuers, and anonymous kidney donors; we devote effort and resources to looking after the poor in our own countries; and we reach across seas, across borders, and even across generations to ease strangers' burdens not only because we want to, but also because we can.

In the upcoming chapter, I begin to lay out this argument by introducing the psychological obstacles that prevent us from taking an intuitive interest in the welfare of strangers. Without conscious, deliberative effort, the research from social and cognitive psychology shows, the human mind is breathtakingly insensate to the welfare of strangers. If natural selection really did design our minds to motivate us to care about strangers, as the stranger-adaptationists assert, then it must have been some pretty shoddy design work.

In Chapters 3 through 6, I will introduce you to those "social instincts and sympathies" that Darwin surmised to be the raw materials out of which our compassion for the distant stranger was fashioned. In the modern language of cognitive science and evolutionary psychology, these instincts and sympathies are the products of evolved cognitive systems—little computational devices that exist in our brains as networks of neurons and synapses—that natural selection fashioned in order to motivate us to care about our friends, our relatives, and our compatriots. I begin in Chapter 3 with a cook's tour of natural selection. In Chapters 4 through 6, I explore what those Darwinian social instincts and sympathies can offer to help us explain the kindness of strangers. As these chapters will show, and contrary to what Darwin seems to have surmised, it's only a small number of our "social instincts and sympathies" that actually make a difference.

From Chapters 7 through 13, each chapter is devoted to one of the seven confrontations with mass suffering that compelled our ancestors to think about the needs of strangers and how they should respond to them. Through this ten-thousand-year history of human compassion, I will show you how our social instincts and sympathies, along with our capacity for reason, interacted to produce the innovations and institutions that we still turn to today.

In Chapter 14, the book's final chapter, I'll weave the natural history and the human history back together, describing the instincts, the reasons, and the progress in technology, science, and trade that have conspired over the past ten millennia to create humanity's compassion for humankind.

## CHAPTER 2

# ADAM SMITH'S LITTLE FINGER

Are humans hardwired to care about strangers? Glancing over my bookshelves, titles such as *Born to Be Good*, *The Compassionate Instinct*, and *The Altruistic Brain* remind me that many of my scientific colleagues answer this question with a resounding yes. Each of these books, in its own unique way, teaches that the animal designated *Homo sapiens* has evolved for compassion. Caring about strangers is just part of who we are. If it doesn't come effortlessly, all it takes is some patience and some practice. Attend a workshop. Volunteer at a homeless shelter, so you can see the face of destitution. Read some fiction, so you can learn how to empathize. Meditate. Compassion is inside of you. You just need to coax it out.

These days, many social scientists are positively exuberant about our innate potential for generosity toward strangers. Their optimistic outlook on the kindness of strangers reminds me of a story I've heard on many occasions. Perhaps you know it as well.

The parable of the Good Samaritan, from the Christian New Testament, is a story Jesus tells after being confronted by one of the local religious scholars who is trying to get the better of him. The scholar challenges Jesus to explain what the biblical commandment to "love your neighbor as yourself" actually means. Jesus replies with a story:

"A man was going down from Jerusalem to Jericho, when he was attacked by robbers. They stripped him of his clothes, beat him and went away, leaving him half dead. A priest happened to be going down the same road, and when he saw the man, he passed by on the other side. So too, a Levite, when he came to the place and saw him, passed by on the other side. But a Samaritan, as he traveled, came where the man was; and when he saw him, he took pity on him. He went to him and bandaged his wounds, pouring on oil and wine. Then he put the man on his own donkey, brought him to an inn and took care of him. The next day he took out two denarii and gave them to the innkeeper. 'Look after him,' he said, 'and when I return, I will reimburse you for any extra expense you may have.'

"Which of these three do you think was a neighbor to the man who fell into the hands of robbers?"

The expert in the law replied, "The one who had mercy on him."

Jesus told him, "Go and do likewise."[1]

The parable of the Good Samaritan is nothing if not optimistic about the human potential for compassion. "Open your mind. Drop your prejudices. Reach out. You can do this."

But there is a competing parable bouncing around out there, of more recent vintage. You've probably heard this one as well. It comes our way from two *New York Times* reporters who wrote about the sexual assault and murder in 1964 of a young woman named Catherine "Kitty" Genovese. The Genovese case became a national sensation not because of Genovese's death, exactly, or even because of the viciousness of the attack, but because of the supposed apathy of the witnesses from a nearby apartment building, who knew something was amiss down at street level, yet did nothing to help her. Two weeks after the murder, Martin Gansberg wrote a *Times* piece about the people who saw the murder:

For more than half an hour 38 respectable, law-abiding citizens in Queens watched a killer stalk and stab a woman in three separate attacks in Kew Gardens. Twice, the sound of their voices and the sudden glow of their bedroom lights interrupted him and frightened him off. Each time he returned, sought her out and stabbed her again. Not one person telephoned the police during the assault; one witness called after the woman was dead.[2]

Later that year, a second *Times* reporter named Abe Rosenthal wrote *Thirty-Eight Witnesses*, a book in which he decried the witnesses' apathy and lamented what it seemed to reveal about human nature:

> She died in the early hours of March 13, 1964, outside the small apartment house in Queens where she lived as neighbors heard her scream her last half hour away and did nothing, nothing at all, to give her succor or even cry alarm. . . . A great many hard things have been said about these thirty-eight, and I am sure they are bewildered, and I know they are resentful. But it is important to say this—that what they did happens every night, in every city. The terror of the story of Catherine Genovese is simply that by happenstance all thirty-eight did that night what each alone might have done any night without the city having known, or cared.[3]

No Good Samaritans showed up to help. At least, that's how the parable of the thirty-eight witnesses goes. In fact, a few of the thirty-eight did try to help. Enough people called down from their windows to scare the assailant away following his first attack. Several people actually did call the police. One neighbor even rushed down to try to help Genovese as she lay dying. The thirty-eight witnesses hadn't been as apathetic as Gansberg and Rosenthal made them out to be. Even so, the Genovese story has been immortalized in books and films as a watchword for human indifference. The parable of Catherine "Kitty" Genovese, if not the actual facts of the case, features in virtually every social psychology textbook of the past half-century.[4]

The parable of the Good Samaritan and the parable of the thirty-eight witnesses could not be more different from each other in what they say about human compassion, but they do have one feature in common: both remind us that there are strangers out there who could use our help. But are we Good Samaritans, or are we unresponsive bystanders? Once we have stripped away any illusions we might be harboring about the basic human potential for kindness, what will we find?

## THE CYNICAL MR. SMITH

Ambrose Bierce, the author in 1906 of the satirical *Devil's Dictionary*, defined a cynic as "a blackguard [dishonorable man] whose faulty vision sees things as they are and not as they ought to be." Bierce's cynic dispenses with comforting, idealistic fictions in order to see reality for what it actually is. If that's a cynic, then few philosophers have been as cynical about humans' intuitive regard for perfect strangers as the Scottish philosopher Adam Smith.

If you're like most people, you are familiar with Adam Smith because of his best-known book, *The Wealth of Nations*, first published in 1776. In his own time, however, Smith was at least as famous for a 1759 book, called *The Theory of Moral Sentiments*, that was dedicated to exploring the psychological foundations of our moral judgments. Much of Smith's impact on the field of psychology today came from the attention he paid to sympathy as the emotion that activates our concern for others' well-being. Without sympathy, Smith thought, humans' natural regard for others—particularly absolute strangers—was ludicrously outmatched by self-love:

> To the selfish and original passions of human nature, the loss or gain of a very small interest of our own, appears to be of vastly more importance, excites a much more passionate joy or sorrow, a much more ardent desire or aversion, than the greatest concern of another with whom we have no particular connexion. His interests, as long as they are surveyed from this station, can never be put into the balance with our own, can never restrain us from doing, whatever may tend to promote our own, how ruinous soever to him.[5]

For his skeptical readers who wanted to hang onto their un-Bierceian illusions about humans' innate concern for the welfare of strangers, Smith offered a thought experiment:

> Let us suppose that the great empire of China, with all its myriads of inhabitants, was suddenly swallowed up by an earthquake, and let us consider how a man of humanity in Europe, who had no sort of connexion with that part of the world, would be affected upon receiving

> intelligence of this dreadful calamity. He would, I imagine, first of all, express very strongly his sorrow for the misfortune of that unhappy people, he would make many melancholy reflections upon the precariousness of human life, and the vanity of all the labours of man, which could thus be annihilated in a moment. He would too, perhaps, if he was a man of speculation, enter into many reasonings concerning the effects which this disaster might produce upon the commerce of Europe, and the trade and business of the world in general. And when all this fine philosophy was over, when all these humane sentiments had been once fairly expressed, he would pursue his business or his pleasure, take his repose or his diversion, with the same ease and tranquillity, as if no such accident had happened. The most frivolous disaster which could befal himself would occasion a more real disturbance. If he was to lose his little finger to-morrow, he would not sleep to-night; but, provided he never saw them, he will snore with the most profound security over the ruin of a hundred millions of his brethren, and the destruction of that immense multitude seems plainly an object less interesting to him, than this paltry misfortune of his own.[6]

Was Smith right to be such a cynic about selflessness? Would we really be more concerned about the integrity of our fifth digits than by news of one hundred million crushed and swallowed-up strangers? To answer such questions head-on, we don't need to content ourselves with waffling, bet-hedging, difference-splitting, parables, or even the psychological insights of a wig-wearing Scotsman from the eighteenth century. Instead, we can evaluate Smith's cynicism in the light of fifty years of scientific research. On the basis of this evidence, I'll argue that Smith got it largely right: each of us is Smith's man or woman of humanity. Our "humanity" (by which he meant something like education and refinement) makes us good at philosophizing about the plights of strangers, expressing our concern, and contemplating the fragility of life, but whatever those deep thoughts and solemn pronouncements are actually about, they cannot be counted on to motivate effective action on others' behalf. The mind, it seems, just doesn't work that way. Better news awaits in later chapters, but let's first try to see human nature through the faulty vision that sees things as they really are.

## THE LIMITS OF ATTENTION

Many people have a story like this one, I suspect: A few years ago, I was having dinner with my family at an outdoor restaurant. After we finished eating, my wife took our kids to get some candy at a store down the street. I chatted with my father-in-law while we finished our drinks. Fifteen minutes later, my wife returned with a pressing question:

"Did the lady get her bag back?"

"Did what lady get her bag back?"

"The lady!"

"What lady?"

"That lady! The one over there who just got robbed?!"

"We didn't see it."

"What do you mean you didn't see it?"

My incredulous wife pointed in the direction of the security guard, not twenty feet from where we sat, who was taking a report from a woman whose purse had evidently been snatched from her shoulder as we were sitting nearby. We had been oblivious to the entire event, including the woman's repeated calls for help, the mall cop on his scooter, and the several bystanders who ran off to catch the culprit.

We missed it all. My father-in-law and I weren't trying to tune out the woman's plight. It's just that our attentional resources were too tied up in talk about craft beers to tune her *in*.

Are you paying attention? It's a wonder if you are because the number of other things you could be paying attention to at this very moment is staggering. The music in the background, the rattling motorcycle outside, every single feature of every single object in your field of view, your tennis elbow, the nail pops in the drywall, the blood pressure in your ears, the aftertaste of coffee, your caffeinated hand tremors, the whispered gossip, the whirring hard drive, your dysfunctional self-talk, the weight of the eyeglasses on the bridge of your nose—is somebody making popcorn? There's simply too much going on out there for our minds to process even a fraction of it. As a result, we unconsciously make hard choices about where to focus our attention. Attention is a spotlight that illuminates a single feature of the world around us to the neglect of all the others.

We can also shine that attentional spotlight inward onto our own mental processes. When we are trying to accomplish a goal, for instance, we focus the spotlight on the goal itself and the tasks we must perform in the service of fulfilling it. As we do so, however, we reduce the reservoir of attention that is available for observing other potentially interesting features of the world. The result is a phenomenon that scientists call inattentional blindness. When people narrow their attention to a single task or goal, they lose the ability to notice even highly unusual features of the environment, including purse-snatchings, clowns on unicycles, dollar bills literally hanging from the trees, and chest-pounding gorillas that walk through the middle of basketball games.[7]

In one revealing demonstration of inattentional blindness, the psychologist Christopher Chabris and his colleagues asked twenty people to take a nighttime run behind another jogger. During the brief run, they were supposed to count the number of times the lead jogger touched his head. Unbeknownst to the participants, the researchers had staged an assault. Just off to the side of the trail, two men faked an attack against another man as the joggers went by. At the end of the experiment, only seven of the twenty participants reported having noticed the incident. For the others, focusing on the lead jogger's head-touches created an inattentional blindness for a stranger in danger.

To make sure it wasn't the nighttime conditions that prevented people from noticing the attack, the researchers repeated the experiment in the daytime. They also varied the difficulty of the head-touch-counting task. One-third of the participants were told to follow along behind the jogger as before, keeping track of the number of times he touched his head. Only 56 percent of people in this condition reported noticing the assault. Another set of participants was told to follow along behind the jogger, but they were not given any further instructions: 72 percent of those participants reported noticing the assault. And in a third condition designed to make inattentional blindness even worse, participants were told to keep two counts: one of the number of times the jogger touched his head with his right hand, and another of the number of times he touched his head with his left hand. Among these participants, only 42 percent noticed the assault. Other researchers have shown that playing with your cell phone causes the same sort of inattentional blindness to others' needs. Our attention is a limited resource: there's just not enough to go around.[8]

It would be a poor design for a human mind if there was not some way for us to monitor the world for important information—even when most of our attention was being devoted to the tasks currently on our plate (such as beer banter or running totals of head-touches). And indeed, scientists have identified a few kinds of information that are good at capturing our attention even when we're wrapped up in something else. Hearing or seeing our own names, for example, tends to capture our attention. Our attention is also readily captured by images of objects that belong to us. The names of friends, family members, and loved ones can also capture our attention (though less often than our own names do). Indeed, so many basic cognitive processes (including attention, perception, awareness, memory, learning, and decision-making) are made more efficient by making them about "me" or "mine" that it's hard to think of one that isn't.[9]

Our attention can also be captured by features of the environment that are relevant to important goals we've had to temporarily put on hold: if you're really hungry, for example, but you are trying to ignore your hunger while finishing up some work, food-relevant stimuli (for example, food-related pictures that might pop up on the Internet) will automatically pull your attention away from the task at hand. Also, like Pavlov's dogs, our attention can be pulled away by a sight or sound that we've come to associate with rewards.[10]

Taking stock of what we know about inattentional blindness and attentional capture, it is fair to say that attention is a self-centered faculty. It loves to illuminate our goals, our needs, our friends, our loved ones, our beers, our conversations, our little fingers. But the welfare of strangers? Not so much. Lifeguards, firefighters, and Superman use their attentional resources to search for strangers in need. The rest of us, for the most part, use our attention to look out for Number One. There's nothing to feel guilty about. It's just the way attention works. I've even stopped beating myself up for my obliviousness to the Case of the Purloined Handbag.

## THE LIMITS OF EMPATHY

In the parable of the Good Samaritan, the Samaritan doesn't just notice the injured stranger and then immediately set to work helping him sort out his problems: between the noticing and the helping, the Good Samaritan "takes pity" on the injured stranger. The Greek word for pity here is *splagchnizomai*, which loosely translates as "to feel compassion in one's guts." It was the Samaritan's gut-sight, not his eyesight, that made him a Good Samaritan.

Like the writer of the Good Samaritan parable, scholars through the centuries have taken it for granted that this emotion plays a critical role in generating the motivation to help others, including strangers. The eighteenth-century Scottish philosopher David Hume thought sympathy converted our *ideas* about someone else's suffering into *impressions* (which today we would call *feelings*) that were consistent with the suffering person's misery:

> It is indeed evident, that when we sympathize with the passions and sentiments of others, these movements appear at first in our mind as mere ideas, and are conceived to belong to another person, as we conceive any other matter of fact. It is also evident, that the ideas of the affections of others are converted into the very impressions they represent, and that the passions arise in conformity to the images we form of them. All this is an object of the plainest experience, and depends not on any hypothesis of philosophy.[11]

According to Hume, the source of our motivation for helping needy others is crystal clear: when you learn of someone's problems, you experience that perception as a mere idea, but when you attach sympathy to that idea, it becomes a feeling. Suddenly, as if by alchemy, you begin to want for the victim what she wants for herself: her purse back, someone to cheer her up, a place to hide from her violent spouse, a high-tech prosthetic limb, the American Dream, justice, a life preserver, a loving home. Sympathy turns lifeless ideas about people's needs into lively desires to help them.

In *The Theory of Moral Sentiments*, Adam Smith developed Hume's theory of sympathy further:

> Pity and compassion are words appropriated to signify our fellow-feeling with the sorrow of others. Sympathy, though its meaning was, perhaps, originally the same, may now, however, without much impropriety, be made use of to denote our fellow-feeling with any passion whatever.

Sympathy, or rather, compassion (which, according to Smith, is just sympathy in its misery-congruent form), was so important to Smith's theorizing about our ability to treat each other kindly that he opened *The Theory of Moral Sentiments* by marveling at its wondrous effects:

> How selfish soever man may be supposed, there are evidently some principles in his nature, which interest him in the fortune of others, and render their happiness necessary to him, though he derives nothing from it except the pleasure of seeing it. Of this kind is pity or compassion, the emotion which we feel for the misery of others, when we either see it, or are made to conceive it in a very lively manner.[12]

Pity, compassion, sympathy, empathy, fellow-feeling—feel free to add your own synonym here—are all regarded by many people as interchangeable terms for a deep reservoir of concern for others' well-being, a precious natural resource that we need to cultivate and protect.[13] Today these terms do have slightly different connotations, but it was not until the early 1900s that scholars began to make really fine-grained distinctions. What Smith and other writers in the 1700s called "sympathy" later became "empathy," particularly as psychologists became more and more interested in the phenomenon. By the 2000s, when a journalist asked the primatologist Frans de Waal what he would change about human nature if he "were God," de Waal offered this answer: "Empathy for 'other people' is the one commodity the world is lacking more than oil. . . . If I were God, I'd work on the reach of empathy."[14]

Like de Waal, the social theorist Jeremy Rifkin sees a broadening of empathy as our best hope for solving humanity's biggest problems:

> The task before the human race is daunting. For the first time, we have to defy our own history as a species and create a new, more interdependent civilization that consumes less rather than more energy, but in a way that allows empathy to continue to mature and global consciousness to expand until we have filled the earth with our compassion and grace rather than our spent energy.[15]

De Waal and Rifkin both have a lot riding on empathy. They're looking to empathy not only to save us from ourselves, but to save the Earth as well. Like our attentional faculties, however, empathy has limitations that caution us not to place too much hope in it.[16]

Empathy, like attention, was almost surely not designed by natural selection to motivate us to take an interest in the welfare of human strangers—never mind nonhuman animals and entire planets. Indeed, virtually every scientist who has thought deeply about the evolutionary origins of empathy understands that it probably evolved to motivate us to assist our nearest and dearest—family and friends.[17] The bright side, as Hume noted, is that we can be quite good at sympathizing with friends, loved ones, and countrymen:

> Accordingly we find, that where, beside the general resemblance of our natures, there is any peculiar similarity in our manners, or character, or country, or language, it facilitates the sympathy. . . . The sentiments of others have little influence, when far removed from us, and require the relation of contiguity, to make them communicate themselves entirely. The relations of blood, being a species of causation, may sometimes contribute to the same effect; as also acquaintance, which operates in the same manner with education and custom. . . . All these relations, when united together, convey the impression or consciousness of our own person to the idea of the sentiments or passions of others, and makes us conceive them in the strongest and most lively manner.[18]

But sympathy—or empathy—has its limits. Natural selection favors the evolution of mental faculties that are highly specialized for solving particular cognitive problems; as a consequence, those mental faculties tend to be rather clunky at solving problems they weren't designed to solve. If empathy exists to motivate us to care for the near and dear, the principle of specialization should lead us to suspect that empathy won't be so great at motivating care for strangers. And the stranger they are, the less empathy we should be able to muster for them. Again, Hume: "Sympathy, we shall allow, is much fainter than our concern for ourselves, and sympathy with persons remote from us much fainter than that with persons near and contiguous."[19]

The implication is that empathy for absolute strangers should be a bit like the Loch Ness monster: much discussed but seldom seen.

Daniel Batson, a social psychologist, is the Sherlock Holmes of altruism. He has spent the past four decades trying to answer a straightforward question: Do people ever help others out of genuine concern for their welfare? Or is the help we extend to others always motivated by selfish concerns, no matter how cleverly hidden (even from ourselves) those selfish concerns might be? Batson suspects that humans really do have the potential for altruistic concern, and that it can be roused by empathy. He formalized his hunch in what he calls the "empathy-altruism hypothesis." Altruism-skeptics disagree with Batson and maintain instead that empathy always motivates helping for selfish reasons rather than altruistic ones. So who's right, Batson or the altruism-skeptics?

Using careful laboratory experiments, Batson has used Sherlock Holmes's own method of investigation to determine whether empathy leads to altruistic motivation: the process of elimination. "Look," one altruism-skeptic might say. "I don't think this thing you call altruistic motivation is altruistic at all. Of course people who feel empathy for someone in trouble try to help, but all they are really trying to do is relieve their own emotional distress. Empathy makes us try to help needy people because needy people make us feel uncomfortable, and we don't like feeling uncomfortable." If our altruism-skeptic is right, then the empathy-helping relationship merely reflects a selfish desire to avoid discomfort, even though it masquerades as virtue.

Batson has dealt with the many self-centered alternatives to his empathy-altruism hypothesis one at a time. If people persist in trying to help a stranger when every conceivable self-centered motivation has been ruled out (for example, the motivation to avoid personal distress, or to avoid guilt, or to appear morally righteous), then the only possible explanation for their persistence, however improbable, is that they wanted to improve the needy stranger's life—that is, that they were altruistically motivated.[20]

Four decades and dozens of experiments later, Batson's empathy-altruism hypothesis appears to be in good shape. Because empathically aroused people persist in trying to help a needy stranger even when they could fulfill a competing, self-centered goal by *not* helping, it would seem that helping the stranger is precisely what they are seeking to accomplish. We can conclude, therefore, that empathy elicits altruistic motivation.[21] Isn't this the evidence we're looking for if we want to know whether human empathy is designed well for motivating us to help strangers in crisis?

Not exactly. In the real world, we don't go around imagining how we might feel if we had to walk a mile in the shoes of every needy stranger we come across, as the subjects in Batson's experiments were instructed to do. If we did, we would never get anything else done: so many people, so little time. In the real world, we rarely bother even trying to feel empathy for anyone unless we value their welfare in the first place. Even Batson conceded as much:

> One often hears lip service paid to valuing all human life or to valuing the welfare of all humanity, but most of us place different value on the welfare of different others. We value the welfare of some highly and some very little. We may even negatively value the welfare of a person we don't like, such as a rival or enemy.
>
> If we place little value on the welfare of someone we perceive to be in need, we aren't likely to think about how this person is affected by the need, except perhaps as a clue to how we might control his or her behavior. We may understand what this person needs, but we don't care. It provides no basis for feeling empathic concern.[22]

And even when we do try to imagine what it would be like to walk a mile in a stranger's shoes—for example, by deliberately trying to see things from her perspective—the mind doesn't always cooperate by producing empathy in response. The psychologist William McAuliffe, who worked in my laboratory during his graduate school years at the University of Miami, spearheaded a project with me and some of the other students in my lab that brings this truth into fine relief. The project involved a systematic quantitative examination, called a meta-analysis, of other scientists' previous experimental evaluations of whether perspective-taking promotes empathy. In each of the experiments we examined, researchers had tried either to encourage people to empathize with a needy stranger, by using explicit instructions to adopt her perspective, or else tried to deter people from empathizing, by using explicit instructions to avoid thinking about the needy person's feelings. Our meta-analysis revealed that deliberately taking the needy stranger's perspective does not, in fact, boost empathy for her. Trying to remain objective while considering the stranger's plight does *reduce* empathy for her, however. We found the exact same result in one of our own experiments, which at the time of this writing is still the biggest experiment on the empathy-altruism hypothesis ever conducted. Evidently, it is easier to squelch our empathy for strangers than it is to boost it.[23]

Recent neuroscientific evidence gets to the heart of the matter. Some researchers think, à la David Hume and Adam Smith, that we come to understand the feelings of others by watching their behavior, simulating that behavior in our own minds as if we were behaving in the same way ourselves, and then monitoring the feelings that arise from that mental simulation. The sequence might work like this: I see that my good friend Harrison (I don't actually have any friends named Harrison, though it might be nice if I did) is sobbing over the defeat of his favorite soccer team, which causes me to mentally simulate the act of sobbing in response to that defeat. Based on this simulation, I formulate a guess about how I would feel if I really were sobbing, and then I use this guess to infer how Harrison himself—the person who actually is crying—must be feeling. It's a nice little hypothesis for how empathy might arise in the human mind. The problem with this hypothesis, however, is that the parts of the brain that enable us to simulate the feelings of others are scarcely even turned on by strangers in need. It's as if we were not even motivated to go to the trouble of trying to simulate their feelings in the first place.

An experiment by the psychologist Meghan Meyer and her colleagues illustrates this phenomenon quite nicely. Meyer and her team had people watch either a friend or a stranger play a computerized game of catch called "Cyberball" with two other anonymous players. (To get a sense of what Cyberball is like, imagine throwing a Frisbee with a couple of people at the park. Now imagine that they simply stop throwing to you after ten turns. That's it—no more Frisbee for you, no reason given. Cyberball has become the standard method for studying social exclusion in the lab.[24]) The researchers found that participants felt more empathy when they were observing a friend who had been excluded from Cyberball than when they were observing a stranger who had been excluded.

What's more, when the friend was excluded, the regions of the brain that are involved in the experience of physical pain were activated; when it was the stranger who was excluded, however, those regions were quiet.[25] Similar studies have shown that this so-called pain network in the brain becomes activated when people observe other people in pain who are from their own racial groups, but not when they are from other racial groups.[26] Taken together, these results indicate that, in a very real sense, we feel the pain of our friends, loved ones, and compatriots much more strongly than we feel the pain of strangers and outsiders.

Depressed yet? Stay with me a little longer because there are two more reasons to be depressed about empathy. The first is this: however difficult it might be to empathize with single, identifiable strangers, it's even more difficult to empathize with *groups* of needy strangers. As the psychologist Paul Bloom put it, empathy is "parochial, narrow-minded, and innumerate"—that is, it does not increase as the number of people involved increases.[27]

The second reason to be depressed is that people tend to actively avoid experiencing empathy for strangers—precisely because they know it will goad them into trying to help people they don't actually care about.[28] I am writing these words around the winter holidays. When I go by the grocery store this afternoon, I will encounter a Salvation Army bell-ringer who will ask me to make a donation to help the poor and needy. If I'm in the right mindset, I will experience this opportunity as a duty, an honor, or perhaps even a pleasure. More likely, though, I'll view it simply as a hassle that I need to negotiate on the way to picking up a quart of egg nog.

One way I might try to negotiate my bell-ringer's dilemma is by preventing myself from experiencing any empathy in the first place. One ingenious group of researchers performed experiments with grocery-store bell-ringers that enabled them to study this form of empathy avoidance. They found that about 14 percent of people made a contribution when directly asked (their average contribution was $1.69). Another 33 percent of people, however, literally went out of their way—by using a much less convenient door into the store—to avoid the bell-ringer entirely. The remaining half of shoppers simply went through the most convenient entry and ignored the bell-ringer's request completely.[29] Far from running toward opportunities to flex our empathy muscles for distant others, most of us choose instead to avoid or ignore them.

Not only is it the case, then, that our "sympathy with persons remote from us [is] much fainter than that with persons near and contiguous," as Hume intuited, but our species' tendency toward xenophobia and groupishness is strong enough to quench empathy before it really even has a chance to get going. None of this is our fault. It's just how empathy works.

## FORMAT ERROR

Formatting problems are a bane of all information-processing systems: if the information you feed into the system isn't in the right format, the system can't read it. This is why Microsoft Word is helpless to read the information contained, say, in a PDF document, unless you specify the PDF document's format (it's written in PDFese, not Wordish) in advance. Likewise, your vision system is helpless to decode the information contained in the air pressure waves that your auditory system decodes effortlessly, and the information in the molecules that hit the taste buds on your tongue is gobbledygook to your auditory system: wrong format.

All cognitive systems, even the more complex ones that govern how we treat each other, are format-sensitive. That format sensitivity allows these systems to excel at extremely specialized tasks, but that specialization comes at a cost: a specialized system cannot read perfectly useful information that comes in a language it wasn't designed to understand. This is a real problem for the kindness of strangers because the people around the world who are in the greatest need of our help are people whom we will never meet face to face, and face-to-face contact is the informational format that these systems evolved to read most proficiently.

The ethicist Peter Singer has a nice thought experiment for demonstrating how finicky our minds can be about the format in which news of other people's plights reaches us:

> On your way to work, you pass a small pond. On hot days, children sometimes play in the pond, which is only about knee-deep. The weather's cool today, though, and the hour is early, so you are surprised to see a child splashing about in the pond. As you get closer, you see that it is a very young child, just a toddler, who is flailing about, unable to stay upright or walk out of the pond. You look for the parents or babysitter, but there is no one else around. The child is unable to keep his head above the water for more than a few seconds at a time. If you don't wade in and pull him out, he seems likely to drown. Wading in is easy and safe, but you will ruin the new shoes you bought only a few days ago, and get your suit wet and muddy. By the time you hand the child over to someone responsible for him, and change your clothes, you'll be late for work. What should you do?[30]

Singer wrote that he regularly poses this thought experiment to his students, and that most of them immediately conclude that they have a moral responsibility to help the drowning child. Singer presses his students: "What about your shoes? And being late for work?" The students stand their ground: "How could anyone consider a pair of shoes, or missing an hour or two at work, a good reason for not saving a child's life?"[31]

It is easy for us to muster the motivation to leap to the drowning child's rescue because the information about the child's need arrives to our evolved caregiving systems in the right format: little arms flail, a thin voice cries for help, two eyes widen with terror, the top of a tiny head disappears below the waterline. When news about another person's dire straits are formatted in flesh and blood, our caregiving responses are speedy, intuitive, and effortless.

But what if the person in need is merely a *notional* stranger, someone whose misery you encounter only as part of a statistic or a viral news story? As Singer pointed out, we make all kinds of excuses for why we're not required to help these notional strangers: We assert that helping far-flung strangers is a matter of personal beliefs rather than duty, or that we have a right to decide how to spend the money we have earned, or that we are already doing enough. We tell ourselves that philanthropic activities delay the political changes that would enable poor countries to solve their own problems, or that helping the poor breeds idle dependency. We think about how one person's contributions are too small to make a meaningful difference. The list of objections goes on and on.[32]

Adam Smith's hypothetical man of humanity, who lamented the earthquake-related deaths of so many innocent people in China, certainly did not learn of China's devastation as an eyewitness. Instead, he learned about the earthquake from an article in the newspaper, or as a rumor from a friend. News about the suffering of notional strangers, whether formatted as column-inches of newspaper print, as banter in a Glasgow coffee house, or as a trending story on social media, cannot activate humans' natural concern for others. As a result, Smith's man of humanity would "snore with the most profound security" after getting the news from China—even when the prospect of a pinky amputation at sunrise would leave him agonizing all night long.[33]

Here's a potential workaround for our format-insensitivity to the plights of notional strangers: What if you try to create a mental picture of

1,000 Wembley Stadiums full of Chinese earthquake victims? Or what if you try to visualize five Wembley Stadiums to hold the nearly 450,000 children who are going to die of malaria this year? Does the picture you created with your mind's eye move you any more than the mere verbal proposition of 450,000 childhood deaths from malaria? Probably. But even this mental picture has a weaker hold on the seats of our emotions than direct experience does. Granted, mental images do resemble the visual perceptions that we generate from direct sensory experience: they draw on some of the same brain regions, for instance, and some of their psychological and behavioral effects are similar. But the effects are much milder for images than for actual perceptions.[34] This is why charities try to put human faces on their causes: simply *knowing* that strangers are suffering, or even visualizing them with the mind's eye, isn't enough. We are most responsive *when we see people's suffering for ourselves.* Anything less turns flesh-and-blood suffering into hollow words and wan imagery.

When the late Janet Reno, US attorney general during President Bill Clinton's administration, wanted to shrug off reporters' "what-if" questions during press conferences, she would reply, "I don't do hypotheticals." The same could be said of our minds' evolved caregiving systems. They don't do hypothetical suffering. They do real suffering.

## AN ENIGMA

Did evolution prepare us to be Good Samaritans, or did it leave us as unresponsive bystanders instead? The characteristic patterns of attention, emotion, and format-sensitivity that we've examined in this chapter are all signs that human minds are, in fact, *not* particularly well designed to motivate regard for distant strangers. Concern for strangers' welfare looks more like a noble aspiration than a natural endowment.

None of this is our fault. Our minds have limitations. We can't imagine ten dimensions. We can't see a bullet fly or grass grow. We can't know what it's like to be a bat. We also possess cognitive limitations that tether our ability to respond eagerly and deftly to the needs of strangers. For all of humanity's wondrous faculties, these limitations conspire to make us reluctant Samaritans at best, and unresponsive bystanders at worst.

This analysis leaves us with a nagging scientific puzzle: How has humanity, outfitted as it is with a mind that is not especially well designed for broadband generosity, managed to create a world in which most of us think we owe strangers some kind of concern? The answer lies in discovering how the mind's evolved faculties, including the Darwinian social instincts that motivate us to care about our relatives and friends, have interacted with the course of human events over the past ten thousand years. But before we can properly understand those social instincts, it is helpful to understand how evolution created them in the first place. It is true that minds make societies, but it is natural selection that makes minds.

## CHAPTER 3

# EVOLUTION'S GRAVITY

Physicists talk about four fundamental forces that govern interactions among the bits of matter that fill our universe. The strongest of these forces, aptly known as the strong force, is so powerful that it can keep the protons inside an atom from ripping the nucleus apart as their positive charges push them in opposite directions. The second fundamental force, electromagnetism, is 137 times weaker than the strong force, but its power to cause bits of matter with opposing electrical charges to attract each other, and those with like charges to avoid each other, is what gives three-dimensional structure to atoms, molecules, and even the proteins that form the cells in our bodies. At only one-millionth the strength of the strong force, the third fundamental force—the so-called weak force—changes quarks from one bizarre "flavor" to another and gives rise to nuclear fusion reactions.

The weak force deserves a better name because it's actually the fourth force—gravity—that's the weakling of the bunch. At only 6/1,000,000,000,000,000,000,000,000,000,000,000,000,000,000 of the strong force's strength, the influence of gravity on the interactions of protons, quarks, and other subatomic particles is, to put it gently, sort of small. When I use the refrigerator magnet that holds up my kid's school photo to lift the ring of keys on the kitchen table, the magnet easily overcomes the gravitational pull of the entire planet. At Subatomic Beach, gravity is the scrawny guy who's always getting sand kicked in his face.

But the only reason gravity looks like such a weakling in comparison to the others is that we haven't yet zoomed out to the scales of mass and distance that reveal its might. For the change of perspective that gives gravity the respect it deserves, we have to use a telescope. When we're studying the interactions of very small things separated by small distances, gravity is the only fundamental force that doesn't matter. But when we're studying the interactions of large things separated by large distances, it's the only one that does.

Every time the mass of an object increases one-hundredfold, gravity's influence increases tenfold. Very massive objects like planets and stars have no net electrical charge because the charges of all their constituent bits cancel each other out, so it's the weakling gravity—acting across huge distances, always attracting, never repelling—that determines their interactions. And when an object gets really large—roughly the size of a hundred Jupiters—gravitation packs all that matter into a sphere so dense that the center of the sphere becomes a nuclear fusion reactor. Gravity is the Charles Atlas of the cosmos. It turns stuff into stars.[1]

Natural selection, one of the fundamental forces of evolution, has something in common with gravity: a public relations problem. From one vantage point, natural selection looks like a chump. When you look up close at the tiny bits of stuff that go into making humans—the sequences of DNA that constitute the human genome—and how they came to be arranged as they are, natural selection doesn't seem to have done much. Other evolutionary processes, such as mutation, migration, and drift, have exerted far more powerful influences on our genomes. For that matter, distinctly nonevolutionary events, including one-off famines, freezes, floods, and fires, can change the fate of a species far more drastically than natural selection ever could.

However, when you zoom out, natural selection is the only evolutionary force that matters at all. That's because it's the only evolutionary force that can produce *design*. Natural selection acts uniformly and consistently, through deep time, to sift genes according to one hard-and-fast criterion: it increases the prevalence of genes that are good at increasing their own rates of propagation. As the evolutionary biologist Richard Dawkins has illustrated so brilliantly in so many different ways, genes contain the recipes for building things. Some of the things they build—first proteins, and then, out of those proteins, specialized organelles, such as mitochondria and ribosomes, and then specialized cells, and then arms, legs, eyes, ears, hearts, lungs, brains, and even beliefs and desires—end up looking like fancy gadgets. The gadgets that facilitate the replication of the genes that construct them get conserved and elaborated. Those gadgets that interfere with the replication of the genes that construct them are shuffled off. As the result of eons and eons of a gene-sifting process that operates according to a single criterion—Does this gene create gadgets that speed up its rate of propagation?—organisms accumulate design. The result of all this relentless gene-sifting and gadget-building is that organisms end up looking like geniuses for thriving in the environments to which they are adapted. They've got the tools for success. Like gravity, natural selection is a star-maker.

## DARWIN'S DANGEROUS IDEA

The apparent design of the natural world has fascinated biologists since Aristotle, and explaining it in purely naturalistic terms is one of evolutionary biology's perennial challenges.[2] For millennia, theologians have marshaled the *argument from design* in their efforts to convince religious skeptics that the Earth's living things were fashioned by an intelligent designer.[3] In *Natural Theology—or Evidences of the Existence and Attributes of the Deity Collected from the Appearances of Nature*, first published in 1802, the British theologian William Paley honed the argument from design to a fare-thee-well:

1. Suppose you're taking a walk through a field and you come across a rock. You wonder how the rock came to be, or what its purpose might be, so you take a closer look. Finding nothing particularly miraculous or improbable about how the rock is put together—a rock is just rock, rock, and more rock—you would probably conclude that the rock *just is*. Rocks don't do anything. They just sit there. They don't appear to be designed for any special purpose.
2. Suppose instead that you come across a pocket watch. You notice its many intricacies, and how all of its gears and springs and other bits cooperate to produce the effect of indexing the passage of time. As you begin to fiddle with the watch, you quickly see

how easily a rearrangement of its parts reduces its effectiveness at indexing the passage of time. You might be tempted to conclude that the watch had been assembled by a designer who envisioned the watch's purpose in advance and then built it with that purpose in mind.

3. It doesn't matter that you haven't met the designer of the watch, or that the watch could be improved, or that it occasionally needs to be repaired, or even that you don't understand what all the parts do: the elegant arrangement of all the watch's pieces, and the improbability of them all coming together randomly, imply that the watch has been assembled by a designer who wanted to make an object to keep time. Even a broken watch appears designed to keep time—a fact that any decent engineer could discover.
4. The creatures and features of the biological world (Paley uses the human eye, with all of its complexly and improbably arranged constituent parts, as an example) are more exquisitely designed than even a watch is. As Paley put it, "the contrivances of nature surpass the contrivances of art, in the complexity, subtilty, and curiosity of the mechanism; and still more, if possible, do they go beyond them in number and variety; yet, in a multitude of cases, are not less evidently mechanical, not less evidently contrivances, not less evidently accommodated to their end, or suited to their office, than are the most perfect productions of human ingenuity."[4]

If you're willing to go along with (1) through (4), then you should be willing to grant that the biological world is caused by the action of an intelligent designer.

It's a tempting argument for God's existence, which is why religious apologists continue to use Paley's argument from design to this day. Ray Comfort, a Christian evangelist from New Zealand, famously argued that the banana was "the atheist's nightmare" because of the way its ridges fit into the grooves of our fingers when we form the standard banana-grip, the way its color signals the moment of perfect ripeness, its easy-open tab, and its high digestibility. The banana's obvious design for human consumption, according to Comfort, is a testimony to God's capacity for intelligent design.

But the apologists' argument from design has a flaw: If an intelligent designer did all of the biological design work that we see around us, then who or what designed the designer? With apologies in advance to fans of the musician Billy Preston, "nothing from nothing means nothing." Throughout history, many thinkers bumped up against this flaw in the argument from design, but nobody could figure out what to do about it (though the Scottish philosopher David Hume came close[5]). Nobody, that is, until Charles Darwin came along.

What makes Charles Darwin special isn't that he noticed the appearance of design in the world of living things. Aristotle, Paley, and virtually every naturalist between the two could see that the living world was chock-a-block with first-rate design work. Instead, what makes Darwin so special is that he identified a purely naturalistic mechanism that could explain all of that design work. That mechanism, of course, is natural selection.

"The prize" of understanding natural selection, according to the philosopher Daniel Dennett, "is, for the first time, a stable system of explanation that does not go round and round in circles or spiral off in an infinite regress of mysteries."[6] I believe that this stable system of explanation is a prize worth possessing as we try to understand the biological basis of human generosity. As I explain natural selection in detail, keep your eyes on the genes: the reason the Earth is so full of elegant biological design is that building well-designed organisms has been a supremely effective way for genes to vouchsafe copies of themselves to the future.

Life on Earth did not begin with genes, but it did begin with things that could replicate. Here's one possible scenario. Before genes, there were certain bits of nonliving matter (perhaps the extremely small silicate crystals that make up tiny particles of clay) that had an interesting property: they attracted molecules from their environments that then assumed the particles' own crystalline structures. Due only to the forces of physics, these early bits of matter ended up making copies of themselves. The copies were also molecules that could replicate, so they, in turn, grabbed raw materials from their environments and then used *those* materials to crank out copies of *them*selves.

Somewhere along the way, the silicon-based replicators might have begun to pull in carbon-based substrates, which could have increased their replication rate even further. Eventually, over an uncountably large number of copy-conscript-copy sequences, the carbon-based add-ons to the silicon-based replicators might have taken on enough structure of their own to leave their silicon-based surrogates behind and go it alone as replicators in their own right.[7]

As these primordial replicators luxuriated in the clay or basked in noxious chemical Jacuzzis—all the while mindlessly conscripting non-living material and then mindlessly making copies of themselves—some of the replicas they synthesized acquired tiny molecular variations, which we now call mutations, that made the replicas slightly different from the molecules that produced them. No physical process is perfect, and molecular replication is no exception. Most of those mutations would have messed up the replication machinery instead of improving it. As Richard Dawkins put it, "*however many ways there may be of being alive, it is certain that there are vastly more ways of being dead*, or rather not alive."[8] On very rare occasions, however, replicators acquired mutations that made them better replicators, perhaps by improving their ability to attract substrates, or perhaps by providing some kind of protection against hostile forces of the universe. The specifics don't matter: all I want you to see is that some mutations turn replicators into better replicators. It is in that sense that we can refer to these mutations as *beneficial* mutations.

When the bearers of these new beneficial mutations started making copies of themselves, those beneficial mutations appeared in their offspring as well. As a consequence, the population of replicators increasingly became characterized by features that increased their rates of replication. That, in a nutshell, is natural selection: the little molecular nips and tucks that improve replicators' effectiveness at making copies of themselves gradually come to characterize the entire population of replicators, and the features that make them worse at replicating gradually vanish from the population.

After so many regimes of replication, mutation, and selection, the Earth's replicators stopped walking around naked. Instead, all of those little molecular nips and tucks that improved their replication rates came to resemble a magic cloak, adorned and lined with nifty design features that helped their wearers—the replicators—to do what replicators do best. This seems like a good time to dispense with the coy talk about "replicators" and "the accumulation of mutations that increase replication rates," and to start talking instead about DNA and the living organisms it builds around itself. From the gene-centered view of natural selection, organisms are vehicles that genes ride around in while they make copies of themselves. In *The Selfish Gene*, Richard Dawkins called them "survival machines":

> Different sorts of survival machine appear very varied on the outside and in their internal organs. An octopus is nothing like a mouse, and both are quite different from an oak tree. Yet in their fundamental chemistry they are rather uniform, and, in particular, the replicators that they bear, the genes, are basically the same kind of molecule in all of us—from bacteria to elephants. We are all survival machines for the same kind of replicator—molecules called DNA—but there are many different ways of making a living in the world, and the replicators have built a vast range of machines to exploit them. A monkey is a machine that preserves genes up trees, a fish is a machine that preserves genes in the water; there is even a small worm that preserves genes in German beer mats [coasters].[9]

When natural selection increases the rate at which a set of genes replicates, it does so by causing the organism—the genes' survival machine—to take on an attribute that surmounts some feature of the environment that is restricting the genes' rate of replication. These rate-limiters can be thought of as *design problems*, and natural selection acts as if it were casting about for engineering solutions to these design problems. In doing so, natural selection retains mutations whose effects make it look as if they were actively and intentionally assisting the evolving organism in clearing its genes' rate-limiting hurdles.

For another example, take the lithe, regal-looking, and completely made-up blue-headed sloozle. Tens of millions of years ago, sloozles were up against a rate-limiting design problem: they could have replicated faster had they been able to adjust their locations in space in response to information about the locations of other things that sloozles cared about (such as food, water, good nesting sites, potential mates, and predators).

One solution to this rate-limiter might have involved exploiting the correlation between the locations of objects in space and the photons that ricochet off of them in sunlight. To use the information contained in those photon showers, sloozles would have needed a sense organ that could capture the photons and wring the information out of them. The sloozles needed eyes. A rate-limiter such as visionlessness, therefore, can be viewed as a design problem for sloozle genes to solve: a design path that caused sloozles to move from a visionless state to a visionful one would have caused the sloozle genes to increase their rate of replication. The problem for sloozle genes is that you can't hop from visionlessness to visionfulness in a single step. You can't build an eye from a single genetic mutation.

Human designers and engineers solve complex multistep problems through purposive, forward-looking cycles of analysis, simulation, testing, and feedback. Explaining how complex design emerges in the biological world is more daunting than explaining how engineers solve problems, however, because our explanation cannot invoke forward-looking intelligent designers. "What Darwin saw" about biological design, as the philosopher Daniel Dennett observed,

> was that in principle the same work [that human R&D accomplishes] could be done by a different sort of process that *distributed* that work over huge amounts of time, by thriftily conserving the design work that had been accomplished at each stage, so that it didn't have to be done over again. In other words, Darwin had hit upon what we might call the Principle of Accumulation of Design.[10]

Taking a page out of Paley's book, Darwin also used the vertebrate eye to illustrate how design accumulates:

> Reason tells me, that if numerous gradations from a simple and imperfect eye to one complex and perfect can be shown to exist, each grade being useful to its possessor, as is certainly the case; if further, the eye ever varies and the variations be inherited, as is likewise certainly the case; and if such variations should be useful to any animal under changing conditions of life, then the difficulty of believing that a perfect and complex eye could be formed by natural selection, though insuperable by our imagination, should not be considered as subversive of the theory.[11]

All that should be necessary for the sloozle to get its eyes, if design really does accumulate as Darwin proposed, is for a vast set of tiny, genetically controlled improvements to arise mutationally during evolution, and for those improvements to accumulate over time. And that's what natural selection does: it keeps the stuff that helps and throws away the stuff that doesn't. But thought experiments involving imaginary blind creatures are one thing; evidence is another. How strong is the evidence that animals evolved from a visionlessness state to a visionful one through a series of extremely small design improvements? Actually, it's very strong.

The design work that went into creating the vertebrate eye, as we now know from research in a branch of biology called phylogenetics (which is devoted to tracing the evolutionary relationships among species, their phenotypic traits, and the genes that give rise to those traits), got started more than 600 million years ago, probably with a sea creature that acquired mutations that caused it to develop a patch of light-sensitive cells. Sunlight catalyzes many chemical reactions, so it's plausible that these light-sensitive cells created chemical changes inside the organism that altered its behavior. For example, the organism might have begun to regulate its behavior on a day-night cycle, making it more active during the day (when food was plentiful) and less active at night (when food was scarce)—or less active during the day (when predators were everywhere) and more active at night (when the predators had gone to bed). Small but progressive design refinements ensued over some 20 million years to make those photocells better at catching photons and then chemically responding to their presence. Although these light-sensitive cells were not yet "eyes," creatures that had them were almost certainly better off than creatures that did not have them.

Natural selection pushed on. By 550 million years ago, new mutations had led to new types of light-responsive cells, and neural connections grew from the proto-eye toward other regions of the brain (including, importantly, a major sensory relay station called the thalamus). A lens also evolved, and as a result, the incoming light rays could be focused into a single image, instead of smearing diffusely across the array. Meanwhile, an iris evolved, allowing depth-of-field control. Muscles also evolved outside the eye, allowing eye-bearers to move their eyes without moving their bodies. By 430 million years ago, the photoreceptor cells had improved further still, and the iris was tweaked. Muscles that appeared inside the eye itself enabled the lens to focus on both close objects and distant ones. Sometime in the past 430 million years, the lens became elliptical in shape, to make it better suited to working in air (rather than in water). As a final flourish, eyelids evolved to protect all of this exquisite design work from the hostile forces of nature.[12]

This highlight reel of vertebrate eye evolution teaches us about more than just our eyes. It shows how vastly long regimes of replication, mutation, and selection can create design pathways that lead to highly improbable destinations. Cells that are sensitive to light are nice things to have, but cells that are even more sensitive to light are even nicer. Connecting those cells up to a sensory relay station means that their outputs can be broadcast to all sorts of behavioral control centers within the brain, which is better than hooking them up directly to only a few. That's nicer still. And using a lens to create a single, detailed image is even nicer than making do with an unfocused blur. With enough replication, enough sifting among variant designs, and enough time, blind evolutionary forces can endow replicating creatures with the ability to see. Natural selection gives us design without a designer.

## EVOLUTIONARY PSYCHOLOGY AND ITS DISCONTENTS

Gene-centered, natural-selection thinking is the best way to understand how the design features of monkeys and fish and German beer-mat worms and sloozles come about, but this isn't a book about such creatures: it's a book about *humans* and our potential for generosity toward strangers. Natural-selection thinking can contribute to this endeavor as well through a field called evolutionary psychology. Some people incorrectly equate evolutionary psychology with the study of sex and violence. Sex sells, and people are so enthralled by violence that news producers follow the rule of thumb "If it bleeds, it leads," so it is not surprising that evolutionary psychologists' research on these topics gets the lion's share of the public's attention. But evolutionary psychology, as it is actually practiced, is a way of thinking about all of psychology, not just the topics that titillate or shock. Evolutionary psychology embraces sensation, perception, psychophysiology, learning, cognition, development, social behavior, psychopathology, and everything else that is relevant to the human mind.

Six decades ago, the evolutionary biologist George C. Williams threw down a gauntlet: "Is it not reasonable to anticipate that our understanding of the human mind would be aided greatly by knowing the purpose for which it was designed?"[13] Evolutionary psychologists have taken Williams up on this challenge. The conviction that inspires their research is a straightforward one: the human brain, like other biological organs, is a product of natural selection, and as such, should accomplish the functions for which it was designed with efficiency, even elegance.

The adaptiveness of many psychological processes is obvious: we get hungry so we'll go look for something to eat, we fear animals with sharp teeth because they could harm us, we acquire aversions to foods that give us food poisoning so we can avoid them in the future. The cognitive systems that produce these responses, however, are often quite unobvious. Even the simplest-seeming behaviors require many computational steps. Those computations, in turn, break down into smaller subcomputations, and those subsidiary computations break down into even more delimited sub-subcomputations. The computations embedded within any single problem-solving system have an almost fractal kind of organization: it's computations all the way down. To use the information that's embedded in the ricochets of photons that reach your eye to create a visual representation of the physical world, for instance, or to store away a memory of a scene you've witnessed, your vision and memory systems execute scores or perhaps hundreds of computations.[14] And it's plainly adaptive to eat when you are hungry—but how do signals from an empty belly eventually send you off in search of a burger and fries? To answer such questions about the operations of the mind, your introspections are going to be useless.

The goal of evolutionary psychology is to identify the mind's adaptations—its evolved computational circuitry. Evolutionary psychologists try to find adaptations not by studying neurons, but instead by studying the end results of our neurons' activities: the perceptions, emotions, decisions, and behaviors that natural selection programmed them to produce. The reasoning behind the evolutionary psychology gambit is simple: the mind is likely to be particularly good at accomplishing the tasks that natural selection designed it to accomplish, and those are the actions that raised our ancestors' fitness while our brain's circuitry was evolving. The things the mind accomplishes only ham-handedly, or not at all, should therefore be things that natural selection did not design the mind to accomplish. That's one way evolutionary biologists and ecologists figure out the functions of various features of nonhuman animals' physiology and behavior, and it's the strategy that evolutionary psychologists use as well.

Evolutionary psychology has drawn its share of criticism over the years: Humans excel at being human not by dint of natural selection, the critics say, but rather by dint of neural plasticity, or their capacity for brute-force statistical learning, or epigenetic effects. Or, say others, the pace of cultural change so outstrips the pace of biological evolution that we can no longer detect the fingerprint of natural selection on human behavior. Or evolutionary psychologists are genetic determinists who ignore development. Other critics indict evolutionary pseudoscience as a tool that privileged white men use to justify their oppression of women and minorities. And don't those dumb evolutionary psychologists realize that not every trait has a function? The belly button doesn't do anything! And we can't test evolutionary hypotheses anyway because behavior does not fossilize. Selfish genes! Sexism! Racism! Reductionism! Genetic determinism! Eugenics![15]

Plenty of evolutionary psychologists have already tackled these objections head-on, so I won't take up your time by responding to them here.[16] Many of the most ferocious criticisms are based on straw men, false dichotomies, or false premises, and virtually every valid critique has already been acknowledged and integrated into the intellectual framework of the discipline.[17] When practiced at its best, evolutionary psychology recognizes and embraces the roles of development, learning, and culture in creating behavior. And many evolutionary psychologists correctly recognize that adaptation, as the biologist George Williams warned, "is a special and onerous concept that should only be used where it is really necessary"—that is, only when the evidence convincingly indicates that a specific trait came into being because its evolution was favored by natural selection.[18] At the end of the day, evolutionary psychology is just psychology, but with an eye toward discovering the species-typical psychological adaptations that evolved because they set our ancestors up for evolutionary success. And it's an outlook that is going to guide our search for the instincts that shape human generosity.

## COSTLY COOPERATION

Adaptations bring evolutionary advantages to their bearers, but they also impose costs. They must be assembled out of bits of matter—molecules, proteins, cells, tissues, organs—during development. They must be fed and maintained once they are up and running. They take up space in the body and in the brain. The resources we use to build, feed, maintain, and house our adaptations are resources that we could be devoting to the development of other adaptations.

Some adaptations are also costly in a different way: they motivate us to transfer valuable resources to others. The kangaroo's pouch, the vampire bat's tendency to share food with hungry neighbors, and humans' willingness to assist each other in too many ways to count: Aren't such actions anti-Darwinian? Natural selection is a relative thing, a competition between rival designs. The benefit I render to you is reckoned as a cost to me. So, how could evolving humans have done better in the eyes of natural selection by diverting resources (such as food, time, and energy) to other people, when they could have been using those resources to support their own survival and reproduction? For decades, this sort of costly cooperation, in which one individual pays costs to provide benefits to another individual, looked like a serious problem for standard natural-selection thinking, so much so that biologists came to call it *the problem of cooperation*. And if we don't understand how natural selection has solved the problem of cooperation, we have scant hope of ever understanding the enigma of generosity in a world of strangers.

The first major breakthrough in solving the problem of cooperation took place in the 1960s. It was a monumental theoretical accomplishment that breathed new life into the scientific study of those social instincts, which Darwin wrote about so long ago, that motivate us to help our family and our friends. It was also a theoretical accomplishment that gave us a useful new way to put the word *altruism* to work.

CHAPTER 4

# IT'S ALL RELATIVE

The standard Darwinian explanation for the adaptive features of the living world is a straightforward one: by causing organisms to reproduce more effectively, adaptations win organisms more offspring. But what about adaptations that cause organisms to leave fewer offspring behind? When the rooster clucks out an alarm call to warn other chickens about the approaching fox, he draws the fox's attention to himself. To the greedy hand, the honeybee's sting is a persuasive deterrent, but it's a death sentence for the bee that delivers it. And, speaking of bees, many social insects (including not only bees but ants and wasps as well) produce workers that do not reproduce at all. These flunkies are specialized for all sorts of fetching, toting, housekeeping, nannying, landscaping, foraging, farming, and soldiering, but they die without ever having known the joys of parenting. And speaking of parenting, when mothers and fathers provide care for their young children, they give up reproductive opportunities out in the wider world in order to stay home and change diapers.

Alarm-calling chickens, kamikaze honeybees, sterile minions, and doting parents are all behaving in ways that are good for other individuals' reproduction, but bad for their own. You can't make new offspring if you're dead, or sterile, or too exhausted by child care to even think about procreating any time soon. Adaptations like these eventually forced biologists to realize that the scope of natural selection was broader than Darwin had realized. Once they adopted a gene's-eye view, they were able to see that natural selection's partiality for any particular gene is determined not only by how the gene affects the reproductive success of the individual organism in whose body it resides, but also by how it affects the reproductive success of copies of itself that are locked inside *other* organisms—namely, the gene-bearer's genetic relatives. Once we begin to think of genes as purpose-driven agents, as Darwin's dangerous idea reveals that we should, it is easier to see that a gene's only agenda is to increase the number of copies of itself that are running around in the world, no matter whose gonads those copies come from.[1]

This profound insight would eventually force biologists, more than a century after *The Origin of Species*, to renovate Darwin's concept of fitness. It was left to a twentieth-century British biologist named William Hamilton to do the heavy mathematical lifting. A gene's *inclusive fitness*, as Hamilton named it, is the sum of the gene's *direct fitness* (its effect on the reproductive success of the individual in whose body it resides) and its *indirect fitness* (its effects on the reproduction of other individuals who also possess copies of that gene).[2] Hamilton showed that a gene that motivates you to help your sisters, brothers, cousins, and offspring can evolve by natural selection even as it reduces your own reproductive success. Hamilton's discovery created a new vantage point for discovering adaptations that evolved to motivate us to care for others—namely, those "well-marked social instincts" that Darwin called "the parental and filial affections." These adaptations have much to teach us about the factors that regulate our concern for others. But, as we will see, these adaptations also tend to be jealous ones, making us eager to help our relatives but reluctant to help nonrelatives. Darwin's parental and filial affections provide a shakier psychological foundation for the Golden Rule than Darwin surmised.[3]

A notion called *relatedness* helps us understand Hamilton's inclusive fitness concept more clearly. From the gene's-eye perspective, relatedness is the probability that a rare gene residing within you also resides within another specific individual. You might recall from high school biology that we use the fraction ½ to represent the relatedness of parents and offspring, ½ to represent the relatedness of two siblings, ¼ for half-siblings, ¼ for aunts/uncles and nieces/nephews, ⅛ for cousins, and so on. To see what these fractions mean, imagine yourself as a mother (recalling that vertebrates all have two copies of each gene), and that you possess one version of that gene that everyone else in the world possesses—let's mark it with an uppercase *A*—along with a mutated copy that no one else has because it resulted from a genetic copying mistake as you were developing from a single cell. Let's mark that rare copy with a lowercase *a*.

Now, if you have a large number of offspring, approximately 50 percent of those offspring will inherit your uppercase *A* version of the gene, and approximately 50 percent of your offspring will inherit your lowercase *a* version of the gene. So what is the probability that any single child of yours possesses your rare lowercase *a* gene? Fifty percent. That is what it means to say that mothers and offspring are 50 percent related. Now imagine things from the offspring's perspective. If you inherited the rare *a* copy from your mother, the probability that any single one of your siblings also possesses that rare gene is also 50 percent. That's what it means to say that siblings are 50 percent related.

When we are thinking about adaptations that cause us to pay direct fitness costs in order to provide direct fitness benefits to others, we use relatedness as a sort of genetic exchange rate: it expresses how much of our own reproductive success a gene locked inside of us might be willing to trade off in order to increase the reproductive success of others. These exchange rates are numbers we can use to estimate a gene's total effects on an organism's inclusive fitness.[4]

Hamilton's major insight, which we now call *Hamilton's rule*, is that a gene that reduces its bearers' direct reproductive success—a gene involved in constructing the bee's stinger, for instance—can evolve by natural selection if the lifetime reproductive benefit (*b*) that gene confers on all of the individuals affected by it, weighted by the average relatedness (*r*) between the gene-bearer and the group of affected individuals, exceeds the lifetime reproductive cost (*c*) that the gene imposes upon its bearer.[5] In other words, genes become more common in the population when $rb - c > 0$. The worse the gene's effects on your direct reproductive success, the better its effects must be on others. If benefactors and beneficiaries are closely related, so much the better for the evolution of that gene. Traits that evolve by causing you to exchange some of your direct fitness for the direct fitness of your genetic relatives can therefore be said to have evolved through an evolutionary pathway that biologists call "kin selection."[6] You can also think of it, as many biologists do, as an *altruistic* approach to organism design.

## THE ALTRUISTIC SCHOOL OF DESIGN

In its short history as an English word, *altruism* has been used to describe several different phenomena. The nineteenth-century British philosopher Herbert Spencer played a major role in establishing a home for *altruism* in the lexicons of biology, social science, and everyday discourse. He used the term in several different senses: as an ethical ideal, as a description of instances in which we help needy persons at our own expense, and as a description of genuine psychological concern for others' welfare.[7] In the twentieth century, evolutionary biologists such as Hamilton also recruited the term *altruism* to characterize adaptations that evolve through kin selection.[8] We can think of those adaptations as altruistic because a design process in which the gene caused its bearers to pay reproductive costs—while simultaneously providing a reproductive advantage to others—is what caused natural selection to favor the gene in the first place.

Some of the best illustrations of altruistically designed adaptations are those that promote parental care. By parental care, I mean the suite of physiological and behavioral adaptations that parents possess for investing in the future reproductive success of their offspring.[9] When a mother furnishes her fertilized egg with proteins and lipids in the form of an egg yolk, or protects it with antibodies, hormones, and antioxidants during gestation, she provides parental care. When a bird sits on a nest full of eggs to keep them warm and safe from egg-eating snakes, she, too, provides parental care. Building shelters for one's offspring is parental care. The same goes for feeding them, helping them in their own breeding efforts, and even helping them as adults.

## MOM'S HOME COOKING

The best example I can think of for illustrating how adaptations for parental care evolve by increasing a mother's indirect fitness, while simultaneously reducing her direct fitness, is breastfeeding. Lactation evolved over more than 300 million years, beginning with a group of vertebrates called the synapsids. The synapsids stood at the head of a phylogenetic trail that eventually led to modern-day mammals, and they were some of the first animals to live their entire lives on land. Their liberation from the water was made possible by the fact that synapsid eggs could survive in the open air. Unlike the ancient egg-layers that led to modern-day birds, whose eggs were hard and watertight, the synapsids' eggshells were soft, permeable, parchment-like affairs. They did a very poor job of retaining moisture, and that was a problem.

The ease with which the synapsids' eggs dried out opened up an opportunity for adaptation, so mutations that contributed to solving this adaptive problem were at a selective premium. In other groups of animals, natural selection solved the dry-egg problem by endowing mothers with behavioral systems that made them bury their eggs underground. In the synapsids, however, natural selection solved the problem by taking advantage of glands on the skin's surface that happened to be good at transferring moisture (as well as lipids that could reduce moisture loss in the first place) onto the newly laid eggs. This is how mammalian lactation got its start.

Transferring moisture from mothers' skin to their eggs was a good thing, but it exposed the developing embryos to fungal and bacterial infections. This vulnerability favored new mutations that added a variety of iron-binding proteins and antimicrobial compounds to the evolving cocktail of secretions. The recipe for those mammary gland secretions took a major evolutionary turn when natural selection added nutrients to the cocktail. Biologists think a small amount of carbohydrate might have been added initially to make it easier for mom's skin secretions to get to the embryo, but once those carbohydrates were inside the egg, additional mutations that enabled embryos to convert those carbohydrates into usable energy would have been extremely advantageous. From there, biologists envision a 150-million-year succession of additional genetic changes that further increased both the quality and the quantity of the evolving cocktail. These changes, in turn, drove natural selection for internal gestation, longer gestation, smaller offspring, and extreme dependence of offspring on their mothers for care and nourishment.[10]

The evolution of lactation, like the evolution of the vertebrate eye, evolved down a garden path of mutations, each of which increased their bearers' inclusive fitness: one mutation after another brought one improvement after another. But there's a crucial difference between the evolutionary pathway that gave us our eyes and the pathway that gave us breastfeeding. The genes that caused better vision evolved simply by boosting our ancestors' direct reproductive success: better seeing led to better living, and better living led to better reproducing. But when we consider the evolution of breastfeeding, we have to consider the fitness of two different classes of individuals—the mothers who pay the costs and the offspring who receive the benefits.

If the mutations that went into the evolution of lactation reduced mothers' direct reproductive success, why did natural selection favor them? Quite simply, because Hamilton's rule rules. A gene will evolve by natural selection if the lifetime direct reproductive costs it imposes on its bearer are less than the total amount of lifetime reproductive benefits it causes its bearer to confer upon a particular class of beneficiaries, discounted by the likelihood that those beneficiaries also possess the gene that causes the benefactor to confer the benefit. Wordy, isn't it? Here's a more concise way to put it: lactation mutations that reduce mom's direct reproductive success by $c$, while also raising her offspring's direct reproductive success by $b$, will evolve by natural selection when $rb > c$. Human children have a 50 percent chance of inheriting a gene from mom, so with breastfeeding, the relationship between benefactors and beneficiaries is $r = 0.5$. With this value for $r$ in hand, we can now be very specific about the minimum cost-benefit ratio required for natural selection to retain a gene that improves lactation. The lifetime reproductive benefit would have to be a tiny bit more than twice as large as its lifetime reproductive cost to the mother. With those conditions in place, $rb > c$, and presto! Lactation becomes one gene better.

## A WOMB WITH A CUE

One factor that favors natural selection's assembly of adaptations for parenting is the high degree of relatedness between parents and offspring. But a second factor is also immensely important, particularly for mothers: their worlds are filled with cues that enable them to see where to direct their help so that it ends up benefiting their own offspring rather than other mothers' offspring. Genes that cause parental care systems to respond to these cues—that is, to discriminate between kin and non-kin—reduce the costs of helping.

A thought experiment will help here. Suppose natural selection built a parental care system that motivated mothers to allocate their assistance according to this dumb rule: "Help anything in your environment that can fit in a bread box." Mothers who obeyed this rule would indiscriminately provide assistance to everything smaller than a bread box, including not only their own infants, but also other mothers' infants, their household pets, other mothers' household pets, toy dolls that looked like infants, toy dolls that looked like unicorns, and actual loaves of bread. These misdirected deliveries of maternal care would be wasted investments because they would cause moms to pay direct fitness costs without yielding any indirect fitness benefits. Here's a better rule, which natural selection appears to readily incorporate into decision-making systems for regulating parental care: "Take care of any squawking thing that lives in your house."[11]

Picture natural selection as a contest between a kin-discrimination system that causes animals to obey the "bread-box" rule and a system that causes them to follow the "feed squawking things in your house" rule. Allow these two rival designs to compete for a few generations of natural selection, and you will find that the squawk-minders did a better job at increasing their inclusive fitness than the bread-boxers did, because most of the squawk-minders' assistance went to their offspring. The bread-boxers, in contrast, will have wasted most of their help. Indiscriminate altruistic designs lose out to more judicious, cue-using designs every time. In fact, many species of birds apply the "feed squawking things in your house" rule so scrupulously that birds such as cuckoos and cowbirds (which are called "brood parasites" for their tendency to surreptitiously dump their gestating eggs into other birds' nests) take advantage of the rule, duping other birds into incubating their own cuckoo or cowbird chicks. The cuckoos and cowbirds of the bird world illustrate just how good the rule must be: brood parasitism is the price you pay for using the "feed squawking things in your house" rule, but it's a price worth paying.

Kin discrimination based on coresidence relies on a cue that is located outside of parents' bodies: if it sleeps where you sleep, feed it. However, many of the kinship cues that mammalian mothers use are located *inside* their bodies. For example, as a soon-to-be-born mammal descends through its mother's birth canal, the stretch of the muscle fibers that line the uterine wall causes nerves in the uterus to send a volley of neural signals to the brain. These neural signals stimulate the brain's pituitary gland to release the hormones prolactin and oxytocin. These two hormones, in turn, hasten labor, stimulate the production of milk, and activate the brain's parenting circuits.[12] Notice that the stretch of those uterine muscles in the context of labor and delivery is actually functioning as a kinship cue—and an almost foolproof one at that: over the course of a woman's life, her first-degree relatives are the only organisms in the universe that will ever cause a sustained and coordinated stretch of her uterus.

## MOTHERS, BROTHERS, AND OTHERS

Because mothers have so many external and internal cues available for identifying their offspring, natural selection solves mother-offspring kin-recognition problems relatively easily. Other kin-recognition problems are harder. How do siblings come to recognize each other as kin? How do fathers figure out which kids are theirs? How do we come to recognize relatives who are further away on the family tree?

Here's one possible answer: When you meet someone for the first time and someone says, "Meet your brother Marvin," or "This is your Uncle Bob," you take these new tidbits ("Marvin's your brother" and "Bob's your uncle") and bind them with other information about Marvin and Bob, such as their looks, sounds, and scents. You then use all of this linked information to create a Marvin database and a Bob database, each of which can be updated as Marvin's and Bob's looks, sounds, and scents change over time.

As tempting as it might be, the idea of an evolved kin-recognition system that's based on kinship labels suffers from two problems. The first problem is that labels such as "brother" and "uncle" are not at all necessary for kin discrimination in the wider world of living things. Organisms from microbes to meerkats manage to recognize kin and then help them in response, and they obviously don't need kinship labels to make the proper distinctions.[13] The second problem is that a kinship recognition system based on labels isn't sufficient either. Consider the term "nephew." Although some languages have words to distinguish my sister's son from my wife's sister's son, many do not. In those more genealogically sparse languages, both people are my "nephews," yet my relatedness to one of them is 0.25, and to the other it is 0. Kinship labels with this much wiggle room could do all kinds of mischief to an evolving kin-recognition system.

So if it's not verbal labels, what cues did natural selection exploit to build kin-recognition systems beyond the mother-child relationship? For sibling recognition, natural selection might have built a system that also follows a coresidency rule. The psychologist Debra Lieberman and her colleagues have hypothesized that the sibling recognition system operates like the odometer in a car: it tallies up the amount of time two people lived together during childhood. Once those two people reach adulthood, the values on their sibship odometers reflect their unconscious confidence that the person they ate, slept, played, and bickered with for all those years is, in fact, a sibling. Part of what justifies Lieberman and colleagues' conclusion is the fact that the more time two people live together as children, the more likely they are to perform favors for each other as adults. In addition, the more time they live together as children, the more disgusted they are by the thought of having sex with each other. The same sibship odometer that regulates our kindness toward our siblings is put to use to prevent us from inbreeding with them.[14]

This sibship odometer doesn't actually *know* whether your putative sibling really is a sibling—at least, not in the sense in which you know the words to the national anthem. The only thing the sibship odometer actually knows is the number of suppers or summers you shared with another child. That's okay. The odometer in your car doesn't know that it measures the distance you travel, either: that information is implicit in the fixed ratio that converts rotations of the car's wheels to rotations of the gears that are attached to the odometer's number display. The one-to-one correspondence between the distance you travel and the rotations of the gears attached to the odometer is what causes the odometer to represent your mileage. Likewise for the sibship odometer: the close correspondence between the number of days we live with our siblings as children and the probability that we really are genetic relatives is what enables natural selection to take advantage of childhood coresidence as a kinship cue.

In the old days, it used to be possible to trick an odometer into displaying a lower mileage by running the car in reverse on a rack. The sibship odometer can be tricked, too—by nontraditional living arrangements between unrelated children, such as those that were implemented on many a kibbutz in Israel throughout the 1970s. Lieberman and Thalma Lobel studied relationships among men and women who had grown up on kibbutzim where the children were reared together. As children, they had eaten, slept, bathed, and spent most of their waking hours not under their own parents' roofs, but instead in communal children's houses with unrelated peers. By interviewing those kids after they had grown up, Lieberman and Lobel discovered that they were treating their childhood housemates like kin: They were most willing to help those with whom they had lived the longest. They were also most revolted by the idea of having sex with the ones with whom they had lived the longest.[15] These findings score major points for the sibship-odometer hypothesis, showing that coresidence doesn't increase your generic attraction to others—it makes you view them as worthy of your goodness, but not your gametes.

Coresidence duration is a nice cue for estimating relatedness, but there's another nice cue to be had: if you see your mother feeding another child day in and day out, then that child is probably your younger sibling. Lieberman calls this cue "maternal perinatal association." In the ancestral world, women gave birth no more than every three or four years (children required two to three years of breastfeeding, during which time their mothers could not get pregnant). You would never have seen your mom breastfeed and coddle an older brother or sister. As a result, the maternal perinatal association cue was available only to older siblings. But this type of cue is so trustworthy for detecting kin that if you see your mother care for a younger child during childhood, then in adulthood you tend to treat that individual kindly—and platonically—no matter how many years the two of you lived together. Indeed, Lieberman and her colleagues have conducted statistical horse races between coresidence duration and maternal perinatal association to see which cue does more work in predicting how adults feel about their grown siblings. They have found that it takes about fifteen years of coresidence for the sibship odometer to generate as much sibling-appropriate behavior as the maternal perinatal association cue can generate on its own. You have to do a lot of hard time with another child before coresidence duration makes a meaningful difference. Maternal perinatal association is the trump card of sibship cues.[16]

Moms matter and kids count, but how do dads figure out who their kids are? One possibility begins with the fact that men need to have sex in order to become fathers (holding aside the wonders of modern fertility medicine). Therefore, a man can rule out the children of all women whom he has never known carnally. For each mother still under consideration, he simply needs to roll back the sex odometer to nine months before the baby's birth. If he had sex with the mom around that time, then the child is probably his.[17]

But only probably. The fact that a man and a woman have sex is hardly a guarantee that any children who appear on the scene nine months later are the man's children.[18] The woman might have had sex with other men. Hamilton tells us that providing parental care to somebody else's child is not so good for one's indirect fitness: each misdirected outlay of parental care increases the fitness costs. Hence the adaptations men deploy in detecting their offspring seem to consider their mates' perceived fidelity.[19]

Natural selection might have found still other cues to help men recognize their children. How about smell, for instance? Many vertebrates detect kin by sensing chemical compounds that are released through the skin. The genes that produce these compounds are located in particularly hypervariable parts of the genome, which in humans are collectively called the human leukocyte antigen complex (or HLA). One of the major functions of your HLA is to discriminate between cells and tissues that belong to you and those that don't. Because of the HLA's effectiveness at identifying stuff that doesn't belong in our bodies, organ donors and the people in line to receive those organs have to be carefully assessed for HLA compatibility to ensure that the organ won't be rejected. And because HLA genes vary so much between individuals, the chemical medleys they release are hypervariable as well. Genetic relatives share more genetic variants than nonrelatives do, which may cause relatives to end up with similar smells.[20] In fact, in blind smell-tests, judges rate identical twins' odors as so similar that it's as if the smell samples (usually T-shirts that have been worn for three consecutive nights) had been taken from the same person.[21] Perhaps we can smell relatedness.

Facial similarity could be another foundation for recognizing kin. The human face is one of the most genetically complex features of human anatomy, so there is a lot of genetic similarity to be detected in facial resemblance.[22] It's unsurprising, therefore, that humans can assess kinship to some degree on the basis of facial resemblance (machine learning algorithms can be trained to do it extremely well).[23] People in ancestral times had no idea what they looked like—In a world without photos, portraits, and mirrors, how could they?—so if we do possess an adaptation for assessing relatedness based on facial resemblance, it would have to work by comparing the faces of new acquaintances to a template of the faces of people we have already judged to be relatives. Odd though it seems, our own faces would have had no role to play.[24]

The psychologist Daniel Krupp and the mathematician Peter Taylor developed a statistical model for how a similarity-based kinship estimator might operate. Their hypothesized estimator first studies the faces (or smells, or some other evolutionarily reliable cue to relatedness) of a key set of relatives, perhaps as defined by a spatial cue such as "all the people who sleep under the same roof as you," or "all the people who eat food that was cooked in the same fire as yours." Second, it builds a kin template that reflects the average of those putative family members' faces (or odors, or whatever). Third, it studies the faces (or odors, or whatever) of everyone else encountered outside of the family and then uses their features, plus some measure of how much those features vary, to build up a non-kin template. With the kin template and the non-kin template ready to go, the kinship estimator would then know how to create a statistical estimate of our relatedness to every new person we encounter. If the person's features are more similar to the kin template than to the non-kin template, the person is tagged as a relative.[25]

But if natural selection was going to go to the trouble of building a cognitive mechanism that could create kin and non-kin templates for facial features or olfactory signatures, why would its curiosity necessarily stop with the scents of people's T-shirts and the size of their ears? Couldn't natural selection create a more open system that could process other kinds of sensory information that distinguish kin from non-kin? Why not height and body mass index, for example—both of which are easy to see and highly heritable?[26] For that matter, why not use psychological attitudes and personality quirks as cues to kinship? These also are moderately heritable.[27] In general, the more cues the kin-discrimination system uses—especially if those cues are controlled by different genes—the more reliable its judgments are likely to be.

And once you've got a system that is open-ended enough to build up a kin template and a non-kin template based on genetically controlled features of our bodies and minds, what would restrain the system from trying to incorporate *any* cue that was reliably correlated during childhood with genetic relatedness—even cues that were not themselves under direct genetic control? After all, some birds learn family-specific songs, and then use those songs to regulate the help they provide to their relatives. What's to stop humans from doing the same sort of thing?[28] How about using family names (also called surnames) as kinship cues, for instance?

It's not as outlandish as it might sound. The evolutionary psychologists Kerris Oates and Margo Wilson sent emails to 6,400 men and women whose names and addresses were listed in a commercial database. The names were either relatively common (according to the US Census), such as "Michelle Smith," or relatively rare, such as "Hugh Morrison." The emails were supposedly sent by a college student who was doing research for a school project, and this fictional student shared either the recipient's first name, the recipient's family name, both names, or neither. Most email recipients ignored the student's request, but recipients who shared a first name and a family name with the student were nearly twice as likely to respond than those who shared neither name. The effect of name sharing was even stronger when the student and the email recipient shared a rarer "Morrison"-type family name than when they shared a common "Smith"-type family name. And the effect was stronger still when the student and the recipient shared a rare "Hugh"-type first name as well as a rare "Morrison"-type family name. A more recent experiment in the spirit of Oates and Wilson's work yielded similar results.[29]

Every McCullough I ever met before leaving home was either a close relative or the spouse of one. Did my kin-detection machinery lock onto the correlation between McCullough-ness and kinship in my childhood world? If so, it would explain why I still find it somewhat charming to meet new McCulloughs. I once went out of my way as a teenager to meet a professional golfer named Michael Earl McCullough (yes, we even share the same middle name), even though I had (and have) zero interest in golf. And I always tried to exchange greetings with my fictive cousin Erica McCullough, who worked at the gym where I used to exercise. One part of my mind knows that my actual relatedness to these two McCulloughs has been diluted to nothing by the global diffusion of McCulloughs over the past few centuries. But another part of my mind, which evolved to track cues to kinship, might not care so much about the history of the Scottish diaspora: perhaps that part of my mind believes instead that I still live in a world in which every single person I care about, and every single person who cares about me, can be reached with a few days of walking.

## FOCUS ON THE FAMILY

The evolutionary story of human altruism is really a story about the human family. Children are seemingly bottomless pits of need, and natural selection made them that way. The children of our hunter-gatherer ancestors lacked the strength and skill to meet their daily needs until their late teens, so they evolved to rely on family members to provide the calories they couldn't obtain for themselves. The children of our own era don't become self-sufficient until they've consumed fifteen, twenty, or even twenty-five years' worth of food, shelter, clothing, and loving care from their mothers, fathers, sisters, brothers, and others. But even adults have a hard time making it without the help of their families: When the money runs out before the rent comes due, or the crops fail, or the car breaks down, or the chicken is stolen, or illness strikes, to whom do most people turn first to loan them some money, give them a ride, fix them a hot meal, or take care of the kids while they recover? Take a guess.[30]

When favoritism for kin leaves the realm of home economics and enters the realm of public economics, as when people in power use their offices to funnel goodies to their own relatives, we give our favoritism for kin another name: nepotism. In the meritocratic West, we regard nepotism as one of the major roots of public corruption, which of course it is. However, throughout most of the world, the ability to find good jobs and fat government contracts for your children, parents, spouses, cousins, nieces, and nephews *is really the point of going into public service in the first place.*

Indonesia's Suharto family, which managed to amass a $40 billion fortune through decades of nepotistic rule, is a case in point. But their nepotism is hardly unusual. Chicago super-mayor Richard J. Daley was once asked to explain why he directed a city insurance contract to a firm that employed his son John (and in doing so, excluded every other company that would have wanted to bid on the contract). The only defense Daley could muster against his critics was a rhetorical one: "If I can't help my sons, then they can kiss my ass. I make apologies to no one. . . . If a man can't put his arms around his sons, then what kind of world are we living in?"[31] It's worth noting that Richard M. Daley, the super-mayor's eldest son, went on to serve as Chicago's mayor from 1989 to 2011. It was the longest mayoral stretch in Chicago's history. In 1999, well into Daley Jr.'s tenure as mayor, the *Chicago Tribune* reported that sixty-eight of his own relatives or in-laws were employed by the city, the county, or some other public institution. Another example of family resemblance.[32]

Across the long arc of history, instances of nepotism such as these have been the rule rather than the exception. As Adam Bellow wrote in his book *In Praise of Nepotism*,

> The press reports [stories of nepotism] in an impotent, exasperated tone and uses terms like "blatant," "flagrant," and "rampant" to describe the extent of the problem. Indeed what seems most striking about the Suhartos and their ilk is their utter lack of shame about their nepotism. Far from being secretive about it, they boast of their nepotistic achievements and are in many cases openly admired for them. We seem to have entered a looking-glass world where our values of efficiency, merit, and fairness are turned upside down. And so we have: for in these societies, nepotism is actually considered a good thing, something to be proud of, not a vice to be concealed.[33]

Democratic ideals and institutions, such as open elections and a free press, help to restrain the nepotistic impulse, but they hardly abolish it.[34] Nepotism is simply too tempting for a species whose very evolution depended on an eagerness to meet the needs of kin. As Hamilton showed, we should expect social creatures—humans included—to evolve both the will and the ways to deliver benefits to their genetic relatives. And because natural selection is a penny-pincher, we should also expect that whenever possible, altruistically designed adaptations will try to minimize their losses by taking advantage of cues that enable us to direct our assistance toward relatives only. It is right to be outraged by nepotism, but it is silly to imagine that the motivations behind nepotism are themselves corrupt. As the anthropologist Helen Fisher wrote, "nepotism, it can be said, is one of our original family values."[35]

## HAMILTON'S RULE AND THE GOLDEN RULE

What, then, should we make of Darwin's claim that our "well-marked social instincts"—particularly "the parental and filial affections"—are the bedrock upon which we have built our modern-day concern for the welfare of strangers? Was Darwin correct to think that these instincts, disciplined by intellect and reason, were responsible for the expansion of our concern? Is Hamilton's rule really the foundation for the Golden Rule?[36] Our use of kinship terminology to express our affinity with nonrelatives might seem like proof for Darwin's speculations. If humans didn't extend their parental and filial instincts to distant strangers, after all, what are we to make of godmothers, *The Godfather*, the International Brotherhood of Electrical Workers, the Daughters of the American Revolution, the Sons of Liberty, the Fraternal Order of Police, and Sorority Row? How else can we make sense of jingoistic speeches about the fatherland or the motherland? What about God the Father, God the Son, and our brothers and sisters in Christ? Doesn't our use of kinship language for members of our own social groups imply that we really are, in some sense, thinking of them as kin and family?

Not at all. Churchgoers might refer to other worshippers as their brothers and sisters, but they don't mistake them for real brothers and sisters. If they did, they'd be too disgusted by the thought of sex with them to ever settle down with them and raise families, and yet people clearly choose their spouses based on shared religious beliefs.[37] Fictive kinship terminology is just a clever tool that the founders and leaders of our modern institutions (such as churches, nations, workers' unions, armies, and terrorist organizations) have exploited to raise the reputational stakes of being selfish toward other members of the collective. When you fail to help a stranger, it's between you and your conscience. When you fail to help a brother or a sister, it's a betrayal.[38]

When Darwin supposed that civilization built its concern for distant strangers on the basis of our parental and filial affections, he wasn't being nearly Darwinian enough. Natural selection idiot-proofed our parental and filial affections precisely so we wouldn't extend them to just anybody. We feel fatherly or motherly or brotherly or sisterly about our children and siblings because of identity-verification procedures that enable our minds to tag our significant others as real, actual, literal, flesh-and-blood relatives. We don't dole out parental and filial regard to others just because we worship at the same temple, fight for the same country, or pay dues to the same workers' union. To understand the evolved features of the human mind that are responsible for our regard for distant strangers, we have to focus beyond the family.

But just how far beyond the family do we need to focus? What about the level of communities, tribes, and nations? For several decades, many scholars have been speculating that we might possess instincts for generosity that evolved not because they serve the reproductive interests of individuals and their kin, but instead because they serve the reproductive interests of the groups to which individual human beings belong. This for-the-good-of-the-group explanation for human generosity comes from a theoretical perspective that is called, appropriately enough, group selection. If the proponents of the group-selection perspective are right, and we care about strangers today because we first evolved to care about our neighbors and countrymen, their insights will be quite useful in our efforts to understand the genealogy of generosity. And if they're mistaken about how much their approach has to offer, it will be good to know that, too.

CHAPTER 5

# FOR THE LOVE OF SPOCK

"Go to the ant, thou sluggard; consider its ways and be wise!" For millennia, people have turned to the social insects—bees, ants, wasps—in search of deep insights about human nature. The writer of the Book of Proverbs thought the ants had much to teach us sluggards about the merits of diligence: "It has no commander, no overseer or ruler, yet it stores its provisions in summer and gathers its food at harvest."[1] The founders of Mormonism so admired the honeybee's diligence that they used the word *Deseret* (a Book of Mormon neologism meaning *honeybee*) to name the territory that later became the state of Utah. Today, Utah's state flag features a honeybee hive with the word *Industry* underneath it. Thomas Hobbes used the social insects to illustrate why seventeenth-century human societies needed absolute monarchs. Karl Marx used such insects to illustrate why nineteenth-century human societies didn't.[2] As polemicists' needs changed, the social insects had to change, too. Occasionally, insect-human comparisons yield new insights about humanity. More often, they lead people to see what they want to see.[3]

For several decades, some evolutionists have been comparing humans and insects to illustrate a trait they credit for humans' generous spirit. The trait is *eusociality*, and the evolutionary process that supposedly created this special characteristic of human communities is called *group selection*. With characteristic verve, the biologist Edward O. Wilson wrote approvingly of these two ideas and their value for understanding human society. "Even by the strictly technical definition as applied to animals," he wrote, "*Homo sapiens* is what biologists call 'eusocial,' meaning group members containing multiple generations and prone to perform altruistic acts as part of their division of labor. In this respect, they are technically comparable to ants, termites, and other eusocial insects."[4] Elsewhere, Wilson wrote, "Of all the primate species that lived across the tropical and subtropical regions for millions of years, only one, an offshoot of the African great apes, an antecedent of *Homo sapiens*, crossed the threshold into *eusociality*."[5]

Wilson's idea of eusociality is generally not what evolutionary biologists mean when they use the term, however. The defining characteristic of eusocial societies is that individuals develop into one or more physically and behaviorally distinct classes of individuals, known as "castes." One caste specializes in producing offspring, and the other castes assist the reproducers by babysitting, getting groceries, and cleaning house. The differences between human societies and actual eusocial societies—those of the ants and honeybees, for instance—could hardly be starker. After all, human communities don't have an irreversibly sterile working class that exists to serve the lucky few who get to reproduce. And although human societies also feature divisions of labor, people are not converted into pin-makers, butchers, brewers, and bakers by chemical signals from their mothers or by the foods they eat as infants (as the eusocial insects are). If you get bored with making pins, you can try your hand at brewing. You can even have a baby in your downtime. Not so with the honeybee. Worker bees don't reproduce and that's that. The idea that humans are eusocial in the traditional sense of the word falls to pieces under the gentlest scrutiny.[6]

## FROM EUSOCIALITY TO GROUP SELECTION

If blurring the line between the social organization of insect communities and the social organization of human communities were the only consequence of characterizing humans as eusocial, there wouldn't be any reason to get uptight. However, this conflation has led some people to believe that humans possess generous instincts that evolved not just because of the fitness benefits of those instincts for the generous person herself, or for her relatives, but instead for the fitness benefits of those instincts for the communities in which the generous person resides. It's a concept known as group selection, and Edward O. Wilson led the charge in promoting this idea as well: "The genetic code prescribing social behavior of modern humans is a chimera," he wrote. "One part describes traits that favor success of individuals within the group. The other part describes the traits that favor group success in competition with other groups."[7]

Some of the advocacy for group selection comes from researchers who have sought to undermine scientists' confidence that Hamilton's inclusive fitness theory, with its *r*, *c*, and *b*, gives an accurate picture of how natural selection generates instincts for cooperation.[8] Wilson wrote:

> The old paradigm of social evolution, grown venerable after four decades, has thus failed. Its line of reasoning, from kin selection as the process, to the Hamiltonian inequality condition for cooperation, and thence to inclusive fitness as the Darwinian status of colony members, does not work. . . . As the object of general theory, inclusive fitness is a phantom mathematical construction that cannot be fixed in any manner that conveys realistic biological meaning.[9]

These are bold claims that threaten to overturn more than half a century of Hamiltonian gene-centered thinking about the evolution of our social sentiments and instincts. Are Wilson's claims correct? Is group selection a better lens through which to see human prosociality? Or is it just a different language for describing the same things that you can describe with inclusive fitness theory? Does it reveal humans to be truly, deeply selfless in a way that inclusive fitness theory doesn't? How do you explain tribes and nations without referring to people taking action for the good of the group? Didn't you read somewhere that group selection doesn't even work? And didn't you read somewhere else that group selection is doing just fine?

*What even is group selection?*

I'm glad you asked.

## A SELECTION OF GROUP SELECTIONS

The term *group selection* has been applied to many different concepts. Some scientists use it to describe how groups, which are composed of individuals, rather than the individuals who make up those groups, are the targets of natural selection's gene-sorting activity. In such instances, group fitness, rather than individual fitness, is the thing that group selection is seen to be maximizing. Other scientists use the term to express how natural selection can work at multiple levels of biological organization: it can change a gene pool by acting on the genetic variation within a single group, or it can do so by acting on the genetic variation between a bunch of different groups. Still other scientists use *group selection* as a proxy for *warfare*: if my group has bashed your group into oblivion, according to this usage, my group (along with the genes of all of the surviving individuals within it) has been group-selected to survive. You and your genes weren't so lucky.

In other uses of the term, *group selection* refers to *nongenetic* processes in which the traits that characterize successful groups (say, the strength of their iron weapons, the speed of their supercomputers, or the quality of the filaments in their light bulbs) come to characterize a population of groups (tribes, research institutes, or light-bulb manufacturers) as a result of competition between the groups. *Group selection* can also refer to processes in which social norms or cultural practices that help people cooperate with each other can serve as selection pressures for genes that make people more responsive to those cooperation-promoting social norms and cultural practices. And the list goes on.[10]

Adding to the pandemonium, scientists who use ideas based on group selection haven't always exercised an abundance of care to clarify how their versions of group selection compare to the others. There's an awful lot one could say about group selection, but in the remainder of this chapter, we'll have our hands full grappling with just three senses in which *group selection* can refer to the natural selection of genes for kindness. Does group selection reveal new evolutionary pathways for generosity to evolve that we couldn't have discovered with the conceptual tools in our Hamiltonian tool kit? Has group-selection thinking led to the discovery of rich reservoirs of generosity and selflessness that we never knew we had? Let's take a look.

## THE NEEDS OF THE MANY

Like many other kids of my generation, my first exposure to group-selection thinking came in the summer of 1982. In *Star Trek II: The Wrath of Khan*, Admiral Kirk, Captain Spock, and the rest of the original *Enterprise* crew matched wits with Khan Singh. Khan, their greatest nemesis from the original television series, was played by Ricardo Montalbán (the versatile Mexican actor and charismatic spokesperson for Chrysler automobiles). In the movie, the Starship *Enterprise* has been called into action because Khan has stolen the Genesis Device, a dreadful gadget with the power to reorganize all of the molecules in the nebula (transmogrifying them, perhaps, into a Chrysler Cordoba, with its soft Corinthian leather). After Khan succeeds in activating the device, the *Enterprise* needs to get as far away from it as possible, and as quickly as possible, but there's a problem: the *Enterprise* isn't going anywhere until someone fixes its warp drive, and the engine room has become flooded with radiation. Whoever fixes the warp drive is going to die. Then again, the entire crew is going to die if no one fixes it.

Mr. Spock takes matters into his own hands. He'll fix the warp drive, even though it means a certain, unpleasant death. He overcomes Dr. Leonard McCoy's resistance to his suicide mission with a standard Vulcan nerve pinch. After the repairs are completed, and Spock has been bathed in a fatal dose of radiation, the warp drive is fired up and the *Enterprise* speeds away from danger. Kirk rejoins Spock in time to watch helplessly as Spock takes his final breaths. Spock comforts his best friend with—what else?—logic:

> Spock: *Don't grieve, Admiral. It is logical. The needs of the many outweigh . . .*
> Kirk: . . . *the needs of the few . . .*
> Spock: . . . *or the one.*

One can only assume that they had riffed on that line a time or two before.[11]

The idea that nature might build organisms to act out of regard for the welfare of their groups, or even their entire species, was an anodyne of wild-animal documentaries and cigarette-break biological theorizing through much of the twentieth century. Why did the lion always catch the weakest wildebeest? Not only because the weakest wildebeest was the slowest of the herd (and therefore the easiest to catch), but also because the worn-out wildebeest had been programmed by natural selection to deliver itself into the jaws of predators so that the stronger and healthier members of the herd might be spared. The needs of the many outweigh the needs of the few. Why did some animals migrate en masse to new locales—even at substantial risk to themselves? Not only to increase their individual odds of finding food, but also to ensure that the species did not drive itself to extinction by extinguishing the food supply in the old locale. The needs of the many outweigh the needs of the few.

For his part, Darwin doubted it was possible to explain the evolution of several interesting animal traits without appealing to the operation of natural selection upon groups of individuals.[12] In *The Origin of Species*, for instance, he confronts the special difficulty of explaining the sterile castes of workers that characterize many insect communities. How could natural selection give rise to individual ants and bees that are destined never to have offspring of their own? In an effort to surmount this difficulty, Darwin invoked a notion we might call *family selection*: "This difficulty, though appearing insuperable," he wrote, "is lessened, or, as I believe, disappears, when it is remembered that selection may be applied to the family, as well as to the individual, and may thus gain the desired end."[13] Darwin solved the sterile castes problem, then, by observing that natural selection could act on parents to cause them to produce some offspring that are sterile: the sterile offspring could help the community carry out its work. He summarized this idea by focusing on the ants:

> We can see how useful their production may have been to a social community of ants, on the same principle that the division of labour is useful to civilised man. Ants, however, work by inherited instincts and by inherited tools or weapons, whilst man works by acquired knowledge and manufactured instruments.[14]

Twelve years later, in *The Descent of Man*, Darwin would turn again to group selection to explain humans' moral virtues. He found it difficult to believe that moral scruples could give anyone a selective advantage over other members of his or her own community, but he did believe that a tribe with many virtuous people could outcompete a tribe composed of people who lacked the same moral fiber. We might call it "tribe selection":

> It must not be forgotten that although a high standard of morality gives but a slight or no advantage to each individual man and his children

> over the other men of the same tribe, yet that an increase in the number of well-endowed men and an advancement in the standard of morality will certainly give an immense advantage to one tribe over another. A tribe including many members who, from possessing in a high degree the spirit of patriotism, fidelity, obedience, courage, and sympathy, were always ready to aid one another, and to sacrifice themselves for the common good, would be victorious over most other tribes; and this would be natural selection. At all times throughout the world tribes have supplanted other tribes; and as morality is one important element in their success, the standard of morality and the number of well-endowed men will thus everywhere tend to rise and increase.[15]

For the century following *The Origin of Species*, the notion that selection occurs at levels of biological organization above the level of the individual—in hives, mounds, families, and tribes, for instance—was heartily endorsed by many first-rate biologists.[16] However, by the 1950s, dissatisfaction with "needs of the many" group-selection theorizing began to coalesce. Many animal behavior researchers were trying to align more closely with the modern synthesis of Mendelian genetics and natural-selection thinking that was revolutionizing all of biology. Group selectionism was viewed by many as an impediment to this goal. After all, the neo-Darwinian synthesis affirmed that natural selection builds adaptations that maximize their bearers' reproductive success. The idea that evolution causes organisms to make sacrifices for the good of their group seemed inimical to that idea. For years, there was little to do about group-selection exuberance other than to write a scientific article or two and to grumble about fuzzy thinking. In 1962, however, a biologist named Vero Wynne-Edwards gave those who objected to group selection a more substantial punching bag: a 653-page book called *Animal Dispersion in Relation to Social Behavior*. The historian Mark Borrello dramatized the debate that ensued in his excellent book *Evolutionary Restraints: The Contentious History of Group Selection*.[17]

In *Animal Dispersion*, Wynne-Edwards proposed that natural selection comprises two distinct evolutionary processes that gave rise to two distinct kinds of adaptations. The first is a Darwinian component that produces adaptations enabling individuals to pursue the maximization of their own reproductive success. Nobody was going to disagree with this statement. But there was a second component to Wynne-Edwards's view of natural selection: "group selection."[18] Group selection, as Wynne-Edwards saw it, produces "group adaptations," some of which create "co-operative and social behavior, and consequently appear to deprive animals of the right to put their own self-interests first, forcing them to make concessions for the common good of their groups instead."[19] Wynne-Edwards explained his position in this way:

> Evolution at this level can be ascribed, therefore, to what is here termed group-selection—still an intra-specific process, and, for everything concerning population dynamics, much more important than selection at the individual level. The latter is concerned with the physiology and attainments of the individual as such, the former with the viability and survival of the stock or race as a whole. Where the two conflict, as they do when the short-term advantage of the individual undermines the future safety of the race, group selection is bound to win, because the race will suffer and decline, and be supplanted by another in which anti-social advancement of the individual is more rigidly inhibited.[20]

Wynne-Edwards's idea here is that natural selection could favor traits that reduce their bearers' reproductive success if those adaptations increase the reproductive success of their communities as a whole. Was it not reasonable to imagine, as Wynne-Edwards did, that the direct competition between breeding populations for really good habitats would favor groups whose members refrained from eating and drinking and reproducing themselves out of a decent place to live? After all, the needs of the many . . .

Wynne-Edwards saw group adaptations everywhere. Sex, as he saw it, was a group adaptation that enabled species to retain genetic variation as a rainy-day fund so that future selection pressures could be more easily surmounted. Animal communication systems were designed not solely for their benefits to individuals, but also for their benefits to groups of animals that needed to coordinate their activities and reproductive efforts. Delayed reproductive maturity at the beginning of life, along with aging and death on the back end, were also group-selected adaptations that kept genetic variance high and the risk of overpopulation low. He even saw group adaptations in animals' aggressive bouts over territories in which the winner gets a breeding site and the loser goes without: "Every step in this elaborate sequence is ritual and conventional," he wrote. "It provides a Utopian formula for selecting the best parental genotypes in the interests of posterity, with the minimum of effort and injury to the competitors. The adaptation is entirely a result of group selection, which sustains viable group gene pools."[21]

To better understand what Wynne-Edwards was arguing for, let's look at how his version of group selection might drive the evolution of reproductive restraint (producing fewer offspring than you could, in principle, raise to maturity) in birds. It's helpful to imagine a physical landscape that is subdivided into a number of discrete spatial units, or patches (as in Figure 5.1). Each patch can sustain five birds and not a single bird more. (It's important to point out that Wynne-Edwards was promoting a form of group selection that involved hypothetical groups of individuals that were not genetically related. Once you allow individuals within a patch to be genetically related, they could easily evolve self-restraint as a benefit to their genetic relatives. Under such a scenario, there would be no need to turn to a group-selection explanation at all: Hamilton's inclusive fitness theory would do just fine.) As soon as the sixth bird settles within the patch, the six birds will collectively eat their favorite prey item out of existence. As a result, the bird population within that patch collapses, and all the birds starve. After a generation of lying empty, devoid

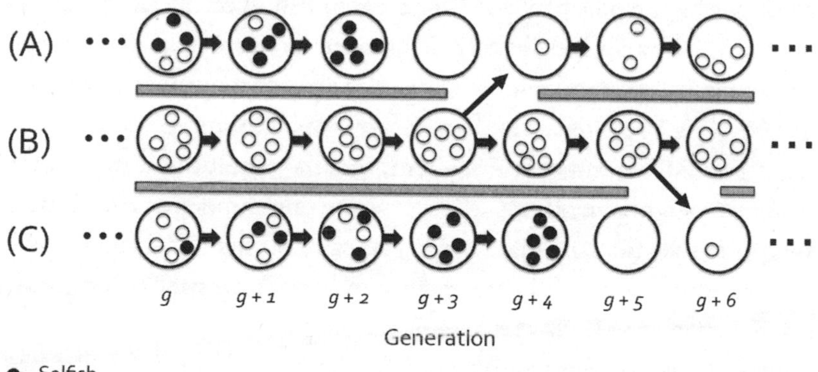


Figure 5.1. An idealized depiction of Wynne-Edwards's group selection.

of birds, the food supply within the patch replenishes and the patch can be recolonized by new birds who migrate into it from different patches. Migration between patches, in Wynne-Edwards's view, is possible only when a patch has been emptied out by a population collapse.

Now, let's imagine that these birds have two different variants of a gene that influences how much they eat. The *selfish* allele (depicted by small black circles in Figure 5.1) causes its bearer to eat as much and as often as possible. The *restrained* allele (depicted with small white circles) causes its bearer to eat with greater temperance. Within any single patch, a gluttonous bird will outreproduce a restrained bird because the glutton will convert its extra food into enhanced reproduction. Thus, within that patch, the selfish allele will increase in frequency as a result of natural selection. A patch with even a single glutton, as in the patches represented by rows A and C of the figure, is destined eventually to crash. By blindly pursuing their individual reproductive success, the gluttons contribute to the deaths of all the individuals within the patch, gluttons and nongluttons alike, including themselves. Too bad.

But now imagine the fate of a patch in which all five resident birds bear the restrained allele, as is the case in the patch represented by row B of Figure 5.1. Because they watch what they eat, they never consume their prey below replacement levels, and so never eat their descendants out of a place to live. Yes, each of these restrained individuals is going to produce fewer offspring than the gluttons in other patches, but so what? There will always be plenty of food for a patch full of restrained birds, so the patch will carry on indefinitely. Moreover, as nearby patches, previously occupied by gluttons who ate themselves out of existence, become replenished with prey items, and therefore available for recolonization, restrained individuals can colonize them, thereby spreading the restraint gene. Given enough generations, the restraint gene can defeat the glutton gene by group selection. In Figure 5.1, you can see this replacement process starting to happen within the top patch, row A, in generation $g + 4$, and in the bottom patch, row C, in generation $g + 6$.

*Animal Dispersion* elevated Wynne-Edwards to minor-celebrity status. Scientists and nonscientists alike took an interest in the book, and invitations to speak at conferences came pouring in from around the world. Two years after its publication, *Scientific American* published a small précis of the book that eventually sold 350,000 copies. The 1960s was the decade when people started to worry about humans' effects on the environment (Rachel Carson published *Silent Spring* in September 1962) and the specter of human overpopulation (Paul and Anne Ehrlich's *The Population Bomb* would come six years later). *Animal Dispersion*, with its idea that animals come into the world, courtesy of group selection, with instincts for reproductive restraint, appeared to offer a woolly sort of moral support for the idea that humanity was up to the task of solving its own existential crises.[22]

The attacks were not long in coming. Two scientists, one at Oxford and the other at Stony Brook University in New York, set to work on book-length rejoinders that, along with several mathematical treatments of group selection, would eventually succeed in taking back nearly all the rhetorical ground that Wynne-Edwards had temporarily captured for group selectionism.

David Lack, a renowned British ornithologist, responded vociferously to Wynne-Edwards's book. Always the individual-level adaptationist, committed to the article of faith that it was *organisms* (or, more specifically, the genes within organisms), rather than groups of organisms, that were the units of natural selection, Lack was the perfect foil for Wynne-Edwards. In a book-length rejoinder to *Animal Dispersion*, Lack reiterated that Darwinian natural selection, acting solely on individuals to build adaptations that maximize their lifetime reproductive success, was sufficient to explain the exact same phenomena that Wynne-Edwards marshaled as proof of group selection in *Animal Dispersion*.[23] No, birds do not delay the onset of reproductive maturity: Lack presented persuasive evidence that they begin breeding as soon as the lifetime reproductive benefits of doing so exceed the lifetime reproductive costs (parenting is draining for everybody). No, birds do not actively restrict the sizes of their broods to safeguard the food supply from overexploitation: Lack showed that they seek to maximize their reproductive success over an entire lifetime, which might mean reducing their egg production when food is scarce. Group selection, as Lack saw it, had nothing to do with these phenomena.[24]

The biggest problem with Wynne-Edwards's group selectionism, in Lack's view, was its lack of parsimony. Why resort to concepts like group selection and group adaptation to explain these behaviors, Lack asked, when individual-level natural selection could explain the same patterns of data with fewer conceptual moving parts? When scientific explanations conflict, conventional wisdom goes, the simpler one is better until someone shows that the more complex explanation can explain phenomena that the simpler explanation can't.[25]

On the other side of the Atlantic, the biologist George Williams had just taken aim at group selection in a book of his own. On page 1 of his *Adaptation and Natural Selection*, Williams depicted the previous century of evolutionary biology as a struggle between two groups of thinkers: those who viewed Darwinian natural selection as the primary driver of organism design, and those who viewed it as but one of many drivers that evolutionary biologists needed to reckon with. In Williams's mind, the debate had been settled thirty years previously when the evolutionary biologists Ronald Fisher, Sewall Wright, and J. B. S. Haldane produced their master works of mathematical population genetics. According to their solution, natural selection's effects on the reproductive success of individuals cause organisms to acquire *design*. As Williams saw it, evolutionary biology's prime directive should be to identify *adaptations*—those features of organisms that natural selection designed because of their contributions to reproductive success.

Williams spelled out two ground rules for identifying adaptations. The first was that the idea of adaptation should be used sparingly. Yes, natural selection builds adaptations, but lots of interesting traits are not adaptations. Hold the term in reserve until a mountain of evidence compels you to conclude that the trait came about by means of natural selection because it executed a function that led to greater reproductive success. The second ground rule was that adaptation should be assigned to no higher a level of biological organization than what the evidence requires. If a trait or behavior can be shown to reflect adaptation for individual reproductive success, then one needn't bother speculating that the trait evolved because of its effects on group reproductive success. Indeed, Williams argued that biologists should adopt the habit of *assuming* that natural selection took place at the simplest level of biological organization, "that of alternative alleles in Mendelian populations, unless the evidence clearly shows that this explanation does not suffice."[26]

Five words in that second ground rule—"alternative alleles in Mendelian populations"—was the proviso in *Adaptation and Natural Selection* that would eventually kill Wynne-Edwards-type group selection. In a chapter specifically devoted to the group-selection controversy, Williams argued that the fact that a population of insects (for instance) is stable does not license the view that the species possesses adaptations *for* group stability. Instead, the numeric stability of a population of insects could be a *side effect* of all the individual-level adaptations operating within a group. And those individual-level adaptations resulted from selective competition among "alternative alleles in Mendelian populations":

> We must decide: Do these processes show an effective design for maximizing the number, rate of growth, or numerical stability of the population or larger system? Any feature of the system that promotes group survival and cannot be explained as an organic adaptation can be called a biotic adaptation. If it does not, if its continued survival is merely incidental to the operation of organic adaptations, it is merely a population of adapted insects.[27]

A population of adapted insects is not, perforce, an adapted population of insects. A herd of fleet deer is not, perforce, a fleet herd of deer. A group of lifetime-fitness-maximizing individuals is not a lifetime-fitness-maximizing group of individuals. As far as Williams was concerned, groups were things that individuals adapted *to*—not units of adaptation in and of themselves. If there were such things as group adaptations, and not just group-level phenomena that arose from the interactions of individuals who possessed individual-level adaptations, Wynne-Edwards had certainly failed—spectacularly, in Williams's opinion—to make a persuasive case for them. "For Williams," as Borrello put it, "one cannot have group selection without biotic adaptations, and the rest of [his] book is dedicated to rejecting their existence."[28]

The nearly simultaneous publication of Lack's book and Williams's book was the radiation-flooded engine room that left Wynne-Edwards-type group selection struggling for its last breaths by the end of the 1960s. The death blow came a few years later from a few mathematically oriented theorists who studied the background conditions that would be required to make Wynne-Edwards-type group selection happen. Yes, they conceded: Wynne-Edwards-type group selection is possible in theory, but the assumptions are so restrictive that its appearance in nature would likely be quite rare.[29] You can get it when the individuals within groups are clonally related, or if competition within groups is completely suppressed. These conditions are satisfied within populations of bacteria (which can be considered to be genetic clones), for instance, or in eusocial insects. But most animal societies don't even begin to fulfill these conditions.[30]

To illustrate, let's return to our birds in their patches. It's easy to see that Wynne-Edwards-type group selection won't work if the gluttons can migrate into patches that contain restrained individuals. Should they begin to do so (as illustrated in Figure 5.2)—for example, when one glutton takes over a nesting site that was occupied by a restrained individual that has died (which is taking place between generation $g + 4$ and generation $g + 5$ in rows C and B)—the glutton will proceed to do what gluttons always do: eat every morsel it can get its little talons on. And as a result, the glutton will have more offspring than his restrained neighbors, and those offspring will also eat more—and reproduce more—than the offspring of their more restrained neighbors. Sooner or later, the gluttons will replace all of the restrained individuals and proceed to eat themselves out of a perfectly nice place to live. A group with even a single glutton is a group that's doomed for extinction.

"Well, okay," you might be thinking, "Couldn't this complication be solved if the group members evolved an adaptation that prevented selfish interlopers from invading the group—for example, an instinct to ostracize gluttons, or to punish them, or to erect a barrier to prevent them from

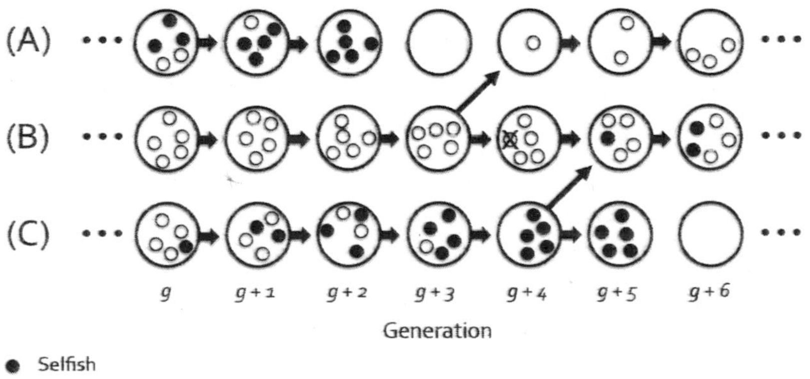


Figure 5.2. A more realistic depiction of Wynne-Edwards's group selection.

entering the patch in the first place?" It is certainly tempting to invoke these sorts of theoretical escape hatches to save Wynne-Edwards-type group selection, but how could you explain how the adaptations for ostracism, punishment, or wall-building evolved? For that, you'd need to invoke Wynne-Edwards-type group selection, which has the same problems here that it has in trying to explain the evolution of the restraint allele. We can't wriggle out of the problems with Wynne-Edwards-type group selection by appealing to other traits that have to evolve by Wynne-Edwards-type group selection.

Toward the end of the 1970s, Wynne-Edwards came to recognize these limitations. "In the last 15 years," he wrote,

> many theoreticians have wrestled with it, and in particular with the specific problem of altruism. The general consensus of theoretical biologists at present is that credible models cannot be devised by which the slow march of group selection could overtake the much faster spread of selfish genes that bring gains in individual fitness. I therefore accept their opinion.[31]

But Wynne-Edwards couldn't let go. His very last scientific article, titled "A Rationale for Group Selection," was more than eleven thousand words long. However, it made no mention whatsoever of the criticisms from Lack or Williams, or the devastating mathematical work from the 1970s.[32] Even so, by that time virtually everyone who understood the field knew that Wynne-Edwards's approach to group selection was, if not exactly dead, then not exactly worth thinking about anymore either.

## MULTILEVEL SELECTION

As Wynne-Edwards's version of group selection was sinking below the horizon, a new version of group selection was in the offing. One of its first proponents was a biologist named William Hamilton. Yes, that William Hamilton—the inclusive fitness one, of Hamilton's rule. Hamilton was not particularly impressed by Wynne-Edwards's "needs of the many" approach to group selection, but he was intrigued by some theoretical work on group selection from the 1930s, and his friend George Price had introduced him to some statistical machinery that convinced him that group selection was indeed viable.[33]

The basic form of "Price's theorem," which had actually been around for years before Price independently derived it, looks something like this:

$$\bar{w}\Delta\bar{z} = \mathrm{Cov}\,(w_i, z_i)$$

Despite the italics, the Greek letters, and the forbidding subscripts, Price's theorem manages to say something that's both simple and profound: generational changes in the average level of a trait (which is $\Delta\bar{z}$ on the left-hand side of the equation; the $\bar{w}$ in front of $\Delta\bar{z}$ is a normalizing constant that's not relevant here) are due to the statistical relationship (technically called a *covariance*) between the number of offspring each individual produces, $w_i$ (we attach the subscripted $i$ to show that each individual has his or her own unique number of offspring), and each individual's value for the trait, $z_i$. If higher levels of trait $z$ are associated with siring more offspring—that is to say, if Cov ($w_i$, $z_i$) is positive—then the average level of the trait among members of the next generation will be higher than its average level in *this* generation. As long as taller individuals have more kids than shorter individuals do, then the average height within the population will steadily increase, generation after generation, until increases in height no longer confer a reproductive advantage—that is, until the covariance between height and fitness goes to zero. Price's theorem communicates this mouthful of an idea in fourteen ASCII characters.[34]

The real value of Price's theorem, however, comes from its ability to illustrate how natural selection can operate in a world in which there are covariances between traits and fitness at intermediate levels of social organization (hives, mounds, families, and tribes) as well as covariances between traits and fitness at the level of individuals (bees, ants, aunts, and warriors). For example, a population of bacteria has an intermediate level of social organization if groups of individual bacteria inhabit different bodies; a population of caterpillars has an intermediate level of social organization if groups of individual caterpillars each live on different plants; and a population of people has an intermediate level of social organization if the population can be subdivided into distinct bands or ethnolinguistic groups.[35] To show how we can extend Price's theorem to multiple levels of biological organization, we will replace the $i$ subscripts with $k$s to keep up with our groups and with *j*s to keep up with the individuals within each of those groups. When we do so, we can create an expanded version of Price's theorem that looks like this:

$$\bar{w}\Delta\bar{z} = \text{Cov}\,(W_k, Z_k) + \text{E}_k(\text{Cov}_k(w_{jk}, z_{jk}))$$

Here, the covariance term, Cov $(W_k, Z_k)$, does not stand for the relationship between individuals' levels of the trait and individuals' fitness, but instead for the relationship between the average levels of the trait *for each of* k *distinct groups* (some of the groups, each of which gets a subscripted value from 1 to *k*, could have members that are taller, on average, than others groups do) and average fitnesses *for each of those distinct groups* (some groups could have members who have higher fitness, on average, than other groups do).

The second term in the right-hand side of the equation—$\text{E}_k(\text{Cov}_k(w_{jk}, z_{jk}))$—is an "Expectation," sort of like an average. But an average of what? The Expectation captures the average *within-groups* covariance between individuals' levels of the trait and individuals' fitness. In the expanded version of Price's theorem, the Expectation captures the relationship between individuals' trait levels and individuals' fitness *within the typical group*: it shows how the simplest version of natural selection—individuals' trait levels changing from one generation to the next as a function of how trait levels are associated with fitness—plays out, on average, within many different hives, mounds, families, or tribes.

This expanded version of Price's theorem brings a valuable perspective to thinking about natural selection in a world in which individuals reside in groups. It gets its value from the fact that a gene could, in principle, reduce its bearers' fitness in comparison to the other members of the groups within which they reside (that is, the Expectation term could be less than zero), while simultaneously raising the average reproductive fitness of the group as a whole relative to other groups (that is, the covariance term could be greater than zero). Selection can act at multiple levels of biological organization, and the expanded version of Price's theorem is a perfect way to understand this sort of *multilevel selection*.[36] It is also useful for showing one way in which cooperation might evolve within a group-structured population—even if cooperation reduces individuals' direct reproductive success. The biologists David Sloan Wilson and Edward O. Wilson summarized multilevel selection for the evolution of cooperation with the following maxim: "Selfishness beats altruism within groups. Altruistic groups beat selfish groups. Everything else is commentary."[37]

To better grasp how multilevel selection might lead to the evolution of a cooperative trait, let's walk through a simple example. For concreteness (for which we will pay by sacrificing biological realism; no primate has a life cycle like what I am about to describe), imagine a population of ten thousand individual monkeys that cluster into one hundred different groups. The life course of this particular species of monkey consists of two distinct stages. In Stage 1, they live in groups. Here is where they will grow and develop and receive their exposure to all the hazards and rewards of life that will affect their survival. If they survive Stage 1, they'll get to move on to Stage 2, during which they will reproduce. Their offspring will then return to the isolated groups of Stage 1 and begin the two-stage process all over again.

For this example, assume that these ten thousand monkeys vary with respect to a gene that has two variants (alleles)—the dominant one represented by an uppercase *A* and the recessive one by a lowercase *a*. Monkeys with the dominant *A* allele engage in a helpful behavior that reduces their reproductive success within groups, but increases the reproductive success of the other individuals in their groups by an even higher degree. Let's imagine, as David Sloan Wilson did, that when predators or enemies approach, monkeys with the dominant *A* genotype produce an *A*larm call. Individuals with the recessive *a* genotype do not.[38] Alarms are valuable to receive, because they give you extra time to hide or flee from an approaching threat. However, alarm calls are costly to produce: predators also have ears, and your alarm calls will tell them exactly where you are.

Let's also assume that the proportions of dominant *A* and recessive *a* individuals differ across the one hundred groups. Some groups are composed almost entirely of dominant *A*s—monkeys that are willing to produce alarm calls—whereas other groups are composed almost solely of recessive *a*s that do not produce alarm calls. Initially, at least, most groups will represent a more balanced mix of the two genotypes.

Using the expanded Price equation, we can show that fitness changes across the entire ten-thousand-individual population can be traced to two independent actions of natural selection operating at the two different levels of social organization. First, natural selection operates at what Hamilton called the *intra-group* (or within-groups) level. Monkeys that produce alarm calls will reduce their chances of survival relative to members of their groups that do not produce alarm calls. With dominant *A* carriers providing benefits to others at their own expense, the dominant allele's effect on its bearer's fitness at the within-groups level is *negative*. Thus, natural selection disfavors alarm calls within groups: you're better off taking advantage of the information you gain from the alarm calls that other individuals produce, while refraining from producing alarm calls yourself.

However, natural selection is also operating at that second, *inter-group* (between-groups) level. Groups with lots of dominant *A* carriers get regular warnings about oncoming predators, so across the entire one-hundred-group population, the survival *rate* will tend to be higher (even for alarm-callers) for groups with lots of alarm-callers than for groups with fewer (or no) alarm-callers. Alarm calls are therefore bad for the survival of the individuals that produce them, relative to the other members of their groups that don't produce them, but good for the *average* survival of groups that have at least some individuals that are willing to produce alarm calls.

At the end of the first stage of the life cycle, when all the growth and development takes place, the surviving monkeys leave their Stage 1 groups and enter Stage 2, when they mate and reproduce. If you survive Stage 1, which is made less likely by possessing the dominant *A* allele yourself, but more likely if you come from a group that has lots of individuals who do possess the dominant *A* allele, then you get to reproduce during Stage 2. After Stage 2, you die, and your descendants move into Stage 1 of their own life cycles, forming groups of their own.

With the help of Price's theorem, Hamilton and Wilson were able to illustrate something important and extremely counterintuitive about evolutionary scenarios like these: if the fitness advantage of living in a group with lots of alarm-callers sufficiently exceeds the fitness disadvantage that alarm-callers incur within their groups, then the *absolute number* of dominant *A* individuals who make it into the Stage 2 migrant cloud can increase over many successive generations, even though their frequency within each of the one hundred groups declines with each successive generation. As a result, natural selection's action at the two different levels of social organization can ratchet up the total number of dominant *A* individuals within the ten-thousand-individual population until it reaches some stable proportion. Selfishness beats altruism within groups. Altruistic groups beat selfish groups. Everything else is commentary.

Now for some commentary.

In principle, multilevel selection can indeed lead to the evolution of cooperative traits. But the initial conditions have to be just right. Whether the two opposing forces of natural selection—selection against alarm calls within groups, and selection for alarm calls between groups—will lead to the evolution of alarm calls in the long run depends on the size of the benefit that an alarm call affords its hearers as well as the size of the cost associated with producing those calls. Hamilton's *c* and *b* matter here, just as elsewhere, for the evolution of cooperation.

Another crucial factor is the proportion of dominant *A* individuals within the groups. Consider an extreme example. If, by chance, the proportion of alarm-callers within a single group drifts its way to 100 percent and stays there, there will be no within-groups selection against alarm calls within that group: each individual within the group will gain the same benefits from receiving the others' alarm calls and incur the same costs from producing them. With the abolition of within-groups selection, which works against the evolution of alarm calls, this pure group of alarm-callers could easily overtake the other groups by outreproducing them during Stage 2 of the life cycle.

However, the tendency for groups with lots of alarm-callers to defeat groups with fewer alarm-callers generates a wrinkle of its own. As groups with lots of alarm-callers begin to fill up the population landscape, the groups will all come to have similarly large proportions of alarm-callers. This is a problem, because multilevel selection requires between-groups variation over multiple generations. Price's theorem makes this much clear: natural selection, whether it is operating at the within-groups level or the between-groups level, runs on covariances between genes and fitness at the level in question. Indeed, this is a fundamental principle of statistics. Men are reliably taller than women, but if you have a group of ninety-nine men and one woman, the covariance between sex and height will be close to zero, because there's so little variance in sex—even if that woman is very short indeed, and even though everybody knows that women on average are shorter than men. Raise the variance in sex (by subtracting forty-nine men and adding forty-nine women, leaving fifty of each sex), and the covariance between sex and height will increase, even though you've done nothing to change the average heights of men and women in the population. Likewise, without between-groups variance in the frequency of the dominant *A* gene, you can't get a between-groups covariance between the dominant gene and fitness.

Let's consider another extreme case. Imagine that between-groups selection has wrung all of the recessive *a*s out of our population of monkeys so that every one of the ten thousand monkeys in the population now carries the dominant *A* allele. At first glance, this seems like good news for the alarm-call gene, but any claim of victory for niceness would be premature. With the abolition of selection between groups, what's left is selection within groups, permitting selfish mutant genes that prevent their bearers from making alarm calls to evolve their way back into the population.

William Hamilton and David Sloan Wilson both discovered a way to iron out this wrinkle: we can stipulate that alarm-callers like to associate with other alarm-callers, and that non-alarm-callers like to associate with other non-alarm-callers. As a result of these preferences, individuals of each genotype end up living with others that share their genotype. This process, known as assortation, could occur for a variety of reasons. Perhaps the gene that causes you to produce alarm calls also causes you to enjoy the sensory properties of alarm calls. Or perhaps dominant *A* individuals like to live at the beach, but recessive *a* individuals like to live in the mountains. It doesn't matter why individuals assort: the math requires only that they *do*.[39]

Assortation during the return to Stage 1 of the life cycle, however it arises, means that the resultant groups will be much more homogeneous than if they had formed by chance. As a consequence, the within-groups covariance between the dominant *A* gene and fitness will be kept to a minimum, which also minimizes within-groups selection against it. At the same time, assortation will maximize the variance in the dominant gene between groups, which increases the between-groups covariance between the dominant gene and fitness, which in turn increases between-groups selection for the gene. The result is that it becomes both less costly to an individual within the group to produce alarm calls and more beneficial to the individual to be associated with a group in which lots of individuals produce alarm calls.

With these conditions in place—behaviors that disadvantage individuals within their living groups, while also advantaging their groups in comparison to other groups, and a demographic process that causes individuals within groups to be more genetically similar to each other than they are to individuals in other groups—we now have a workable recipe for the evolution of cooperative behavior through natural selection's work at two distinct levels of biological organization. Selfishness beats altruism within groups, but altruistic groups beat selfish groups. The needs of the many outweigh the needs of the few.

Bear in mind, however, that ideas like these can also be expressed in the Hamiltonian language of direct fitness and indirect fitness. Once individuals have assorted themselves into majority-dominant-*A* and majority-recessive-*a* groups, after all, the alarm-callers' actions will tend to benefit genetic relatives—that is, other dominant *A* individuals (even though some nonrelatives, who bear the recessive *a* genotype, could exist within the alarm-callers' groups as well). Whether you use the Price equation, with its variances and covariances, or Hamilton's rule, with its costs, benefits, and relatednesses, you end up predicting the same thing: natural selection can favor a gene that reduces its bearer's fitness relative to other members of its group if that gene also raises the fitness of others who possess that same gene. Multilevel selection and inclusive fitness give exactly the same answers, always. Price's theorem, $\bar{w}\Delta\bar{z} = \mathrm{Cov}\,(w_i, z_i)$, says nothing with its fourteen ASCII characters that Hamilton's rule, $rb - c > 0$, can't say in six. Despite the hype in the social sciences around the concept of multilevel selection over the past few decades, it has not—and, indeed, it could not—yield new insights into the origins of human concern.[40]

## KILLING THEM WITH KINDNESS

Over the past couple of decades, group-selection enthusiasts have been looking into the possibility that our concern for the welfare of strangers emerged through, of all things, inter-group warfare. The scenario they envision has a few moving parts. First, they speculate, our ancestors' willingness to make sacrifices on behalf of their comrades turned their tribes into more efficient war machines: as a result of their improved fighting ability, they became better at exterminating rival tribes whose members weren't so willing to make such sacrifices. And because the nonsacrificial tribes got slaughtered, the instinct for self-sacrifice on behalf of the tribe became a universal human trait.[41]

The conjecture that human generosity emerged from a coevolutionary union of hatred for our enemies and love for our comrades has been given a winsome twenty-first-century name—*parochial altruism*.[42] But guess who made the same conjecture in 1871:

> When two tribes of primeval man, living in the same country, came into competition, if (other circumstances being equal), the one tribe included a great number of courageous, sympathetic and faithful members, who were always ready to warn each other of danger, to aid and defend each other, this tribe would succeed better and conquer the other. . . . A tribe rich in the above qualities would spread and be victorious over other tribes: but in the course of time it would, judging from all past history, be in its turn overcome by some other tribe still more highly endowed. Thus the social and moral qualities would tend slowly to advance and be diffused throughout the world.[43]

Darwin didn't know what a gene was, so he couldn't take a gene's-eye view on anything, much less on the role of intertribal warfare in the evolution of compassion. But we do, so we can. And when we take that gene's-eye view, we discover that parochial altruism—"individual payment of fitness costs in order to improve the group's success in warfare"—is also, in Hamiltonian inclusive fitness lingo, "individual payment of direct fitness costs that also increase one's indirect fitness, thereby increasing one's inclusive fitness." There's nothing about parochial altruism that can't be explained (and more revealingly, in my opinion—but that's just an opinion) with the *r*s, *b*s, and *c*s of inclusive fitness theory.

Suppose you possess a gene for parochial altruism that causes you to provide benefits to your comrades at some cost to yourself. As a result of your sacrifices, your tribe becomes a more effective fighting unit than it was before, which boosts its success in vaporizing its enemies. Once you've defeated an enemy tribe, your tribe can either grow larger (by annexing the losers' territory and resources) or splinter off a new group that colonizes the extinct group's territory. Either way, your gene for parochial altruism spreads across the population landscape even though that same gene puts you at a fitness disadvantage in comparison to members of your tribe who lack it: you make sacrifices on behalf of your tribe, but they don't.[44] That's okay: selfishness beats altruism within groups, as we well know, but altruistic groups beat selfish groups. It's multilevel selection at its finest.

But it's also regular old inclusive fitness maximization at its finest. If my parochial altruism gives my tribe a big enough competitive edge, then the gene responsible for my parochial altruism has a survival edge over the non-parochially-altruistic genes of the tribes we defeat: after all, they won't be able to reproduce after we've killed them. Additionally, if warring tribes are genetically isolated, so that they avoid intermarrying (and bringing non-parochially-altruistic genes into tribes that have lots of parochially altruistic genes already), then tribes with parochial altruists will become ever thicker with parochial altruism genes as the generations tick by. As a result, parochially altruistic sacrifices will increasingly benefit *other individuals who also possess the parochial altruism gene*, thereby fetching *indirect* fitness benefits for the parochially altruistic gene by promoting the direct reproductive success of its relatives.

The notion of natural selection for self-sacrificial traits that promote success in battle is not a far-fetched one, at least not from a mathematical point of view. The big problem with the idea of parochial altruism is not mathematical to begin with, however: it's behavioral. The "sacrifice for the good of the group" that scientists have in mind in their models of parochial altruism is just a mathematical entity that stands for some sort of behavior that promotes combat effectiveness at some cost to the individual performing the behavior. What is that behavior, exactly? The models don't say. It could be a warm and fuzzy trait, such as sharing food, or helping injured comrades on the battlefield. But it also could be something more directly combat-related—nerves of steel, for instance, or a strong stomach for gore, or a lust for blood and mayhem, or a seemingly irrational willingness to rush headlong into a volley of enemy arrows. The only thing the mathematical models tell us is that the parochially altruistic behavior creates a fitness cost within tribes while creating an even larger fitness benefit between tribes. The altruism could be something nice, but it's just as likely to be something nasty.[45]

I'm placing my bet on "something nasty." To be sure, sharing food and helping injured comrades will help your tribe defeat the enemy (while also, possibly, getting you killed), but bravery, ferocity, and an appetite for risk have the same effects. Yet there is an important difference between parochially altruistic traits of the menacing variety and those of the compassionate variety: the menacing traits might also help you collect the spoils of war once the enemy tribe has been destroyed; the compassionate traits won't. Wouldn't natural selection, miser that it is, use the same traits for the two jobs, instead of fashioning a set of compassionate traits to support the killing and a set of menacing traits to support the looting and pillaging? Perhaps it depends on what the spoils of war actually are. In fact, there are only two spoils of war that directly translated into improved reproductive success for our ancestors, so these are the resources that natural selection designed them to care about. The first was the enemy's territory. The second was their women.[46]

When a group of warriors arrogate their enemies' women for themselves, it looks an awful lot like rape and sexual slavery. That's because it is rape and sexual slavery. When hunter-gatherers go on raids against other groups, they often do so for the express purpose of raping the enemy camp's women or taking them as "wives." Rape and sexual violence in small-scale societies, particularly in the context of inter-group conflict, is such a banal form of evil that the anthropologist Donald Brown was forced to include it among his "human universals"—traits so common across societies that they can be considered species-typical.[47]

As an example of the close connection between war and rape, consider an 1870 interview between two anthropologists and a man from Australia's indigenous Kurnai people, as the latter recounted a nighttime raid against an enemy camp:

> Two spies went first. Two other spies who had gone on now met them. "Where are they?" They reply, "Just here." The dawn was coming. Then all rapidly painted themselves with pipeclay—red ochre is no use, it cannot frighten an enemy—and divided, so as to surround the camp. The spies whistled like birds, to tell when all was ready. Then all ran in;

> they speared away, they speared away! They only speared the men, and perhaps some children. Whoever caught a woman kept her himself.[48]

Young, unmarried men around the world tend to be restless, belligerent risk-takers. It's a perverse set of traits that increases their risk of dying in combat while also increasing their appetite for going into combat in the first place. Why would young men possess traits that tend to get them killed at the end of a spear? Israeli military scholar Azar Gat has argued that the risks associated with possessing such traits are vastly overshadowed by the potential rewards that come from success in combat. The young men left standing at the end of the battle would have been more than repaid for their risk-taking by capturing territory, livestock, or, horrifically, one or more mates from the decimated tribe. "The interconnected competition over resources and reproduction," Gat wrote,

> is the *root* cause of conflict and fighting in humans, as in all other animal species. Other causes and expressions of fighting in nature, and the motivational and emotional mechanisms associated with them, are derivative of, and subordinate to, these primary causes, and *originally* evolved in this way in humans as well.[49]

Wartime rape and sexual slavery aren't just faint echoes from our deep ancestral past. They are also scrawled all over the earliest written accounts of war. After examining the history of wartime rape, the literary scholar Jonathan Gottschall was left with

> the distinct impression that whenever and wherever men have gone to war, many of them have reasoned like old Nestor in the *Iliad*, who concludes his pep talk to war-weary Greek troops by reminding them of the spoils of victory: "So don't let anyone hurry to return homeward until after he has lain down alongside a wife of some Trojan."[50]

And as rare or as common as you believe rape to have been throughout human history, the fact that the frequency of rape skyrockets during war even today is beyond dispute.[51] In the twentieth century alone, military or paramilitary groups in more than forty nations committed millions of rapes.[52]

We don't know what sorts of behavioral traits got created by an evolutionary process that involved parochial altruism, and we never will. For that matter, we can't even know whether parochial altruism was a vehicle for natural selection in the first place. Recall that parochial altruism is a concept that may exist inside mathematical models and nowhere else. But assuming that parochial altruism really was a feature of our evolutionary history, it seems as likely as not that it outfitted us with behavioral traits that make us cruel, not kind.

## ANOTHER DEAD END

What should we make of E. O. Wilson's assertion that group selection turned humans into a eusocial species? Not much, actually. The model of group selection that Vero Wynne-Edwards championed so vigorously in the 1960s and 1970s is incorrect under most biological scenarios. And when it is correct, it manages to be correct only because some force of nature has eliminated reproductive competition within groups. Individual bacteria from a single bacterial species can't compete because they are genetic clones of each other: their reproductive interests are perfectly aligned. And the eusocial insects eliminate reproductive competition by assigning the community's reproductive tasks to one or more queens. Needless to say, humans aren't genetic clones of each other, and we don't assign reproduction to a small minority. Human societies just don't work that way.

What about the type of group selection that can take place in organisms with intermediate levels of social organization—bacteria within bodies within a population of bodies, for instance, or caterpillars on different bushes within a population of bushes, or monkeys within different colonies within a population of colonies? The operation of natural selection in communities like these can be helpfully visualized in terms of multilevel selection, but the multilevel perspective doesn't identify new resources for understanding the evolutionary basis of human generosity. How could it? "Selfishness beats altruism within groups, but altruistic groups beat selfish groups" is a pithy thing to say, but there's nothing behind that trope that couldn't also be said using the Hamiltonian language of inclusive fitness and kin selection. When a gene motivates us to assist the members of our group at a cost to ourselves, what it is really doing is boosting the reproductive success of its genetic relatives. Multilevel selection at its finest is also Hamiltonian kin selection at its finest. And as we saw in Chapter 4, kin selection doesn't get us very far in explaining our modern concern for distant strangers.

Models of group selection that are based on head-to-head competition between groups aren't helpful, either. Despite Darwin's speculations, it's not clear that natural selection endowed us with an appetite for self-sacrifice so that we could rape and pillage more effectively. Maybe it did, maybe it didn't. And even if it did, there's no telling whether those self-sacrificial tendencies look like compassion for one's fellows or eagerness to become cannon fodder.

If the various models of group selection prove anything, it's this: the hypothetically group-selected genes that are responsible for all the hypothetical things that group-selection theorists have attributed to them are merely doing things that favor their own replication. They do these things either through the bodies in which they reside or by helping the bodies of other people who bear exact copies of those same genes. It couldn't be otherwise: promoting their own replication is what genes do. Go to the ants all you want, thou sluggard. You still won't learn why humans (sometimes) take care of strangers.

CHAPTER 6

# THE BIG PAYBACK

Tit for tat. Quid pro quo. An eye for an eye. Scratch my back and I'll scratch yours. One good turn deserves another. Turnabout is fair play. Payback is hell. I owe you one. You'll pay for this. Right back atcha. Measure for measure. A taste of your own medicine. One hand washes the other. Payback. Getback. The feeling is mutual. It takes one to know one. Requite, repay, revenge, reward, retaliate. As you give, so shall you receive. Sow the wind, reap the whirlwind.

Did you get all that? Of course you did. Everyone speaks the language of reciprocity. It is a language that's so widely spoken that the anthropologist Donald Brown could declare reciprocity a "human universal."[1] It's so useful, so easily mastered, and so vital to how we survive and thrive in this world that children and even chimpanzees can understand it.[2] Reciprocity also has some insights to offer us in a search for the origins of humans' regard for the welfare of strangers—more than the evolutionary forces of kin selection and group selection, at any rate—so we will be rewarded for taking the time to understand it more deeply.

Darwin was quite impressed by many animals' tendencies to help unrelated individuals, observing that "the social instincts lead an animal to take pleasure in the society of its fellows, to feel a certain amount of sympathy with them, and to perform various services for them." But how did these social instincts for "performing various services" for one's fellows evolve in the first place? He suspected that reciprocity had some kind of role to play. In *The Descent of Man*, he casually observed that as humanity became better fitted for group life, "each man would eventually learn that if he aided his own fellow-man, he would commonly receive aid in return."[3] He was a bit more explicit in a short footnote, in which he quoted the Scottish philosopher Alexander Bain on the subject:

> Mr. Bain states that, "sympathy is, indirectly a source of pleasure to the sympathizer"; and he accounts for this through reciprocity. He remarks that "the person benefited or others in his stead, may make up, by sympathy and good offices returned, for all the sacrifice."[4]

But that's about it. Other than a couple of short asides and a conscientious footnote, Darwin wrote little or nothing on the topic of reciprocity. The world would have to wait a century for a proper evolutionary treatment of reciprocity and the psychological faculties with which it endowed us. The credit for this achievement is due to an evolutionary biologist named Robert Trivers. In 1969, while still a PhD student, Trivers set out to determine whether humans could obtain direct fitness benefits—and not merely the indirect fitness benefits that Hamilton had discovered—as a consequence of rendering benefits to nonrelatives. Specifically, Trivers wanted an answer to this question: Is there a way for natural selection to favor the evolution of a psychology that causes me to render a favor to you, even when I incur an immediate cost by doing so? Trivers thought an answer in the affirmative was possible, and that it lay in the notion of reciprocity. He published a systematic treatment of this idea in 1971—exactly one hundred years after Darwin wrote his one hundred words on the subject.[5]

To understand Trivers's ideas, it's helpful to start with a thought experiment. Suppose I possess a mutant gene that causes me to dispense benefits to my neighbors indiscriminately, without regard to how they treat me in return. This gene is obviously going to be an evolutionary dead end because it will motivate me to deliver benefits to others who have no corresponding tendency to provide benefits to me in return. In such a case, a stingier gene wins out, and no propensity to deliver benefits to nonrelatives will evolve. Likewise, imagine that the population starts out with the gene that motivates benefit-delivery toward neighbors, but a mutant gene arises that causes its bearer to take benefits from neighbors without dispensing them. In this scenario, the stingy mutant will take over the population. From this gene's-eye view, a generalized willingness to provide benefits to one's neighbors, irrespective of kinship, looks like a nonstarter.

This depressing conclusion reminded Trivers of the *prisoner's dilemma*, which derives its name from another thought experiment. Imagine that the police have arrested two thugs—Tom and Jerry—on suspicion that they committed a burglary. The police don't have enough evidence to charge either of them, however, so they question them individually, in hopes of getting one of them to implicate the other. If Tom betrays Jerry, Tom will be let free and Jerry will receive a three-year sentence. If neither suspect betrays the other, both will be charged with a lesser crime—trespassing, perhaps—for which they will each get a one-year sentence. If each implicates the other for the burglary, they'll both get two-year sentences (Table 6.1).

| | **If Jerry denies Tom's involvement** | **If Jerry implicates Tom** |
|---|---|---|
| **If Tom denies Jerry's involvement** | Tom gets a 1-year sentence<br>Jerry gets a 1-year sentence | Tom gets a 3-year sentence<br>Jerry goes free |
| **If Tom implicates Jerry** | Tom goes free<br>Jerry gets a 3-year sentence | Tom gets a 2-year sentence<br>Jerry gets a 2-year sentence |

Table 6.1. Prisoner's dilemma outcomes for two accused burglars.

If Tom and Jerry both stay quiet, they're better off, on average (each gets a one-year sentence), than if one betrays the other (one is let free and the other receives a three-year sentence) or if they betray each other (each gets a two-year sentence). So what should they do? Should they hang together or risk hanging separately? It's a tough choice—that's why it's called the "prisoner's dilemma" and not the "prisoner's easy decision."

Game theorists' analyses of the prisoner's dilemma in the 1950s had led to an unsettling conclusion: assuming you are a rational agent, and your primary interest is in maximizing your own welfare, you should always defect. It's the decision you cannot regret, no matter what your accomplice chooses to do. As the left column of the table shows, Tom goes free, instead of serving a one-year sentence, by implicating Jerry, assuming Jerry denies Tom's involvement. And as the right column shows, Tom knocks a year off a three-year sentence by implicating Jerry, assuming Jerry chooses to implicate him as well. Either way, Tom saves a year in prison by cooperating with the police rather than with Jerry. And Jerry, being no less rational and no less self-interested than Tom, undoubtedly comes to the same conclusion. As a result, they both end up serving two-year sentences. Each ends up with a worse deal than was hypothetically available—that is, if they had hung together—but they got the best deal possible in a world of rational actors. It's a dismal conclusion, but too bad. The math doesn't lie.[6]

Subsequent research on the prisoner's dilemma produced a more hopeful insight, however. When flesh-and-blood human beings actually play the game with a partner, they're irrationally nice: they cooperate with each other most of the time. In fact, on the afternoon of the very day in 1950 when the game theorists Merrill Flood and Melvin Dresher invented the prisoner's dilemma, they had two human subjects play the game for one hundred rounds (to earn real money instead of to avoid "prison time"). Each player chose to cooperate with the other player on about seventy of those rounds, which is seventy rounds more than a purely rational actor should.[7]

To the game theorists, humans' irrational tendency to cooperate was an interesting curiosity, but hardly noteworthy from the point of view of Capital-S Science: So what if people aren't rational? But the biology graduate student Trivers thought he saw something in the prisoner's dilemma game that had profound evolutionary significance. Trivers recognized that natural selection tallies up organisms' wins and losses at the end of their lives, not somewhere in the middle. Thus, an organism's "score" in any single prisoner's dilemma in the real world—measured, say, as the amount of food it shares in a social interaction with a nonrelative—is meaningful from an evolutionary viewpoint only to the extent that it influences the final outcome: total lifetime inclusive fitness.

In other words, Trivers realized that natural selection doesn't care whether you cooperate with a noncooperator on a single occasion, or two, or twenty, or twenty thousand. All it cares about are the *lifetime* costs a gene imposes upon its bearers and the *lifetime* benefits the gene accumulates for its bearers. A gene that promotes defection in the prisoner's dilemmas of life, even when one's interaction partners possess a gene that promotes cooperation, might get the upper hand over the cooperative gene for a single moment in time, but it could still be an evolutionary loser if it subsequently reduces its neighbors' willingness to bestow *additional* benefits in the future. Perhaps people are so irrationally cooperative in the prisoner's dilemma, or in real-life situations that look like prisoner's dilemmas, because their irrational levels of cooperation enable them to coax benefits out of others over longer time scales.

Trivers proposed that a gene for generosity might implement the following strategy: Help your neighbor when it is cheap for you to do so. Ask your neighbor for help when you need it. Continue providing benefits to neighbors who return favors, and stifle your generous impulses toward neighbors who don't. Trivers postulated that genes that implement such a strategy could evolve even in a population that was already dominated by nonreciprocators—because the reciprocators would get the benefits that other reciprocators extended to them, and they would keep their own costs down by responding with stinginess toward individuals who did not reciprocate when called upon. The game theorists had already discovered this strategy before Trivers, and they had named it "tit for tat."[8] To Trivers, it seemed like a system for benefit-delivery that could evolve by natural selection. The math looked right, at any rate. But would a formal evolutionary treatment bear out his intuitions? He didn't even know if a formal evolutionary treatment was possible.

## THE DIGITAL EVOLUTION

While Robert Trivers was puzzling about the prisoner's dilemma and its relevance for the evolution of generosity, another Robert was wondering about optimal strategies for the *iterated* prisoner's dilemma—a series of repeated games with the same partner. What's the best way to play a game like that? An infinite number of strategies are possible: you could cooperate in every encounter, defect in every encounter, defect in every hundredth encounter, alternate between defection and cooperation, or cooperate when your partner does certain things and defect when your partner does other things. Which strategy wins against all others? Robert Axelrod, a political scientist, resolved to find out. He organized a prisoner's dilemma tournament and invited other experts in game theory to submit computer scripts that would implement their favorite strategies. (Trivers was invited to submit a strategy, but he declined, pleading computer illiteracy.) Fourteen entries were submitted in total. In the tournament, each strategy was pitted against each of the others in iterated games consisting of two hundred prisoner's dilemmas. Each strategy's performance was based on the average number of points it scored in all the matches it played.

In the end, tit for tat emerged victorious. Like Trivers's ideal reciprocator, tit for tat cooperated in Round 1, and from there on out, it cooperated on each successive round as long as its partner had cooperated on the previous round. If its partner defected, then tit for tat defected on the next round. If the defector ever returned to cooperation, then tit for tat returned to cooperation as well.[9] The secret to tit for tat's success, then, was that it started out nice, retaliated against defectors, and returned to cooperation with defectors who mended their ways. Axelrod publicized the results, and then he solicited entries for a second tournament. This time around, sixty-two entrants submitted strategies. Tit-for-tat won again.[10]

Axelrod had become aware that evolutionary biologists such as Trivers were thinking about the prisoner's dilemma and its evolutionary implications, so after the results of the second tournament were in, he wanted to examine whether tit for tat could outcompete other strategies when playing for resources that translated directly into reproductive success. To do so, Axelrod teamed up with William Hamilton, the discoverer of inclusive fitness, to turn the tournament into a simulation of gene-frequency change in a population—not exactly a model of natural selection, because they failed to build in a mutation process, but a baby step in that direction. Once again, tit for tat came out on top.[11] Nice guys, it seemed, really could finish first.

## LESS SPHERICAL COWS

Before long, however, researchers who were interested in applying game theory to social evolution realized that cooperation in the real world possessed some important features that the prisoner's dilemma models had been ignoring. Scientists' tendency to remove the complexities of reality from their models is so typical that engineers have a joke about it. A farmer comes to a theoretical physicist for advice on how to boost his cows' milk production. After working on the problem for a few weeks, the physicist calls the farmer to tell him that she has found an answer. She asks the farmer, "Can you come over to the university next Wednesday to hear me present my discovery at a departmental seminar?" He'd be honored. The next Wednesday, the farmer shows up to find a packed seminar room, with his physicist friend at the chalkboard. As the farmer takes his seat and the crowd settles down, the physicist turns to the board to draw a bold, jaunty circle. "First," she pronounces, "we assume a spherical cow."[12]

Which assumptions really matter and which ones don't? Does it matter that the model cow is a sphere-shaped thing and not a cow-shaped thing? It's not always easy to know. We do know, however, that Axelrod and Hamilton made a simplifying assumption about reciprocal cooperation that did matter: they neglected a fact of life that game theorists call "noise."[13] It's unrealistic to assume that organisms will perfectly implement their strategies and perfectly perceive the intentions of their interaction partners. Instead, we must assume that their actions and perceptions have some noise in them: actors sometimes defect when they mean to cooperate, and the perceivers of their actions sometimes mistake their partners' cooperative overtures for defections. If you ignore the possibility of noise when you use the prisoner's dilemma to model reciprocal cooperation, you get misleading conclusions. Even when errors in implementation and perception are rare, two players who are playing a tit-for-tat strategy will eventually encounter a round of play in which Tom defects when he intended to cooperate, or else Jerry misperceives Tom's genuinely cooperative action as a defection. When these errors occur, Jerry, who is also playing a tit-for-tat strategy, will defect in response. And because Tom is implementing a tit-for-tat strategy, Tom will subsequently respond to Jerry's defection with a reciprocal defection. What follows, of course, is another defection from Jerry. And then another defection from Tom. And so on and so on, ad infinitum. This annoying descent into madness was named the *echo effect*.[14] It's an assumption about the nature of reciprocal cooperation that any reasonable model simply cannot set aside.

Once scientists recognized the need to build the echo effect into their models of the prisoner's dilemma, they set off in search of strategies for the prisoner's dilemma that were better at handling noise. One pair of mathematical biologists found evidence that the problem could be solved with strategies that forgive some of their partners' defections, but they had to change their minds after they discovered an even better strategy that shifted between cooperation and defection purely in response to its payoffs from the previous round.[15] Axelrod challenged these researchers' results in a subsequent project, concluding that cooperative strategies that occasionally forgive really are best.[16] All of these different simulations had their own spherical cow assumptions, however, so it's hard to know what to make of any of them.

In the ensuing decades, the model livestock that grazed their way through these simulation experiments came in ever-weirder shapes. Assume cube-shaped cows and you get a fancy reciprocal strategy called "firm-but-fair."[17] Assume pyramid-shaped cows that like to roam with a few good friends and you get a tendency toward hyper-generous cooperation.[18] Change from spherical cows to okra-shaped sheep and guitar-shaped goats that engage only in one-shot interactions with their adjacent neighbors, and you get mixed herds in which pure cooperators and pure defectors can exist side by side.[19] The stranger the ruminants got, the stranger the conclusions became.[20]

No matter. These models are supposed to be tools to help us think straight, not reconstructions of our natural history. Their most enduring contribution, in fact, is a simple mathematical inequality that works as a sort of Hamilton's rule for reciprocal altruism. (I'm tempted to call it Axelrod's rule, in honor of its discoverer.) For mechanisms for generosity to evolve by reciprocity (that is, by virtue of a generous act's ability to lure neighbors into being generous in the future), Axelrod's rule states that the cost to the donor must be less than the benefit to the recipient, discounted by the probability that the donor and recipient will encounter each other again in the future (which Axelrod called "the shadow of the future": the higher the probability of a re-encounter, the longer the shadow).[21] To express Axelrod's rule algebraically, we just start with the inequality from Hamilton's rule ($rb > c$) and replace Hamilton's relatedness coefficient, $r$, with a coefficient that represents the shadow of the future, which Axelrod represented with the letter $w$:

$$wb > c$$

When the benefit to the donor, multiplied by the length of the future's shadow, exceeds the cost to the donor—that is, when $wb > c$—the stage is set for the evolution of direct reciprocity. If we want to find real-life examples of generosity that evolved via reciprocity, the inequality in Axelrod's rule tells us where to look: in species in which a donor can cheaply produce a benefit that a recipient will find highly valuable, and in situations in which the individuals involved are likely to have many opportunities to help each other as they go through life. For his part, Trivers recommended looking for reciprocity-based helping in long-lived species with many opportunities for re-encounter over the life course—for example, those that live in stable groups, those that rely on high levels of parental care, and those that tend to stay close to their places of birth.[22] Under conditions like these, the shadow of the future can be a long one indeed.

In the wake of the initial work by Trivers, Axelrod, and others, naturalists began to fill the pages of evolutionary biology journals with examples of helping behaviors that arguably evolved via the benefits of reciprocity. Nominees included fish that work in pairs to inspect predators, birds that reciprocally squawk out alarm calls when predators arrive, lionesses that look after each other's cubs, and monkeys that take turns grooming each other.[23] After a proper scientific vetting, though, some of these nominees turned out to be adaptations for delivering benefits to kin or potential mates. Other nominees were shown to be mere mutualisms, in which individuals pursue their own interests and create benefits for others only as an unintended by-product. Still others looked like cases of coercion in which one individual harasses or threatens another individual into delivering a benefit.[24] First the party, and then the hangover: reciprocity looked to be rarer in nature than the early theorists might have hoped.

## WHAT HAVE YOU DONE FOR ME LATELY?

Even so, a few examples of reciprocity in nonhuman animals have withstood closer scrutiny. The common vampire bat is the poster child for reciprocity. Each night, vampire bats leave their roosts in search of food, which they obtain by sneaking up on drowsy livestock (or slumbering humans), chewing a hole in them, adding in some anticoagulant saliva, and then lapping up the flowing blood. For relatively inexperienced young bats, however, it's not uncommon to taste nothing but failure on these nighttime excursions, which is a problem: vampire bats have high metabolic rates and their furry little bodies store almost no energy in the form of fat. They die after three nights without food.

In the late 1970s, the biologist Gerald Wilkinson noticed that when the vampire bats he had been studying in Costa Rica returned with full stomachs, they would occasionally regurgitate some of their partially digested food into the mouths of roostmates that had failed to find meals for themselves. Most donors directed their blood donations to their own offspring, but occasionally a bat would regurgitate a small amount of its dinner into the mouth of a nonrelative. Wilkinson thought he had perhaps stumbled on an example of reciprocity. The economics of blood sharing certainly make blood seem like the ideal resource for reciprocal sharing: a well-fed bat that regurgitates 5 percent of its body weight into the mouth of a hungry neighbor might hasten its own death by a few hours (assuming no future meals are forthcoming), but in doing so, it pulls the hungry bat back from the brink of death by eighteen hours. So the ratio of benefits to costs exceeds 1: a promising start.

Wilkinson also found that he could induce reciprocal sharing experimentally: if he removed a bat from its home on, say, a Tuesday (so that it went hungry Tuesday night), one or more neighbors would share food with the hungry animal after it was returned home on Wednesday. Moreover, Wilkinson found that when Wednesday's donor bat was removed for twenty-four hours and then returned to the roost on Thursday, Wednesday's beneficiary would try to return the favor. Bats donate blood to hungry neighbors, Wilkinson proposed, because as the species was evolving, those that did so were able to get favors in return when the bat shoe was on the other cute little bat foot.[25]

Wilkinson's conclusion did not go unchallenged. Some researchers wondered about an alternative explanation they called the miscalibrated kin-recognition hypothesis: if genetic relatedness is generally very high in bats' natural living groups—but happened to be very low in the experimental groups Wilkinson had studied—perhaps his bats had been unable to build up high-fidelity kin templates early in life, which might have impaired their ability to discriminate between kin and non-kin later in life. Skeptics also wondered about an alternative they called the group-level altruism hypothesis. This hypothesis suggests that vampire bats are so highly related to each other in their natural living groups that natural selection didn't even bother hooking up their blood-sharing instincts to their kin-recognition system in the first place. After all, if everyone shares with everyone else, so that all the bats obtain fitness benefits both from helping and from being helped, reciprocation is beside the point. A third group of skeptics hypothesized that hungry bats use coercion or begging to harass well-fed neighbors into trading a little food for some peace and quiet. This alternative explanation was called the harassment hypothesis.

The three rival hypotheses sat in the annals of science, largely unaddressed, for more than two decades. In 2010, however, another scientist named Gerald—this one a Carter—joined Wilkinson. Together, the two Geralds performed an experiment that put all of these hypotheses into a war of all against all. Carter and Wilkinson got access to an artificial colony of twenty-five mostly unrelated vampire bats (there were a few mother-offspring pairs) that had been living together for two full years before the study began. Consequently, they had had plenty of time to learn to recognize each other by sight, sound, and smell. If these bats were to share blood with each other, it would not have been due to failures of kin recognition. Over the course of the experiment, Carter and Wilkinson removed individual bats from the group and put them on a 24-hour fast, just as Wilkinson had done in the 1980s. Afterward, when the hungry bats were returned to the group, Carter and Wilkinson could monitor who shared food with whom.

They found that the bats did tend to share more with relatives than with nonrelatives, as Hamilton's rule would lead us to expect. However, their results were not kind to the miscalibrated kin-recognition hypothesis: on average, each reintroduced bat received offers of donations from four other bats, virtually all of whom were nonrelatives. This is not what one would expect from a group of well-acquainted bats who regulated their helping on the basis of kinship alone. In fact, the bats' tendency to share reciprocally was even stronger among kin than among non-kin. Relatives asked, "What have you done for me lately?" even more insistently than nonrelatives did.

The results were also bad for the group-level altruism hypothesis. If bats shared with all roostmates indiscriminately, one would expect a correlation between one bat's donations in a single encounter and the amount of total aid she received from all of her roostmates. Carter and

Wilkinson's data did not yield even a hint of such a correlation. Instead, bats shared specifically with bats who had previously shared with them.

This experimental result left the harassment hypothesis as the only viable competitor. Were donors merely trying to avoid harassment so they could get back to sleep after a night of fine dining? Au contraire: on 62 percent of the occasions when a donor shared food, it was the donor, not the recipient, who initiated the interaction. As a matter of fact, donors were so eager to share that they appeared to be in a competition with other aspiring donors for the opportunity to do so. For lack of a better hypothesis to entertain, most biologists now believe that vampire bats do share food with each other on the basis of reciprocity.

Vampire bats are now the textbook example of reciprocity in action, but there are other examples as well. A European bird called the fly-catcher, for instance, will help a neighbor mob a predator (an owl, for instance) that has approached the neighbor's nesting site—but only if the neighbor provided assistance the last time the helper needed help.[26] And many primates, including monkeys, chimpanzees, and bonobos, reciprocally share both food and grooming with unrelated individuals.[27] Indeed, some researchers, including Gerald Carter, have argued that systems of helping that evolved due to reciprocity are actually quite abundant in nature. We fail to recognize them as such, Carter and others argue, because the preoccupation with spherical cows made it difficult for researchers to spot reciprocity in actual cows.[28]

## MOVABLE FEASTS

It's a lot easier to spot reciprocity in nature by focusing on human nature. The Hadza people of Tanzania are one of the last groups of humans on Earth who live more or less year-round as nomadic hunter-gatherers. It's a lifestyle they have been practicing for fifty thousand years, give or take. For this reason, the Hadza provide one of the best portraits of how humans lived before agriculture, writing, money, cars, houses, the Internet, and all the other affordances of the modern world. It's a portrait with a lot of meat in it.

Hadza men regularly kill birds, jackals, warthogs, baboons, and pretty much any other animal they can shoot with an arrow. It's the large animals, however—giraffes, zebras, impalas, buffaloes—that really grab their attention. The hunter who shoots the decisive arrow gets to take the best share of the dead animal for himself and his family. The remainder is either butchered on the spot or taken back into camp to be shared with others. Nobody would object if a Hadza hunter killed a small animal and kept it for himself, but it's a sin to be stingy with the large ones. Successful hunters are expected to share. It's just the way things are done. But why isn't it okay for the man who shoots the giraffe to keep the entire animal for himself? Wouldn't a "finders keepers" norm work just as well as a "Take the best, share the rest" norm? It's a question that has interested anthropologists for decades.

It's tempting to think that the Hadza share meat with each other simply because a successful hunter has no incentive to monopolize an entire giraffe carcass, most of which would go bad before he and his family could eat it all. This explanation was the received professional wisdom for many years, but that can't be quite right. The Hadza are willing to eat meat from animals that have been dead for a week. They also know how to preserve meat by drying it under the Tanzanian sun. Alternatively, they could trade their leftovers with non-Hadza neighbors for other stuff they like, or they could sell it in a trading village.[29] All to say, even if a hunter can't monopolize eight hundred pounds of giraffe meat, he could be getting more profit from it than he does.

A more plausible explanation for meat sharing among the Hadza relies on two facts about human evolution and human behavior. The first is that humans evolved to depend on meat. The second is that no hunter is good enough to get a sufficient amount of meat by relying only on his own wits. First things first: Meat is at the base of the hunter-gatherer food pyramid. In small-scale societies, about 60 percent of people's daily calories come from meat.[30] (For chimpanzees, it's about 2 percent.) Although the details of humans' evolutionary transition to a meat-centered diet remain murky (some anthropologists believe that the switch to meat enabled hominins to colonize a grassland environment that lacked the vegetarian buffet of the African rain forest), the nutritional benefits of meat created selection pressures for our ancestors to get better at obtaining it. The first great leap forward came in the form of new skills and technology for scavenging: *Homo habilis*, one of our early hominin ancestors, used two-sided hand axes that were perfect for cutting hunks of meat from animal carcasses. But as evolving humans came to depend more and more on meat, selection pressure mounted for approaches to bringing the meat supply under better control.

The principal difference between the meat that our pre-*sapiens* predecessors scavenged and the meat that *Homo sapiens* hunts is that dead meat—the kind you scavenge from the body of a creature that someone or something else already killed—does not care one way or another whether you eat it. Live animals, in contrast to dead ones, very strongly prefer that you do not eat them, so they do their best to frustrate your efforts. The live animals to which meat is attached can hide in plain sight, outrun you, or kill or maim you with horns, tusks, and hooves. So hunting is hard: on average, a Hadza hunter needs twenty-nine days of hunting to kill a large game animal.[31] Overall, daily failure rates for hunter-gatherers range from 40 percent to 96 percent.[32] You can't evolve into a solitary meat specialist if you can obtain meat only once a month.

However, there is a way to increase your access to meat without increasing your success rate: if hunters share with others what they can't eat themselves, they can smooth out their own daily consumption by calling in the future on the repayment of their previous meat-sharing favors. It's a simple matter of risk-pooling, not unlike modern health-insurance schemes, in which everybody pays, even when they're healthy, so they can all receive medical care when they're ill. Like the vampire bat, a Hadza hunter banks some of today's protein windfall in the only rainy-day fund that ancestral humans ever had: the bellies of grateful neighbors who were ready to return the favor when the need arose.[33] And once our ancestors came to possess instincts for sharing meat in this reciprocal fashion, "the propensity to truck, barter, and exchange one thing for another" (as Adam Smith put it) would have been a mere evolutionary eyeblink away.[34]

## IMAGE-CONSCIOUS

Biological markets—including vampire bats' market for blood from large living animals and the Hadza's market for meat from large dead ones—are possible because of an economic reality called diminishing marginal utility: the value of the next packet of a resource you can obtain depends on how many packets you already possess. A few milliliters of blood are more valuable to a starving bat than to a well-fed one. Likewise, ten pounds of giraffe meat is worth less to a hunter who has already claimed the best forty pounds for himself than to a hunter who hasn't dined on giraffe in a week.[35] Diminishing marginal utility helps to fulfill one of the most important conditions for the evolution of reciprocity: the thing I am giving you must benefit you more than it costs me to part with it. For this reason, we actively consider the urgency of our potential beneficiaries' needs. The needier you are, the more valuable the benefit to you. The more valuable the benefit to you, the more $b > c$. All else being equal, we ought to be more inclined to share with needy people, and we are.[36] What looks like tenderhearted charity could be a simple case of buying your friends at an affordable price.

It's not enough that $b > c$: Axelrod's rule has that third term, $w$, that we can't ignore. The only way the donor gets her investment back is if she meets back up with the recipient at some point in the future. For this reason, we should expect that our instincts for reciprocity, along with the feelings they engender and the services they motivate us to render, are sensitive to the likelihood that we will meet up again with the person we've helped—and that he will be in a position to return the favor when that time comes.

How do you make a good guess about the probability that you'll run into your beneficiary again in the future? One thing worth considering is whether you knew each other before the helping encounter. If you knew him, odds are good that you'll keep knowing him.[37] It might also be useful to consider whether he is a member of your own cultural group. I imagine this insight was just a pen-stroke away from Darwin when he observed that "with all animals, sympathy is directed solely towards members of the same community, and therefore towards known, and more or less beloved members, but not to all the individuals of the same species," and, a couple of pages later, when he pointed out that "a savage will risk his own life to save that of a member of the same community, but will be wholly indifferent about a stranger."[38]

There is, of course, another sense in which a low likelihood of re-encounter might prevent someone from requiting a favor, and that is the sense of being dead. The closer you are to death, the lower the chances you will live long enough to return my favor. By this reasoning, we should expect people to be eager to help the healthy, the strong, and the young, and averse to helping the sick, the weak, and the old. And they are.[39]

But what if someone's likelihood of reciprocating is not an issue of "Can he or can't he," but instead an issue of "Will he or won't he?" Some perfectly able-bodied people might require your assistance not because of bad luck, but because of their own laziness or negligence. If it's the latter, you have to wonder whether they will ever work hard enough to repay their debt to you. Others might be unwilling to return your assistance because of inattention, ingratitude, or sociopathy. A psychology of helping that is not sensitive to these considerations is at an evolutionary disadvantage: from the viewpoint of natural selection, assisting people who refuse to fend for themselves or to repay their debts is like throwing your resources into the campfire. For this reason, we judge people who've asked for our help on the basis of what the political scientist Michael Bang Petersen and others have called the "deservingness heuristic." When we apply this rule of thumb, we use people's past behavior to quickly estimate the amount of effort they will invest in trying to repay their debts to us.[40] If we think they'll work hard to even things up, then we judge them as deserving of our help.

But what if you haven't had any direct experiences with the person in need? How are you supposed to apply the deservingness heuristic to someone you've never dealt with before? Easy: Just ask others about their experiences with the person in question. We ought to care, therefore, about the would-be beneficiary's own history of generosity with others, and we do. And because we care about others' reputations, they ought to care about their reputations as well. And they do.[41]

Darwin, dyed-in-the-wool Victorian that he was, had no trouble grasping how we judge the moral character of other people on the basis of their past kindness and generosity:

> We may therefore conclude that primeval man, at a very remote period, was influenced by the praise and blame of his fellows. It is obvious, that the members of the same tribe would approve of conduct which appeared to them to be for the general good, and would reprobate that which appeared evil. To do good unto others—to do unto others as ye would they should do unto you—is the foundation stone of morality. It

> is, therefore, hardly possible to exaggerate the importance during rude times of the love of praise and the dread of blame.[42]

Indeed, Darwin believed that our love of praise and dread of blame was strong enough to motivate us to care for others even in the absence of any genuine concern for their well-being:

> A man who was not impelled by any deep instinctive feeling, to sacrifice his life for the good of others, yet was roused to such actions by a sense of glory, would by his example excite the same wish for glory in other men, and would strengthen by exercise the noble feelings of admiration.[43]

More than a century after Darwin noted humanity's "love of praise and dread of blame," the American evolutionary biologist Richard Alexander proposed a new form of reciprocity, which he named indirect reciprocity, that relied heavily on the power of reputation. Indirect reciprocity differs from regular old Trivers-type reciprocity in one important way. Trivers-type reciprocity requires only two people—one person who needs help and another person in a position to provide it. For indirect reciprocity, it takes at least three: a person in need, a helper, and a bystander who helps the helper. An indirect reciprocator is a helper-helper. And when *fourth* parties learn of the helper-helpers' help, they reward them for their helper-helping, making them helper-helper-helpers. And so on and so on, ad infinitum. Like a contagion that jumps from one person's need to another person's generosity to a third person's moral judgments, generosity can spread through an entire community, creating self-reinforcing virtuous circles. According to Alexander's model of indirect reciprocity, kindness really is contagious.

Alexander envisioned three scenarios for how a helper's reputation for generosity might yield fitness benefits.[44] In the first scenario, the helper is directly compensated by third parties who learn of his generosity: the helper helps, and a great cloud of witnesses step forward to reward him. As a result, the helper is encouraged to help again in the future, and the group that rewards him enjoys the benefit of having a helpful person in their community. In Alexander's second scenario, a helper receives Hamilton-type indirect fitness benefits by helping others. The helper helps, perhaps at the cost of her own life, and the beneficiaries of that help take actions that raise the fitness of *the helper's children and other kin*.

Alexander's third scenario is the cleverest of the three: by helping, the helper later becomes "engaged in profitable reciprocal interactions by individuals who have observed his behavior in directly reciprocal interactions and judged him to be a potentially rewarding interactant (his 'reputation' or 'status' is enhanced, to his ultimate benefit)."[45] In the scenario Alexander envisions here, I pay attention to your generosity toward other people because it conveys information about your inclination to be generous toward other people *in general*. "People in general," of course, includes me. Under this scenario, I'm interested in how you treat others not so I can reward you, but instead, so I can figure out whether you are the sort of person I want as a friend.[46] Notice here that indirect reciprocity and direct reciprocity merge into one. When I, as an indirect reciprocator, help you after you have helped someone else, it might look like I am rewarding you, but what I'm really doing is wooing you.

Altruism has Hamilton's rule, with its relatedness coefficient ($r$), and direct reciprocity has Axelrod's rule, with its shadow of the future coefficient ($w$). Indirect reciprocity has a rule of its own. One might be tempted to call it Alexander's rule if hadn't been two mathematicians—Martin Nowak and Karl Sigmund—who discovered it. Fine: let's call it Nowak and Sigmund's rule. It relies on the following inequality:

$$qb > c$$

Nowak and Sigmund's rule says that indirect reciprocity can evolve when the benefit to the donor, discounted by the coefficient $q$, exceeds the cost to the donor. Here, $q$ doesn't represent the likelihood that the beneficiary and benefactor share the gene that motivates the helping (which is the meaning of the $r$ in Hamilton's rule), or the probability that the person helped will reciprocate in the future (which is the meaning of the $w$ in Axelrod's rule). Instead, $q$ stands for the probability that a helper's behavior will be observed by an indirect reciprocator, who will then help the helper. As $q$ approaches a value of 1, the likelihood that our generous act will become known to others approaches 100 percent, which is how we get repaid for the original act of helping. Because $q$ determines the likelihood that we'll get repaid, we have evolved to drive $q$ as high as it can go. Thanks to Alexander, and to Nowak and Sigmund, we finally have an evolutionary explanation for humanity's "love of praise and the strong feeling of glory, and the still stronger horror of scorn and infamy," as Darwin put it in *The Descent of Man*.[47] We love others because we love our reputations, and we love our reputations because they motivate others to love us.

But if humans care so much about their reputations, why do we sometimes donate anonymously to needy individuals and charities? Wouldn't we want to grab the reputational benefits that come along with grand gestures and random acts of kindness? Don't our occasional efforts to conceal our generosity (or, to paraphrase Jesus, to prevent our left hand from knowing what our right hand is doing) prove that we really do sometimes act out of love for others rather than from a love of praise? Not so fast. I'm sure there are people in the world you would be happy to impress with your generosity, but there almost certainly are other people you would just as soon not impress. After all, you probably don't want the entire world beating a path to your door for handouts.[48]

And anyway, concealing your generosity from the hordes of people you don't care about impressing doesn't prevent you from advertising your generosity to the people you do want to impress. In fact, the only people you might want to impress with your magnanimity are people whose capacity for generosity equals or exceeds your own. These are, after all, the people who might be in the best position to give you things you can't easily obtain for yourself. And you can let *those* people know about your $10 million donation to your alma mater even if you don't insist on having your name placed on the building.[49] What's more, if you conceal your generosity—as though your goal really is to help others rather than to advertise what a mensch you are—it's likely that you'll receive even more praise if your generosity is ever discovered.[50]

Once glory-loving organisms begin to advertise their value as cooperation partners through their generosity, a community of people seeking out the best possible friends begins to resemble a mating market in which people seek out the best possible lovers.[51] In social markets such as these, everyone is both a buyer and a seller: people try to buy the best partners they can, and they do so by putting their own social value up for sale. As a result, each person's friendship goes to the highest bidder. Bob, who everyone agrees is the best possible friend to have, tries to establish a friendship with Sarah, because she is second best. Theresa, who is third best, would really like to make friends with Sarah, but Sarah is already tied up with Bob, so Sarah makes do with Fred, who is the fourth best person to befriend. And so on, and so on, ad infinitum.[52]

This tendency for like to associate with like, called homophily (literally, "love of the same"), is the inevitable result of people's efforts to acquire the most valuable friends their own social value enables them to afford.[53] Among hunter-gatherers, homophily manifests itself as friendships between people of similar physical strength, height, and body weight, all of which are good indicators of one's economic value in a world of hunting and foraging.[54] In modern societies, homophily shows up as friendships between people of similar income, educational attainment, and occupational prestige, all of which are good indicators of one's economic value in a nine-to-five world of pay stubs and hedge funds.[55] From how we design society's moral systems to how we fill our address books with friends and associates, reputation matters.

## STONE AGE SAMARITANS?

For 95 percent of our time on Earth, humans lived in nomadic bands of hunter-gatherers consisting of a few dozen or perhaps a few hundred individuals. Dozens or scores of these small foraging bands made up a cultural group consisting of hundreds or thousands of people who spoke the same language, shared the same cultural practices, and intermarried.[56] When they weren't fighting over territory, resources, women, or old grudges, our hunter-gatherer ancestors were often prepared to dispense kindness to other members of their cultural groups—even those whom they might not have met before. Here's how a German anthropologist named Viktor Lebzelter, who spent lots of time with the San people of South Africa during the 1920s, characterized their typical responses to strangers in need: "Allowing a stranger to partake of a meal, as well as giving him some of the available water, is wholly matter-of-course with them."[57]

One of Lebzelter's close associates, an anthropologist named Martin Gusinde, similarly characterized how the Yahgan people of Tierra del

Fuego (at the southern tip of South America) treated hungry strangers who happened to pass through camp around suppertime:

> Just as every visitor is readily received, so that one scarcely notices a difference in the treatment of friend or stranger, so is he generously provided with food. The owners of the hut spontaneously and regularly offer the best they have. They hold nothing back, for they aim to please and satisfy their guests, and they are not sorry if their entire supply is consumed. Even a person who has come for the express purpose of eating his fill for once—perhaps he saw his host returning home with rich spoils—is not turned away, but can freely indulge his appetite. It is not customary for the visitor to ask for anything. The owner of the hut already knows the other's intention and serves him everything he has on hand.[58]

This sumptuous hospitality sounds like an ancient version of universal human concern, but it's not. Ancestral humans didn't think of strangers the way we do. Today, you can travel to another continent on business and have lunch in a café with dozens of strangers you would never possibly meet again. That's what "stranger" means to us. Not so in the Paleolithic world in which our minds evolved. In that world, a stranger was usually someone who spoke your language and dressed like you did. And if you talked with a stranger long enough (assuming you weren't too terrified of him or too interested in killing him), you would eventually discover a mutual acquaintance: your brother-in-law might be a friend of his son; your half-sister's daughter might be his nephew's wife. Strangerdom among people inside your own cultural group was a temporary state, a momentary social blip that was quickly surmounted with a bit of ice-breaking chit-chat and name-dropping.

Armed with this understanding of what strangerdom actually meant within hunter-gatherer societies, we can easily appreciate the self-serving logic of their generosity toward strangers. Why were the San people so eager to help strangers? To Lebzelter, the answer was obvious: because "anyone who does the Bushmen a kindness may be sure that they will manifest their gratitude by deeds to the point of self-sacrifice."[59] And why did the Yahgan lavish hospitality on the unannounced visitor? This, too, was a no-brainer: because "at the next opportunity, he will visit the one who is now his guest. Thus a person may eat first in one hut and then in another, but eventually he also must show that he is a generous host. . . . Since everyone wishes to avoid being called stingy or mean, he offers his guest everything he has with unmistakable generosity."[60]

The rule in small-scale societies, then, was that what comes around goes around. When you help a stranger, you are almost surely investing in your own welfare by burnishing your reputation as a warm and attentive host, or convincing your guest to take better care of your relatives who live in his camp, or securing a warm welcome for yourself next time your travels take you in his direction.

## RECIPROCITY, REPUTATION, REASON

What are reciprocators like? They're smart investors, searching for good returns by purchasing shares in others' lives. They're wary of con men. They discriminate based on health, strength, skill, and success. They're show-offs, advertising their beneficence in order to attract the best possible friends and allies. They delight in gossip. They love praise and dread blame. They admire virtue and shun callousness. They're wary of false friends, and they treasure true ones. In short, if you want to see what a reciprocator looks like, take a look in the mirror.

Do our ancient instincts for reciprocity undergird our modern regard for strangers? To some extent, yes. Probably. Reciprocity certainly has more to offer for an understanding of our attitude toward strangers than Hamilton-type altruism and group selection do, at any rate. As we have seen, our appetite for establishing Trivers-type reciprocal relationships is sensitive to cost-benefit ratios. To people in dire need, assistance that seems trivial to us can take on tremendous value. Because of diminishing marginal utility, a loaf of bread is more valuable to someone who's starving than it is to someone with three other loaves of bread already in the freezer. Under such cost-benefit ratios, helping can seem like a good investment. Everybody likes a bargain.

Even so, our instincts for reciprocal sharing can deter generosity almost as much as they can help it. Investments in someone's welfare might not seem worth the hassle unless there's a reasonable chance of receiving a favor in return. Consequently, our evolved instincts for Trivers-type reciprocity restrain us from helping the weak, the sick, the old, and the isolated—precisely the people who tend to need the most help, tragically—because they are the people from whom we are least likely to recoup our investments. Our instincts for reciprocity also make us reluctant to help people we're unlikely to meet again, so how strongly could they possibly motivate us to help people we've never even met in the first place? And then there's the deservingness heuristic: in the absence of information to the contrary, it's tempting to assume that a stranger in need is responsible for her own problems. And if she's too irresponsible to look after her own interests, why should we think she's responsible enough to pay back her debts?

As an evolutionary explanation for modern generosity toward strangers, Alexander-Nowak-Sigmund-type indirect reciprocity may hold even more promise than Trivers-type direct reciprocity. We value magnanimity in our friends, our mates, our business partners, and our leaders. And we want them to view us in precisely the same way. "A good name," as the author of the Book of Proverbs noted, "is to be chosen rather than great riches, and favor is better than silver or gold."[61]

In the next seven chapters, we'll incorporate what scientists have learned about the evolution of human generosity over the past two hundred millennia to better understand the revolutions in human generosity over the past ten. As we proceed, we will pinpoint seven distinct confrontations with mass suffering that left permanent marks on how we, in our modern societies, strive to meet the needs of the poor and downtrodden. In examining these seven hinges of history, we will see that the major revolutions in how we regard strangers in peril have been underwritten by a surprisingly small set of evolved psychological faculties (three, to be precise). The first is our keen nose for good reciprocal investments, and the second is our love of a good name. As we have seen in this chapter, reciprocity and reputation matter. But there's a third R in the genealogy of generosity that we mustn't neglect: reason. Darwin didn't take reason for granted, and he encouraged us not to take it for granted either.

## CHAPTER 7

# THE AGE OF ORPHANS

For the first eight million years of hominid evolution, our capacity for compassion advanced at a snail's pace as gene-by-gene alterations to our ancestors' minds and bodies slowly crafted their social instincts and sentiments. In comparison, the evolution of humans' concern for the welfare of strangers over the past ten thousand years has proceeded at light speed. In the next seven chapters, we will carve that ten-thousand-year arc into seven historical ages, each of which featured a confrontation with mass suffering that forced our forebears to revise their beliefs about human suffering and to devise new solutions for addressing it. As we go, we will see how three of our instincts—our knack for reciprocity, our desire for a good reputation, and our capacity for reasoning—shaped our ancestors' responses to these encounters.

The historical era we will explore in this chapter, which I call the Age of Orphans, coincides with humans' transition from a nomadic lifestyle that revolved around hunting and gathering to a sedentary lifestyle that revolved around farming. During the Age of Orphans, wealth inequality shot up to unprecedented levels, creating a kind of hardship that humans had never before faced: becoming so deeply indebted to one's creditors that the only ways to dig oneself out were to sell everything one owned or to consent to a lifetime of indentured servitude. In the face of so much inequality and oppression, the kings of the archaic world came up with a new idea: by protecting the society's most vulnerable people from oppression, they would be repaid with their subjects' loyalty and with a strengthened reputation for goodness and wisdom.

## FARM FOLK

Poverty is fundamentally about food, so it is fitting that the first stop on our seven-stop tour of the history of human generosity involves a big change in what people ate. The stage for these changes was set approximately 12,500 years ago by a group of Middle Eastern hunter-gatherers whom we now call the Natufians. To supplement their traditional hunter-gatherer diet of meat and foraged foods, the Natufians began to take advantage of wild grains, which had become more plentiful as climate changes made the region warmer and wetter. Armed only with stone sickles with wooden handles, the Natufians slowly transitioned to a grain-based diet.[1]

The precise details of this transition are lost to time, but here's what we do know from the archaeological data: Within two millennia, the Natufians went from full-time mobile hunter-gatherers to full-time farmers organized into sixty-person communities on half-acre sites. They used large communal mortars to grind their grain into flour, and the grain they didn't eat immediately was stored as surplus in communal granaries.[2] Though their diets had changed, the Natufians continued to pool their labor and to share the fruits of that labor, just as hunter-gatherers always had.

By 8000 BCE, however, the Natufian experiment came to an end, perhaps because a long drought forced them to migrate elsewhere. However, it didn't take long for another group of enterprising prehistoric farmers to reoccupy the old Natufian settlements. The Natufians' successors, the settlers of the so-called Pre-Pottery Neolithic era (or PPN), made several important advances in agriculture, the most momentous of which was a selective breeding program that yielded an energy-dense strain of wheat that required so much love and care that it could not even grow without direct human intervention. As the cereals became dependent on the humans, the humans became dependent on the cereals.[3]

There was another innovation as well: rather than pooling the labor of the entire village to harvest grain, as the Natufians had done, the villagers of the PPN began to pool their farming labor within families.

Where feeding the village was once the responsibility of the entire village, it was now every family for itself. The communal mortars and granaries of the Natufians fell into disuse, and their Neolithic successors began building mortars and specialized grain storage rooms within their own homes.[4] Communal work and communal sharing were out; private property small family businesses were in.[5]

What happened to the old hunter-gatherer esprit de corps? It seems strange that a few thousand years of plopping seeds into holes, and then eating the plants that grew out of those holes, could disrupt a way of life that humans had practiced for two hundred thousand years. But disrupt that way of life it did, and with only three mean strokes of the sickle.[6] Agriculture's first fatal blow against communal sharing was a new approach to avoiding hunger. It has been said that our hunter-gatherer ancestors managed the risk of starvation by taking advantage of the Two Fs: fat and friends. If you obtained a surplus that you couldn't immediately turn into energy for fueling your body, you could store it as fat (which your body could convert into energy during leaner times), or you could give it to a friend (who would likely reciprocate in the future). But with the advent of farmed cereals, a third F for managing the risk of starvation arose: a fast food.

Unlike most other foods, which lose their value over time due to spoilage, grain is fast—in the sense of durable, stable, and not easily corrupted. All you need to do is store it somewhere cool and dry and keep the pests out of it. Cared for properly, grain is a fast friend that will be worth as much in six months as it is today. The same was true for the agricultural pioneers of the PPN. By storing their surpluses away in private granaries, they could protect their future selves against the risk of starvation. No need to gorge on those surplus calories all at once; no need to share them with needy friends to build up favors. To take advantage of this new opportunity for hedging against risk, the PPN farmers of the Levant (where modern Israel, Jordan, Lebanon, and Syria are located) began to cache their grain away, far from the prying eyes of others, in up to a dozen little storage rooms within their own homes.[7]

Cereals have another feature that discourages sharing: their size. Hunter-gatherers tend to share foods that come in big packages (think gazelles, zebras, and baby giraffes). They're much less likely to share foods that come in small packages (like fruits and tubers), presumably because they intuitively grasp that any able-bodied person can pick ripe fruit and dig yams out of the ground.[8] As a result of this aversion to sharing easily acquired foods that come in small packages, the PPN farmers of the Levant, who still possessed hunter-gatherer minds, may have found the idea of sharing grain sort of unbecoming.

Agriculture's second fatal blow against communal sharing was the transition to private ownership of farmland, which alters how people think about the relationship between effort and reward.[9] Hunting food is a risky way to put it on the table because it's not enough to be good at hunting—you also have to be lucky.[10] For even the best hunter, most hunting trips are failures. To mitigate against the very real risk of coming back with nothing to eat, therefore, hunters pooled their risk by sharing their kills with other members of the community.

Farming is quite different. With hunting, you get a return on your investment as soon as you or one of your friends kills something made of meat. With farming, the payoffs are separated from the labor by weeks and months. Also, farmers don't have to be just lucky: they have to be extraordinarily diligent. The more time and effort you put into acquiring land, clearing it, removing stones, fertilizing the soil, pulling weeds, planting, watering, and saving seed for replanting, the bigger your harvests will be. Compared to the least industrious farmers, the most industrious will work longer hours in their fields and spend less of their time socializing.[11]

In a world in which some people were more motivated than others to take part in such work, farming became increasingly organized around families rather than around communities. Traditional foragers are fiercely egalitarian—nobody has the right to boss anybody else around—so any industrious and hardworking farmer who tried to coerce a lazier peer into working harder on a communal patch would have been ignored or ridiculed.[12] For this reason, communal farming simply wouldn't work. Family-based organization of farming provided an easy solution to the problem of recruiting farm labor: "management" (parents) could easily coerce "labor" (children) into working long hours in unpleasant conditions.

In his *Second Treatise of Government*, John Locke argued that land becomes private property when an individual devotes his own labor to the task of improving the land. "As much land as a man tills, plants, improves, cultivates, and can use the product of," Locke wrote, "so much is his property. He by his labour does, as it were, inclose it from the common."[13] Humans everywhere seem to share Locke's intuition that labor transforms material things into property. On the basis of this basic human intuition, I am guessing that the PPN farmers of the Levant came to the following understanding of land ownership: the land you work belongs to you and your family, and the land I work belongs to me and mine. To this very day, 95 percent of farms in the United States and Europe are controlled by single families.[14]

Agriculture's third and final blow against communal sharing came through the new forms of wealth that farming created. Before farming, success in life was built in roughly equal measures of so-called embodied capital (physical strength, skill, and smarts), social capital (strong relationships with other people), and material capital (stuff you owned). But with the transition to farming, success came to have much less to do with your embodied capital and social capital and much more to do with how much land and how many animals you possessed. Land and livestock are also much more readily passed down to one's heirs than are strength, skill, smarts, and social connections.[15] You can't take it with you, the old saying goes, but if "it" is farmland and goats, you can leave it to your kids.

The increased materiality and heritability of wealth during the PPN ushered in a profound reorientation of social life. The egalitarianism and communal sharing that governed life in hunter-gatherer bands lost some of their value. Family and lineage came to mean more and more. The ethnographic record even suggests that parents stopped promoting liberal, unconditional generosity through the moral lessons they taught their children.[16] Families got bigger, houses got bigger, and social life came more and more to revolve around the family hearth rather than around the community campfire.[17]

Proving your ancestry was how you defended your land rights, and the Neolithic villagers of the Levant probably proved theirs in two ways. First, they arranged marriages between their children and their cousins' children, as is common in small-scale farming societies around the world.[18] Cousin-marriage is a great way to make sure that wealth stays in the family.

They also kept their ancestors close. Really close. As in "bury-the-body-under-the-floor-of-the-kitchen" close. The farmers of the PPN interred their dearly departed right under the floors of their own homes, taking care to note the locations of the skulls by etching bright red circles into the plaster. After a couple of years, the plaster would be broken up, the defleshed skulls would be removed, and skilled artisans would be hired to plaster and paint faces back onto them. Many of these exhumed skulls eventually made their way out into the community, but only to be buried again—often in public spaces, in multi-skull caches.[19] This ritual practice could have meant many things, but the philosopher Kim Sterelny interprets it as evidence that people were using their ancestors' remains to assert property rights: "One could hardly make the claim to be the continuants and heirs of the departed more vividly," Sterelny ventured, "than by living on top of their bodies and by removing, preparing, and displaying their skulls (in all probability) as props in ritual recitals of genealogy and connection."[20]

With these three blows—fast food, privatized farmland, and new sources of wealth—family-based agriculture displaced communal sharing as the world's primary economic model. Which is not to say that the Levantine farmers of the PPN suddenly stopped caring about anyone outside of their family and lineage: they surely would have continued to help their friends and neighbors with everyday tasks and with the construction of mutually valuable assets, such as fences, walls, or irrigation technology. And then, just as now, friends and neighbors were relied upon when the crops failed or they really needed to borrow an onion.[21] But as ever more cooperative effort was organized within lineages, concern for the welfare of the extended family crowded out concern for the welfare of the families across town.

As people began to concentrate their labor and wealth within extended families, another seismic cultural shift got under way: families and lineages started sorting into rich ones and poor ones. The wealth associated with agriculture (houses, land, animals) could be retained within lineages, so lucky breaks (an especially good breeding season, for instance), devastating setbacks (a father's untimely death), and wise investments (buying the house and farm from the poor widow next door, or hiring an extra hand to help with all of those extra cows) began to influence the long-term fortunes of lineages. Over time, positive feedbacks caused the rich to become richer and the poor to become poorer. Consequently, high levels of income inequality became ever higher.

Within five thousand years of the Natufians' experiments with farming, the PPN villagers who succeeded them were living in huge communities. The Pre-Pottery town of Jericho, for instance, was home to as many as three thousand people. For the first time in this region of the world (and perhaps anywhere), people began to live among throngs of other people with whom they shared no ties of obligation or affection. Their world had become a world of strangers.

## FROM INDIFFERENCE TO INEQUALITY

Around 6200 BCE, the PPN farmers of the Levant left home, probably driven south by abrupt climate changes. By then, the Neolithic experimentation with farming had spread eastward from the Levant to Mesopotamia. (Similar agrarian explosions would also soon take place in Egypt, South Asia, China, and the New World.[22]) Like the Pre-Pottery Neolithic villages of the Levant, societies of Upper Mesopotamia with names such as Hassuna, Samarra, and Halaf pivoted away from the classic hunter-gatherer pattern of communal sharing. The remains of these societies also show the telltale signs of rising inequality: some families had very big houses; others had tiny ones. These clues and others suggest that the Mesopotamian agricultural explosion killed communal sharing much as it had in the Levant, leaving inequality, and indifference to the welfare of strangers, in its wake.[23]

Just how high did inequality soar in Mesopotamia? It's anyone's guess, but what we know about wealth inequality in the world's existing small-scale societies offers some clues. Among those that focus on hunting and gathering, inequality tends to be just a little higher than in the modern Scandinavian nations, which is to say, very low indeed. In small-scale agricultural societies, conversely, inequality is on par with modern Venezuela and the Democratic Republic of Congo, which is to say, very high indeed.[24]

## FROM INEQUALITY TO OPPRESSION

By 4000 BCE, as embryonic Sumerian cities such as Uruk became more stratified and complex, infrastructure became more important. Canals and ditches needed digging. Earthworks needed to be built and repaired. Water rights needed to be negotiated and enforced. Wars of defense and expansion needed to be planned and executed. None of this stuff was going to build itself, or pay for itself. Thus, the world's first systems of taxation were born.[25]

In a world without money, taxes came in the form of meat, produce, supplies, and corvée labor (corvée comes from the Latin *corrogare*, which means "to requisition"). Like the weekend warriors of the United States National Guard, who serve "one weekend a month, or two weeks a year," the poor and middling farmers of the world's ancient city-states left their own fields several days out of each month to work as farmers or soldiers in the fields and armies of the kings, priests, and other elites.[26] Because the farms of the royal house, the divines, and the aristocrats needed to be tended to at exactly the same time that everybody else's farms did (at planting and harvest times, for example), the corvée requirements were a real hardship.

With such a system of taxation in place, one might imagine that the kings of the ancient Near East would set up distribution systems that enabled them to feed society's most vulnerable people during hard times. After all, that is part of what our governments do for us with our taxes. However, there is no evidence that they did anything of the sort. True, the rulers of the ancient world did store up lots of grain—in some cases, enough, in principle, to supply the city's needs during famines—but it's at least as likely that those granaries and warehouses were simply tools, like banks, for managing the institutional wealth of the state. What seems probable, given the evidence at scholars' disposal, is that those giant caches of food were rarely, if ever, used to buffer the poor against simple food scarcity.[27] Even if we decide to read the history of the ancient Near East as "avowed Marxists," the Assyriologist Benjamin Foster wrote, "we find little direct evidence for social discontent, popular movements, charismatic crusaders, or other familiar forerunners of social reform."[28]

Instead, citizens of the world's earliest cities were fatalists who viewed poverty and hunger as escapable facts of life, as illustrated by this Sumerian proverb:

> Let the poor man die, let him not live. When he finds bread, he finds no salt. When he finds salt, he finds no bread. When he finds meat, he finds no condiments. When he finds condiments, he finds no meat. When he finds oil, he finds no jar. When he finds a jar, he finds no oil.[29]

The poor didn't expect their sorry lots in life to become less sorry, and they certainly didn't expect their governments to do anything about it. Even so, some kings eventually began to take action to protect society's most vulnerable people from oppression. Consequently, ancient documents, such as the Reforms of Uruinimgina (2300 BCE), the Code of Ur-Nammu (2050 BCE), and the better-known Code of Hammurabi (1750 BCE), suggest that the world's first kings began to place some sort of priority on shielding society's most vulnerable people—the orphan, the widow, and "the man of one shekel"—from the most flagrant forms of oppression.[30]

There's some radical stuff to be found in those reform documents. At turns, the reformer-kings reduced the number of years one could be enslaved to service a debt, capped interest rates and modified repayment terms on onerous personal loans, and allowed people to repurchase land and property they had previously sold in order to settle debts or buy food. Hammurabi prevented army officers from stealing their subordinates' property. Uruinimgina outlawed extortion and other Mafia-style methods to pressure poor people into selling their property at far below asking price. Another king even bragged that he had reduced the corvée labor requirement to *only* four days per month![31]

The reformer-kings also shielded orphans and widows from exploitation by old laws that disqualified women from inheriting the property of their deceased husbands and fathers. Uruinimgina, the king of ancient Lagash, gave some teeth to his new protections by promising the city's patron deity that he would not allow powerful men to commit injustices against orphans and widows.[32] Hammurabi acknowledged similar concerns in both the prologue and the epilogue of his code, and his Law Number 177 allowed widows to stay in their deceased husbands' homes. It also prevented creditors from liquidating widows' household goods to settle debts. For their part, the rulers and high officials of Egypt happily accepted epithets such as "defender of the orphan," "husband to the widow," "rescuer of the fearful," and "savior of the distressed."[33]

## A COMPASSION PLAY

It is tempting to see the reforms instituted by the kings of the ancient Near East as proof of their intrinsic concern for the welfare of their subjects, just as it is tempting to imagine that they used their riches to establish robust programs for redistribution. But before drawing such hasty conclusions, it is instructive to consider how Uruinimgina and the others stood to benefit by protecting the orphan, the widow, and the man of one shekel. I see two possibilities.

First, they stood to benefit by winning their love. The archaic kings defended their legitimacy by presenting themselves as a god's representatives on Earth, or as gods in their own right. Any god or god-proxy who couldn't protect his people from ruin had a major public relations problem. The gods themselves were often portrayed as shepherds, and their subjects as the livestock under their care. In Egypt, the people were the "cattle of God." Among the Babylonians and Hebrews, the people were flocks of sheep and the kings were shepherds. It is a sorry shepherd who can't keep his sheep from starving or becoming the wolf's dinner, and if you can't keep the god's cattle healthy and well-fed, pretty soon you're out of meat and milk. Compassion for the poor helped to prevent all that.[34]

Compassion benefited the god-kings in a second way as well: by implementing reforms that limited oppression (capping interest rates and canceling debts, for instance), they could limit the power of their aristocratic rivals. Debtors often settled their debts by selling off their land or subjecting themselves to a fixed term of slave labor, so the most ambitious elites were able to assemble huge farms that required many hundreds or even thousands of workers, not unlike the Spanish haciendas of colonial South America, or the massive Sicilian *latifundia* of ancient Rome. Such massive land appropriations were terrible for regular folk, and they increased landholders' power and prestige considerably.[35]

The last thing an archaic king wanted was competition, especially in a society in which the elites operated as autonomous centers of power.[36] Reforms were an easy way to keep those competitors in check. In fact, canceling debts, reducing the terms of slave labor, and returning property to its ancestral owners enabled an ancient king to kill two birds with one stone: winning his sheep's devotion and limiting his rivals' power.[37] In Kanesh, for example, where several trading families had begun to grow obscenely rich around 1875 BCE, the king began to tax their profits. He also forced them to donate much of their wealth to religious foundations that ostensibly served the common good.[38] The result? The rich continued to get richer (though less quickly), the poor were better off, and the king managed to keep the favor of the common people and stay one step ahead of his competitors.[39]

Well played, Your Highness. Well played.

By the end of the twelfth century BCE, the Age of Orphans came to an end in a conflagration of war and natural disaster. In the centuries that followed, the entire political and cultural map of the ancient Near East would be redrawn. Many of the new societies to emerge from the chaos, however, retained the conviction that the poor and the powerless should be shielded from oppression. This conviction would become stronger, more extensive, and more popular in the coming Axial Age. New innovations for caring for the poor were on the horizon as well.

CHAPTER 8

# THE AGE OF COMPASSION

By 1200 BCE, Egypt, Mycenaean Greece, and the other great kingdoms around the Mediterranean had broken up, each having suffered its own unique combination of internal political strife, natural disaster, mass migration, and marauding by the still-mysterious sea peoples. Historians refer to this upheaval as the Bronze Age collapse. Similar disruptions of the political status quo were under way at roughly the same time in China and India.[1]

Within four hundred years, new societies had begun to replace the old ones. The new ones were different. For one thing, they put the relationship between rulers and the ruled on a more egalitarian footing: no more divine shepherds, no more cattle of God. For another, they produced many of the religions and philosophical systems that continue to inform our intellectual and ethical concerns today. It was the age of Second Temple Judaism and of classical Athens. It was the age of the Buddha; of Confucius, Laozi, and Mencius; of Socrates, Plato, and Aristotle. The religions that arose and grew during (or just after) this period, including Buddhism, Hinduism, Judaism, Christianity, and Islam, are today collectively practiced by five out of every seven people on the planet.

The six-hundred-year period from 800 to 200 BCE was a time of profound cultural change. Some scholars have gone so far as to view it as a critical turning point—an *axis*—when people began to think in new ways, believe new things, and adopt new views of the universe. Few scholars have seen this moment in history as a more momentous turning point for human consciousness than a German psychiatrist-turned-philosopher named Karl Jaspers, who named it the Axial Age. In his book *The Origin and Goal of History*, Jaspers wrote:

> It would seem that this axis of history is to be found in the period around 500 B.C., in the spiritual process that occurred between 800 and 200 B.C. It is there that we meet with the most deepcut dividing line in history. Man, as we know him today, came into being.[2]

Despite Jaspers's enthusiasm, the historical evidence does not exactly support his idea of a single moment in history that cleaves human consciousness into two distinct eras. Experts can't even agree on when the Axial Age began and ended (or even if it really "existed" at all). Some choose 1400 BCE as a beginning; others pick 400 BCE. Some scholars say it lasted for two hundred years; others say two thousand. What's more, many Axial Age ideas had been circulating well before the classic Axial Age societies began. The Zoroastrians borrowed thousand-year-old ideas from the Indo-Iranians, the Greeks borrowed from the Akkadians, and the Jews borrowed from the Persians, who had liberated them from the Babylonian exile. On top of that, Christianity and Islam—religions that bear many Axial Age features—came along centuries after the official close of the Axial Age.[3]

Even so, the centuries that straddled Jaspers's "deepcut dividing line in history" really did bring some important conceptual changes. Laws were woven together into overarching visions of justice, for instance, and the gods became nosier about humans and their moral affairs. The Greek writers of the Axial Age even increased their use of the word *theos*, which means "god," and *theos* became more closely linked with words related to power and morality.[4]

In some societies, there were even efforts to level inequalities between the social classes, along with a new emphasis on prosocial concern for others.[5] Perhaps most critically (for our purposes here, anyway), spiritual concerns became tightly yoked to concerns for others' well-being. As the writer Karen Armstrong described this particular feature of the Axial Age,

> the only way you could encounter what they called "God," "Nirvana," "Brahman," or the "Way" was to live a compassionate life. Indeed,

> religion was compassion. . . . Further, nearly all the Axial sages realized that you could not confine your benevolence to your own people: *your concern must somehow extend to the entire world.*[6]

Concern for the entire world? How, exactly, are you supposed to pull that off? Each Axial Age society came up with its own approach. The Hebrew scriptures of Second Temple Judaism, for instance, directed the wealthy to set aside some of their crops for the poor. They also called for the cancellation of onerous debts and the restoration of land to its original owners during the Jubilee year (every fiftieth year).[7] The Second Temple books of Tobit and Sirach introduced almsgiving to the West.[8] Several centuries later, Christians adopted these Jewish ideas and institutions for their own communities. Islam, another religious tradition to grow out of Axial Age Judaism, held almsgiving (zakat) in such high regard as to extol it as one of Islam's five pillars.[9]

Axial Age Buddhism placed a similar emphasis on compassion and generosity. One sutra in particular addresses the topic:

> There are three fundamentals to all kinds of giving: (1) giving compassionately to the poor, (2) giving to foes without seeking rewards, and (3) giving joyfully and respectfully to the virtuous. . . . If one can teach others before giving them material things, one is called a great giver. . . . If a wise person is wealthy, he should give like that. If he is not wealthy, he should teach other wealthy people to practice giving. . . . If he is poor and has nothing to give, he should recite curative mantras, give inexpensive medicines to the needy, sincerely take care of the ill for recuperation, and exhort the rich to provide medicines; if he knows medical remedies . . . he should provide treatment according to the diagnosis.[10]

Greece didn't view generosity toward the poor as a matter of moral character in the way the other Axial Age societies did. Instead, the Greeks encouraged civic generosity—philanthropy—in the form of liturgies (in later Rome, *munera*) through which wealthy Athenians supplied money, goods, and services to support the essential institutions for running a state. Out of the city's wealth came an *obol* or two each day in *poleos argurion* (city money) for disabled people to buy food for themselves. For the veterans of Athenian wars, there were public pensions. For the unemployed, there were grand public works projects, jobs aboard the ships in the city-state's fleet, and the opportunity to settle in the newly conquered territories. In times of plenty, all Athenians received small disbursements of money as a perk of citizenship. In times of want, all Athenians received disbursements of grain. Even without turning concern for the poor into a personal virtue, the Athenian social welfare program managed to become the most comprehensive of its time.[11]

## THE CAUSES OF AXIAL AGE GENEROSITY

It would be valuable to know why the ideologies of the Axial Age, with their attention to the needs of strangers (the strangers who were citizens of the *polis*, at least), came about, but until very recently, scholars could only speculate. For his part, Jaspers was struck by the confluence of interstate conflict and trade-fueled increases in wealth in all of the regions that gave rise to Axial Age ideology: "There were a multitude of small States and cities," Jaspers wrote, "a struggle of all against all, which to begin with nevertheless permitted an astonishing prosperity, an unfolding of vigour and wealth."[12] The Italian historian Arnaldo Momigliano noticed that all of the Axial Age civilizations "display literacy, a complex political organization combining central government and local authorities, elaborate town-planning, advanced metal technology and the practice of international diplomacy."[13] The sociologist Robert Bellah thought literacy was of paramount importance because it enabled people to move away from understandings of the world based on myth and allegory and to move instead toward explanations based on analysis, symbolic logic, and evidence.[14] So who's right? It's hard to know.

Frustrated by the decades of data-free speculation about the causes of the Axial Age, the psychologist Nicolas Baumard and his colleagues assembled a data set that included century-by-century measurements of material affluence and political success in eight different ancient societies. Of the eight (Egypt, Mesopotamia, Greece, China, India, Mesoamerica, the Andes, and Anatolia), only three would go on to become Axial Age societies (Greece, China, and India). With these data in hand, Baumard and colleagues conducted statistical tests comparing the societies and concluded that it was material affluence (measured, for example, as the number of calories the average adult could extract from the environment through a hard day's work), and not political success (measured by the size of the state), that best distinguished the three societies that became axial from the five that did not. Put plainly, the axial societies were richer.

Even more impressively, Baumard and colleagues found that the axial societies experienced significant gains in affluence right around the time of the axial transition in 500 to 300 BCE.[15] In the five hundred years leading up to the Axial Age, vigorous trade and improved technology raised the average adult's daily economic output in the three Axial Age societies by five thousand calories—more than enough to sustain two adults. Houses became bigger, too: In 800 BCE, the average Greek home measured about six hundred square feet. At the height of the Greek Axial Age, the average home was five times that size.[16] These improvements in comfort and affluence were not shared equally among all citizens: by modern standards, economic inequality continued to be very high. But for society's elites, at least, it would not have been hard to see that theirs were the best of times.[17]

Baumard's conclusion that people became more "axial" (and, by extension, more concerned about the welfare of strangers) after they became richer is bolstered by other research teaching the same lesson, albeit in very different ways. In one study, for instance, the economists Jon Bakija and Bradley Heim studied the link between income and charitable donations by analyzing data from the federal income tax returns of more than sixty thousand anonymous individuals and couples. They found that every 1 percent increase in persistent income (for example, the income that would come from a pay raise, which one would expect to enjoy permanently) corresponded to a 0.5 percent increase in charitable contributions.[18] As people became more prosperous, Bakija and Heim found, they become more generous. Other work reveals that changes in Americans' charitable contributions track the year-over-year changes in several other indicators of material prosperity, including the stock values of America's five hundred largest corporations, the gross domestic product (GDP), and the employment rate.[19] Finally, the economists Chau Do and Irina Paley found that charitable giving tracks home equity: the more equity people have in their homes, the more they donate to charity.[20] Taking all of these facts into account, it is reasonable to surmise, as Baumard and his colleagues have, that "concern for the entire world" became easier for our Axial Age forebears because their lives had become easier.[21]

Not everyone agrees with that conclusion. The sociologist Stephen Sanderson argued, based on his own research, that it was not wealth that ushered in the Axial Age, but two other factors. The first was urbanization: cities were getting bigger. The second was the intense interstate warfare that the new Iron Age weapons made possible. Whereas Baumard and colleagues attributed Axial Age compassion to Axial Age prosperity, Sanderson attributed it to Axial Age angst:

> The enormously disruptive effects of large-scale and rapid urbanization and the intensification of warfare during the second half of the first-millennium BCE created new human needs for ontological security, anxiety reduction, and release from suffering. People's existing social attachments were being undermined by the altered circumstances they faced. The old pagan religions of the ancient world were not up to the task of meeting these new challenges. As a result, people began to create new kinds of religion.[22]

## THE AXIAL PRIME DIRECTIVE

Though scholars continue to disagree sharply about the causes of the Axial Age, they agree about its most important ethical innovation. The Golden Rule—the mandate to do unto others as you would have them do unto you—did not get its start with a 1961 Norman Rockwell painting: it shows up in many of the Axial Age's most influential texts.[23] In his Analects, for example, Confucius wrote, "What you do not want done to yourself, do not do to others."[24] The Mahabharata of Axial Age Hinduism gave similar guidance: "Knowing how painful it is to himself, a person should never do that to others which he dislikes when done to him by others."[25] The Book of Leviticus in the Jewish scriptures features an Axial Age Yahweh who commands his followers, "Thou shalt love thy neighbor as thyself." Centuries later, Jesus took the "love thy neighbor" idea even further with his parable of the Good Samaritan. Aristotle didn't leave us with any pithy sayings about doing unto others, but he based his theory of friendship on the notion that friends wish for each other what they would wish for themselves.[26]

Today, the Golden Rule may seem quaint, but 2,500 years ago, it would have sounded outré, perhaps even revolutionary. After all, the ethical ideals that people had relied upon before the Axial Age implied that the only people worthy of real moral consideration were those within one's own home, lineage, or hometown. The Golden Rule pushed against those boundaries. We get a hint of how this new idea sounded to naïve listeners from the response of Severus Alexander, who ruled the Roman Empire from 222 to 235 CE, as written down by the author of the *Historia Augusta*:

> If any man turned aside from the road into someone's private property, he was punished in the Emperor's presence according to the character of his rank, either by the club or by the rod or by condemnation to death, or, if his rank placed him above all these penalties, by the sternest sort of a rebuke, the Emperor saying, "Do you desire this to be done to your land which you are doing to another's?" He used often to exclaim what he had heard from someone, either a Jew or a Christian, and always remembered, and he also had it announced by a herald whenever he was disciplining anyone, "What you do not wish that a man should do to you, do not do to him." And so highly did he value this sentiment that he had it written up in the Palace and in public buildings.[27]

Confucius, Moses, Jesus, and Severus Alexander didn't teach the Golden Rule because they thought it was cute: they taught it because they believed it made for better people. Not everyone has agreed, then or now. Some contemporary philosophers argue that the Golden Rule is hopelessly flawed by its implication that we can figure out how to treat others by simply considering our own wants and needs. Others worry about a rule that seems to entitle masochists to treat others sadistically—after all, that's how masochists would like others to treat them. Other philosophers object that looking inward to discover what is right breeds ethnocentrism and perpetuates the moral status quo.[28] Some would even have us imagine a judge who uses the Golden Rule to justify why she decides to let a convicted mass murderer go free: If the shoe were on the other foot, the judge would want to avoid prison time, so shouldn't she extend the same consideration to the killer? Because the Golden Rule supposedly has problems such as these, the ethicist Kwame Anthony Appiah has called it "fool's gold."[29]

These worries are quite silly, though, unless you assume that the person attempting to live by the Golden Rule lacks the moral judgment of any neurologically intact fifth grader. An actual masochist who got sexual pleasure from abuse at the hands of others, yet sought to live by the Golden Rule, wouldn't follow it so literally as to assume that it obligated him to abuse other people; instead, he would know that others usually had tastes and preferences different from his own, and treat them accordingly. Likewise, a judge seeking to follow the Golden Rule in her professional decisions would not need to vacate the sentences of mass murderers; instead, she would also consider her obligations to the law-abiding people who would not want convicted murderers running around free.

The philosopher Harry Gensler is the world's leading exponent of the idea that the Golden Rule can, when read properly, withstand close ethical scrutiny. As he explained in his book *Ethics and the Golden Rule*, many philosophical objections to the Golden Rule vanish once we understand how to implement it intelligently. Toward this end, Gensler recommends we use an algorithm with four steps represented by the acronym KITA (Know-Imagine-Test-Act). In the Know step, we take time to discover what will help a specific person and what will harm him. A conscientious Golden Rule follower does her homework. After learning about a specific person's basic needs and desires, a conscientious student of the Golden Rule will then implement the Imagine step by trying to picture how her possible courses of action will affect others. Gensler isn't talking about idle, millisecond flashes of intuition. He's talking about a concerted effort to work through the possible consequences for everyone who might be affected. The judge has to consider not only how her sentence will affect the convicted criminal, but also how it will affect the community.

Next, a conscientious Golden Rule follower proceeds to KITA's third step, which is the Test for consistency: she must ask whether the action she has in mind is consistent with how she would want to be treated in exactly the same circumstances. Finally, she is ready to execute the fourth and final step: she can Act.[30] Applying the KITA algorithm responsibly, Gensler maintained, leads us to what we would surely want out of a truly effective Golden Rule for human conduct: it shows us how to "treat others only as you [would] consent to being treated in the same situation."[31]

The Golden Rule does have a real weakness, however, and it is a weakness that it shares with all explanations for human behavior that rely on moral principles to explain why we do what we do: Which moral principle gets to control your behavior at any given time? After all, there are many other very fine-sounding chestnuts besides the Golden Rule that might cross your mind when you encounter a stranger in need. There's "Be wary of strangers," for instance, and "God helps those who help themselves," and (my personal favorite) "A failure to plan on your part does not constitute an emergency on my part."

Because there are so many principles one could apply in any given situation—including some that might stir up our compassion and others that might stifle it—it seems unlikely that people's endorsement of the little maxim that we know as the Golden Rule could fully explain how our Axial Age ancestors came to care about the welfare of strangers as they did. As the social psychologists Bibb Latané and John Darley observed,

> like the King of England, norms may reign, but not rule. That is, they may exist, but have so little application to the complex real-life situations in which ethical considerations arise as not to be useful in explaining the actual variations in help-giving.[32]

Latané and Darley may have overplayed their hand a bit, but they have a good point. Even so, it's possible that Golden Rule reasoning can generate compassion through a more circuitous route. Rather than simply calling on us to reason our way to a decision about whether to help each new stranger we encounter (where the Golden Rule might conflict with other perfectly sensible norms), perhaps the Golden Rule works hand in hand with our ability to form new habits. Once you have used Golden Rule reasoning to decide on a course of action toward any particular person—say, someone whose car has broken down on the side of the highway—your policy for responding to the next stranded motorist you come across becomes simpler: just think about doing what you did the time before. As people repeatedly apply their new policy, which they originally acquired through the sort of Golden Rule reasoning that might please Harry Gensler and KITA practitioners everywhere, they eventually acquire a Golden Rule habit.

Darwin was a committed enough evolutionist to conjecture that our social instincts and sentiments, designed to motivate us to help family and friends, "naturally lead to the Golden Rule," but he was also a good enough psychologist to recognize that those social instincts and sentiments could lead to the Golden Rule only "with the aid of active intellectual powers and the effects of habit."[33] In their book *The Power of Ideals*, the developmental psychologists William Damon and Anne Colby elaborated this important point about how moral reasoning can lead to ingrained habits that support moral behavior:

> Most of the routine habits we rely on to guide us honorably through daily life have been worked out during childhood and do not need to be thought about once they've been solidly acquired. Mature people don't need to make a conscious choice about whether or not to grab some candy they might like from a store shelf if no one is looking. They pay for it without thinking twice about it. But at age four this basic rule of conduct may have been less obvious. For the young child, taking candy without paying for it may be a tempting option. The choice to pay for the candy needs to be learned, perhaps in the course of stern inputs from nearby adults. If learned well, the choice eventually becomes habitual and non-conscious. In general, choices that become habitual and automatic later in life often begin as decisions requiring learning and reflection. This is a basic principle of development: Once a skill is learned, we can perform it quickly; but the process of learning can be laboriously reflective and far from instantaneous.[34]

## FROM REASONING TO RHETORIC

So far, I have concentrated on the roles of effortful deliberation and the force of habit to help Golden Rule reasoners reach satisfactory ethical decisions. However, the Golden Rule's impact on humanity's regard for strangers wasn't only through its effects on individuals in isolation. Much of its impact (if not most of it) came instead from its rhetorical force. Imagine a group of Axial Age elites—teachers, politicians, elders, and leading citizens—as they tried to decide how best to help the poor residents in their communities (along with poor travelers who were just passing through). In arguments about matters such as these, the Golden Rule may have sounded persuasive because of its folksy sensibility, its consistency with other concerns of the Axial Age, and its inexorable logic.

Opponents of Golden Rule reasoning would have put themselves at risk of seeming hard-hearted and hypocritical.

Even today, political leaders press the Golden Rule into rhetorical service when it suits them. On June 1, 1963, for instance, the Alabama National Guard was sent to the University of Alabama at the direction of the United States District Court for the Northern District of Alabama. The Guardsmen were tasked with ensuring that two African American students who had been admitted to the university were able to attend their classes despite hateful invective and defiant rhetoric that had been pouring in from all corners of the South. Later that evening, President John F. Kennedy delivered a televised speech in which he invoked the Golden Rule to explain the federal government's determination to keep the students safe: "Every American ought to have the right to be treated as he would wish to be treated, as one would wish his children to be treated."[35]

Likewise, when asked in 2016 to defend a federal order requiring public schools to allow transgender kids to use any bathroom they preferred, President Barack Obama appealed not only to his understanding of what the law required, but also to his understanding of what the Golden Rule required:

> My answer is that we should deal with this issue the same way we'd want it dealt with if it was our child. And that is to try to create an environment of some dignity and kindness for these kids. . . . Look . . . I have profound respect for everybody's religious beliefs on this. But if you're at a public school, the question is, how do we just make sure that children are treated with kindness. That's all. And you know, my reading of scripture tells me that that golden rule is pretty high up there in terms of my Christian belief.[36]

To any list of the Golden Rule's positive attributes, I would therefore add the following: it sounds good when you say it loud and with conviction.

And here, I think, is one of the neglected ways that the Golden Rule changed history. Teachers, priests, and other elites used it to make their cases for how institutions should be organized to best meet the needs of the poor and disadvantaged while also preserving the welfare of the community at large.[37] And after those institutions were up and running, individual observance of the Golden Rule—either through active attempts to abide by it or through the power of habit—became less central to the enterprise. Once a society has well-functioning institutions, Golden Rule reasoning and even Golden Rule habits can take a back seat to responsible administration and savvy fundraising.[38]

## JEWISH CHARITIES IN THE GOLDEN RULE ERA

For a case study in the power of the Golden Rule to transform a community's approach to generosity, especially when it is combined with strong institutions, we can look to the Jewish communities of the Axial Age and the first few centuries thereafter. Through most of the first millennium BCE, Jewish law provided rather little guidance on how to support the poor, aside from the same sorts of laws about gleaning, burdensome interest rates, and protections for orphans and widows that had been on the books for centuries in other Near Eastern societies. There wasn't much else to draw from. And in the absence of strong institutions with clear rules and clear plans for enforcement, people were motivated to help mainly by the tugs of their own consciences, their appetites for approval in the eyes of their peers, and the prospect of obtaining God's blessings.

This approach had changed by the first century of the Common Era. By then, Jewish laws that required farmers to set aside portions of their fields for the poor had finally been given some teeth. Infractions of the gleaning laws came to be regarded as theft from the poor. According to the new laws, the poor could take you to court and sue you for damages. If found guilty, your thievery could even get you publicly flogged.

Later, the Jews set up institutions such as community soup kitchens, called *tamhu'i*; poor-relief funds, or *kuppah*; and a charity fund, the *kis* (or *arnak shel tzedaka*, which means "charity purse"), that helped orphans with their dowries (which were essential for good marriage matches). To fund these efforts, charity wardens collected taxes from all households; additional collections took place in the synagogue on Sabbath days. Those who did not contribute to these fundraising efforts were regularly threatened with shaming, ostracism, flogging, and even court-ordered seizures of property. In time, responsibility for meeting the needs of the poor was shifted almost entirely away from voluntary initiatives and to community institutions instead.

Jewish experimentation with institutional support for the poor did not end there. They made laws to protect workers from exploitation: they forced employers to pay their workers promptly, they limited the length of the workday, they devised methods for handling disputes between workers and employers, and they set up rules to protect slaves from life long indenture. By the first century CE, synagogues throughout Judea had also established hostels to provide poor travelers with room and board. These hostels were a big departure from the Greek hostels, which provided hospitality (*xenia*) only to visitors with the demonstrated ability to reciprocate once they made it home. Poor travelers to Greek towns had to content themselves with sleeping in the city square.[39] By the time of the Talmud, Jewish law required every town to maintain a school for children and a physician (even though medical care at the time consisted mostly of keeping the patient clean and dry, fetching food and blankets, and the occasional bloodletting). By the second century CE, Jewish communities had opened these charitable institutions to *all* residents—including non-Jews.[40] These advances became the new ethical standard in just a few hundred years.[41]

## A DEEP BREATH

"The Axial Age," Karl Jaspers wrote, "can be called an interregnum between two ages of great empire, a pause for liberty, a deep breath bringing the most lucid consciousness."[42] It was a cosmopolitan time, an era when it was cool to be engaged in the wider world. Writing was maturing from its former existence as a tool for bean-counting and imperial crowing into a robust, high-capacity data-storage system that could support abstract thought and advances in knowledge. When the cosmopolitanism and rising literacy of the era intersected with its urbanization and rising affluence, scholars began to ask deeper questions about how to live a good life—questions they couldn't answer very well, it turned out, without also figuring out how to live an ethical life.

Enter the Golden Rule, that target of so much clucking from professional philosophers. The Golden Rule was taught far and wide—in India,

China, Egypt, Persia, Greece, and Israel—as one of the great advances in ethics, and it became vitally important as our Axial Age forebears argued about how best to help people in need. Most importantly, on the basis of Golden Rule reasoning, the Axial Agers experimented with a wide variety of social institutions for meeting the needs of people in dire straits. If people wanted to live by the Golden Rule, the charitable institutions of the Axial Age made it easier by freeing them from the cognitive burdens of reasoning through the dilemmas of every new stranger they encountered. All they had to do was support the institutions that had been created to implement the Golden Rule on their behalf.

The Axial Age, for all its "lucid consciousness," didn't turn the deserts of antiquity into ethical Gardens of Eden. Life remained poor and precarious for most people, just as it always had been, so the idea that poverty could somehow be eradicated on a large scale would have been unimaginable—even when Axial Age visionaries were at their most lucidly conscious. The writers of the Book of Deuteronomy reminded the Jews that the poor would never cease to be in the land. Several centuries later, Jesus had to remind his students of the same hard truth. The Hindu religion normalized poverty by conceptualizing middle adulthood as a time of satisfaction and abundance bookended by two stages of penury—the first as a poor young student of a guru, the second as a street beggar who, made wise by a full life, would finish that life in renunciation of the world's pleasures and treasures.

It is also worth remembering that the selfish benefits associated with kin altruism, direct reciprocity, and indirect reciprocity remained as alluring as ever. Donors to hospitals, guesthouses, and food distribution centers sometimes earmarked their donations with particular sets of beneficiaries in mind: family and lineage first. The general public got what was left.[43] Helping people in need also remained a trusty method for buying friendship and loyalty—often at very favorable rates of exchange.[44] Perhaps it is no coincidence that mouse-and-lion-type fables—in which a weak creature gains a lifetime of grateful indebtedness from a much stronger one by helping her in a time of extreme crisis—originated in the literature of Axial Age Greece and India.[45] Finally, because grand public gestures of generosity toward strangers say something about the generous person's character (and bank balance), conspicuous giving would surely have helped people obtain new business contacts, new friends down in

City Hall, and better marriage matches for their kids. The Greeks provided hospitality to out-of-towners not only because they pitied them (though perhaps they did), but also because they knew their hospitality marked them as men of good standing.[46] When people helped the poor, they were not unaware that they were helping themselves as well.

All of that is to say that Axial Age generosity was not governed exclusively by Golden Rule reasoning. It was also motivated by a mix of fatalism, evolved appetites for the material benefits that generosity can bring, and the promise of spiritual benefits—peace of mind, an encounter with God disguised in a beggar's cloak, the forgiveness of sins, or another rung up the ladder to Nirvana.[47]

Finally, we should keep in mind that Axial Age generosity was rarely inspired by the concerns about efficiency that preoccupy us today: "Because our focus is so singularly set on these sorts of factors," Gary Anderson wrote in *Charity: The Place of the Poor in the Biblical Tradition*,

> one often misses the crucial variable that drove much of Christian history for its first fifteen hundred years: the promise that Scripture provides that one could meet God in the face of the poor. Charity was, to put it briefly, a sacramental act. That is, an act that established a contact point between the believer and God. To think of poverty as a social problem that could be solved was not really imaginable in the mindset of premodern man.[48]

To think of poverty as a social problem that could be solved, or at least brought to heel, the world would require new ideas, new institutions, and a new approach to financing it all. The poor would have to wait another 1,500 years to see these innovations come to fruition.

## CHAPTER 9

# THE AGE OF PREVENTION

Through most of history, people thought about poverty the way we think today about death, taxes, and thinning hair: as just another of life's unpleasant inevitabilities. The Jewish teachers of the Axial Age complicated that ancient view of poverty with this counterintuitive idea: of all his children, despite appearances to the contrary, God loved his poor ones best. Jesus's words, "Blessed are the poor in spirit, for theirs is the kingdom of heaven," sum up the Axial Age attitude toward the poor quite nicely. It was a theological viewpoint that lasted all the way through the Middle Ages.[1]

That said, you rarely ran across long lines of people signing up for lives of poverty: most people were happy to let this particular blessing pass them by. This is true despite the fact that Christian traditions, texts, and teachers warned that wealth, if not handled responsibly, could be hazardous, and eternally so. Even in the Middle Ages, when most people were illiterate, every European knew Jesus's parable about the rich man and Lazarus the beggar:

> There was a rich man who was dressed in purple and fine linen and lived in luxury every day. At his gate was laid a beggar named Lazarus, covered with sores and longing to eat what fell from the rich man's table. Even the dogs came and licked his sores. The time came when the beggar died and the angels carried him to Abraham's side. The rich man also died and was buried. In Hades, where he was in torment, he

> looked up and saw Abraham far away, with Lazarus by his side. So he called to him, "Father Abraham, have pity on me and send Lazarus to dip the tip of his finger in water and cool my tongue, because I am in agony in this fire." But Abraham replied, "Son, remember that in your lifetime you received your good things, while Lazarus received bad things, but now he is comforted here and you are in agony."[2]

Eternal torment in a lake of fire was an afterlife fate that most medieval Europeans were eager to avoid, so they raised memorials to Lazarus and the rich man in carved wood, wrote about them on parchment, and depicted them in stained-glass windows and in the frescoes that adorned the cathedral walls. Flesh-and-blood encounters with the poor did an even better job of alerting people to the perils of wealth. In return for their services as living reminders of these hazards, wealthy people hosted banquets, distributed alms, and left a little something for the poor in their wills. Municipal authorities and religious orders ran almshouses for widows and the disabled, along with hospitals for orphans and foundlings.[3] Although formalized poor relief never amounted to more than 1 or 2 percent of gross domestic income, the churches and monasteries devoted as much as 7 or 8 percent of their incomes to poor relief.[4] Working peasants also formed cooperatives, called "fraternities" and "gilds," that collected contributions to support members during hard times. An additional benefit of membership in these cooperatives was that they gave people a structured opportunity to build up treasures in Heaven.[5]

By the 1500s, however, new tidal waves of poverty began to outstrip the existing institutions for care of the poor. Agrarian economies were giving way to economies based on global trade. Wealthy landowners bought up as much land as they could to raise sheep for wool and to grow crops that could be turned into high-priced goods to sell on international markets.[6] As a consequence, many peasants turned to work as laborers in building construction and manufacture, but population growth was suppressing wages: there were too many workers for too few jobs. Veterans from Europe's many wars, lacking good options for making a living, also swelled the numbers of the vagrant poor. Meanwhile, food prices kept rising because the land that remained for domestic food production was becoming less productive. Low wages and high food costs worked as a sort of vise, squeezing most fifteenth-century Europeans' incomes below what they actually needed to meet the needs of a small family.[7] Without enough land to grow their food, and without enough money to buy it, the working poor in England and in Europe were being pushed into destitution.[8] The rampant poverty and hunger were making Europe's cities intolerable, as one visitor to Vincenza in northern Italy (about fifty miles west of Venice) confided to his diary in 1528:

> If you give alms to two hundred paupers, you will immediately be surrounded by two hundred more; if you walk down the street, or cross a square, or enter a church, a crowd of beggars will rush at you. Hunger is written on their faces; their eyes are like empty holes and their bodies seem to be no more than bones covered with skin.[9]

The throngs of indigent people who were surging into Europe's major cities in search of work were also bringing the worst effects of mass poverty—malnutrition and disease—into bold relief. One diarist's account of a Parisian epidemic in the summer of 1522 gives a sense of just how devastating such outbreaks of disease could be:

> A plague reigned in Paris at this time, dangerous and wondrous strange, and so deadly that a hundred and twenty people died within three days so it was said in the Hôtel-Dieu of this city. At the cemetery of the Holy Innocents more than forty were buried in one day, and common burial was given each day to twenty-eight or thirty people, so that after two months there were dead in great number, not counting those buried in the cemeteries of other churches. It was said that death descended mainly on the poor, so that porters, the pittance-workers of Paris, of whom before there had been a great number, seven or eight hundred perhaps, were reduced to a small handful after this disaster befell them. As for the Petits Champs quarter, it was quite cleansed of paupers, who had lived there in great number.[10]

Europe's crisis of poverty sent politicians scrambling for solutions. Even if poverty was inevitable, they did begin to wonder whether society might be better protected from its worst effects with the right sorts of interventions. Some of Europe's finest minds began to ask surprisingly modern questions about the causes of poverty, its consequences, and the policies that were best suited to protecting society from them. In doing so, they ushered in the Age of Prevention.

In the imaginary land that the English humanist Sir Thomas More invented for his 1515 book *Utopia*, the obligation to work was universal and absolute. In Utopia, "strong and lusty beggars that go about pretending some disease in excuse for their begging" were nowhere to be found.[11] Likewise, the Dutch humanist Erasmus argued in the 1524 edition of his *Colloquia* that city officials could do more to reduce public begging by finding work for the poor than they ever could by simply providing alms.[12]

Emboldened by *Utopia* and other humanist writings, secular officials in many European cities began to play with a new philosophical outlook on poverty. According to this new viewpoint, poverty was not to be regarded primarily as one of the universal facts of life (as our Neolithic ancestors viewed it), or as a sad inevitability for widows and orphans (as the fatalists of the Bronze Age viewed it), or as an opportunity for the rich to buy their way out of a fiery afterlife (as the Christians of medieval Europe viewed it), or as a primarily spiritual condition (as the priests and monks viewed it). Instead, it was to be regarded as a social ill that harmed not only the poor, but also the secular state and its citizens. A profound change in how we think about poverty was in the offing.[13]

## VIVA VIVES

No one did more to lay the groundwork for the Age of Prevention than the humanist scholar Juan Luis Vives.[14] In 1525, the senate of the city of Bruges, northwest of Brussels in Belgium's Flemish region, asked Vives to design a program to help solve the city's poverty problem. The resulting book, *De Subventione Pauperum* (Concerning the relief of the poor), included a comprehensive plan for identifying the people whom the city should help, a rational approach to allocating that help, and a scheme to pay for it all.

Vives has rightly been called the "Reformer of Charity," but his ideas weren't exactly new: Erasmus and More had been flirting with similar ideas for at least a decade. However, Erasmus and More weren't trying to win a policy war. Vives was. In a classic example of speaking truth to power, Vives laid blame for poverty and its consequences not at the feet of the poor themselves, but at the feet of the people who ruled them:

> For they have no conception of the duty of government who wish to limit it to the settling of disputes over money or to the punishment of criminals. On the contrary, it is much more important for the magistrates to devote their energy to the producing of good citizens than to the punishment and restraint of evildoers. How much less need would there be to punish, if these matters were rightly looked after beforehand![15]

Vives leavened his argument with appeals to common sense. "Why is it not true," he wrote,

> that, just as everything in the state is restored which is subject to the ravages of time and fortune—such as walls, ditches, ramparts, streams, institutions, customs, and the laws themselves—so it would be suitable to aid in meeting that primary obligation of giving, which has suffered damage in various ways?[16]

As Erasmus before him had done, Vives appealed to his readers' sense of Christian duty ("Surely it is a shame and disgrace to us Christians, to whom nothing has been more explicitly commanded than charity . . . that we meet everywhere in our cities so many poor men and beggars").[17] Even so, Vives wasn't writing a work of theology. The major rhetorical force for his argument came from his enumeration of the many social ills that a comprehensive reform of charity could *prevent.* Poor relief needed to be reformed because poverty caused misery and civic disorder *now, in this world.* Poverty is a risk factor for disease, Vives asserted, so mass poverty turns the city into a sink for epidemics. When the masses are miserable, while the privileged few live in comfort, the city becomes vulnerable to civil unrest. When the poor can't meet their basic needs through honest work, they turn to crime. Poverty should be reduced, Vives believed, not chiefly because God wanted the poor to be comforted, but because poverty was bad for public health, bad for social order, and bad for business.

The Vives plan went straight to the heart of things. Anticipating the modern business consultant's mantra that "You can't manage what you don't measure," Vives called for the city to comprehensively assess the extent of its poverty and to catalog its many manifestations. He also advised the city's magistrates to identify every existing charity so they could coordinate their efforts: "Let the governors of the state realize that all these institutions are a part of their responsibility," he wrote.[18] He also called for a medical exam and a sixteenth-century version of a social work assessment for every poor person. Poverty had many causes, Vives observed—the death of a family breadwinner, child abandonment, addiction, lack of job skills, mental illness, blindness, laziness—and he recommended specific remedies for each one.

Food, shelter, clothing, and the basic necessities of life were to be provided to all eligible poor people: enough to sustain them, but not enough to make them content to stay idle. For Vives, education and work were the most important antipoverty measures, so he advised the adoption of a comprehensive program of education and job training. He also recommended that prospective aid recipients' talents and abilities be taken into account when arranging their job-training opportunities: more measurement still.[19]

No one was to be permitted to wander around aimlessly. "Let no one among the poor, therefore, be idle," Vives advised, "provided of course he is fit for work by his age and the condition of his health."[20] Even blind people were to be trained in the workshops and artisans' studios:

> Nor would I allow the blind either to sit idle or to wander around in idleness. There are a great many things at which they may employ themselves. Some are suited to letters; let them study, for in some of them we see an aptitude for learning by no means to be despised. Others are suited to the art of music; let them sing, pluck the lute, blow the flute. Let others turn wheels and work the treadmills; tread the wine-presses; blow the bellows in the smithies. We know the blind can make little boxes and chests, fruit baskets, and cages. Let the blind women spin and wind yarn. Let them not be willing to sit idle and seek to avoid work; it is easy enough to find employment for them. Laziness and a love of ease are the reasons for their pretending they can not do anything, not feebleness of body.[21]

It would be a mistake to view the Vives plan as merely a blueprint for improving charity. In fact, it was a comprehensive plan for determining the factors that led people into poverty, for identifying the measures that would get them out of poverty, and for giving them skills that could prevent them from falling into poverty in the first place. For all its carrots, the Vives plan had its share of sticks as well. Anyone who had previously shown a preference for drinking and gambling, Vives counseled, should be assigned the most menial and unpleasant jobs: "But to them more irksome tasks should be assigned and smaller rations, that they may be an example to others, and may repent of their former life and may not relapse easily into the same vices."[22] Any foreign beggar was to receive travel money and a one-way trip back home. Anyone feigning disability was to be punished. Gamblers, prostitutes, and heavy drinkers were also to be rounded up and punished, if that was what it took to transform them into productive members of society. Any administrator caught trying to cheat the system was to be punished severely.[23]

Big municipal plans require big municipal purses. How to pay for it all? Vives thought most of the expenses could be covered simply by using existing resources more efficiently. More vigorous efforts to solicit donations and end-of-life bequests from individual citizens, along with the new income that the poor would generate in their new jobs, would pay for the rest. Vives expected that some readers would be skeptical of his financing plan. He simply reminded them to have faith: "Wherefore, in pious undertakings it is sacrilegious to consider how much you can do; consider rather how much faith you have in Him to whom all things are possible."[24]

The senators of Bruges responded to Vives's plan decisively: First, they printed it up and distributed it so the city's stakeholders and decision-makers could give it the attention that such an important document deserved. Then, they dutifully buried it.[25] Bruges was not yet quite ready to enter the Age of Prevention.

Church officials weren't ready either. To some leaders in the Catholic Church, the demoralization of poverty and the regulation of begging seemed like a sort of violence against the poor. Others resisted Vives-style plans because they threatened the status quo for some of Europe's most influential monastic orders (including the Franciscans, the Dominicans, and the Augustinians), which had specialized for centuries in providing poor relief. Many church leaders also suspected that the assignment of poor relief to the secular authorities was a Lutheran power-grab. Vives and several of his reform-minded contemporaries in Belgium, France, and Spain had to tread lightly to sidestep formal charges of blasphemy and heresy.[26]

Progressives within the church and the government, however, saw the wisdom in reform. Cities began to centralize their charitable efforts, beginning with a few German cities in the 1520s that had begun to institute secular reforms (owing to Martin Luther's influence) even before Vives came up with his proposal. Paris put its own extensive reforms in place in 1525, and a string of Flemish cities (including Bruges) followed suit in the 1530s.[27] An imperial edict that Charles V issued in 1531 led to Vives-style reforms throughout the Holy Roman Empire. In most cities, begging was either outlawed outright or severely restricted: in certain parts of Spain, for example, it became illegal to beg without a license. In some cities, violations of these ordinances were punishable by hard labor, flogging, a few years of indentured servitude, or even public hanging. All in all, more than sixty European cities reworked their systems of poor relief within just a few decades.[28]

Even the church gave way as the tide of learned opinion moved in the direction of Vives. By the end of the sixteenth century, as the historian Bronisław Geremek put it,

> the great controversy over charity had burnt itself out and the conflicts abated. This is not to say that they came to a definitive end, for the following century saw a continuation of the debate on both the practical and on the theoretical plane, and the subject was still a prominent one in religious and ethical literature. The reform of charity ceased, however, to be perceived as the "municipal heresy," a threat to the interests of the Church. It became incorporated into the ideology of the modern State, and accepted as the State's prerogative.[29]

We can learn a great deal about the Age of Prevention and how it unfolded by looking at how it manifested itself in two different European countries: England and the Dutch Republic.

## ENGLISH REFORM

England's sixteenth-century efforts illustrate just how deliberately some countries worked to put the ideas of Vives and the other Christian humanists to the test. The English members of Parliament (MPs) enacted a long list of Vives-style laws for checking the growth of poverty. They experimented with registering and licensing beggars. They tweaked the punishments for vagrancy. They tried to figure out how to distribute aid more rationally. Frustratingly, these measures did little to improve the lot of the poor. Most of the lawmakers' earliest efforts were reduced to "false starts, parliamentary compromises, and half-measures."[30] In England, as elsewhere throughout Europe, populations continued to grow, food prices continued to rise, and real wages continued to plummet.[31] Even fully employed people had a hard time staying above the poverty line, so life had the potential to be truly miserable for the unemployed. In the face of this continued misery, along with the appearance that most reform efforts were only making things worse, the poor in many European cities responded with resistance and riots.

King Henry VIII's eagerness to reduce the Catholic Church's influence made things worse still. His Protestant fervor, along with his bitterness over the pope's refusal to nullify his heirless marriage to Catherine of Aragon, compelled him to declare himself head of the new Church of England in 1534. In due course, Henry closed England's Catholic monasteries—and along with them, the hospitals they operated for the benefit of the poor.[32] The consequences were disastrous. In 1500, England's monasteries were responsible for more than 20 percent of all care for the indigent poor.[33] With the dissolution of the monasteries, the poor relief they had been providing dissolved as well. By the end of the sixteenth century, nearly half of England's institutions for care of the sick and the aged had been shuttered. The same fate awaited thousands of farmers' cooperatives. Henry ransacked virtually all of them, pocketing their assets.[34]

Although Henry managed to do away with most of the existing institutions for poor relief in the country, a few new institutions did come online in the meantime (including several new large-capacity royal hospitals, outfitted with workshops in which the poor could perform unskilled labor to offset the costs of their room and board). Most of the new institutions were smaller than the ones they replaced, however. Meanwhile, England was undergoing a population boom. The decline in institutional resources, in tandem with the expanding English population, created a scissoring effect that halved the per capita number of beds available for care of the indigent poor by the end of the century.[35] Private philanthropy was almost surely too sluggish to make up for what was lost.[36] The sixteenth century, for all its efforts at reform, was not a good time to be both English and poor. Of course, it wasn't a great time to be poor anywhere else, either.

The Vives-inspired reforms of sixteenth-century England didn't work out as many had hoped, but they were a dress rehearsal for the improved policies that came after them. In response to more bad harvests and the food shortages that followed, in 1598 Parliament passed a raft of new laws designed to finally bring England's poverty problem to heel. Collectively, these acts came to be known as the Elizabethan Poor Laws, and they would remain in effect for the next two centuries.

Some of the Elizabethan Poor Laws solidified government policy on the punishment of "rogues, vagabonds, and sturdy beggars." People caught begging or hustling were to be whipped upon the first offense and then returned to their homes or places of birth. Second offenses were punished with banishment (although this penalty was later reduced to branding). Third offenses drew the death penalty. People seeking aid were forced to do so within the parishes in which they were born. Still other legislation, which was crystallized a few years later in the Charitable Uses Act of 1601, cut the red tape required for private philanthropists to found hospitals, almshouses, and so-called houses of correction, which provided the indigent with rudimentary work in exchange for rudimentary food, shelter, and clothing.

But the most important difference between the Elizabethan Poor Laws and the reams of ineffectual legislation from earlier in the sixteenth century was that the new laws regulated how poor relief was to be distributed and paid for. The new laws mandated that England's ten thousand parishes (a level of English political organization below the county) regulate their own poverty through their own charitable institutions and their own fundraising efforts. Each parish was required to provide food to those who were too young, too old, or too sick to work. The new laws also required the parishes to provide jobs for those healthy enough to work, to create apprenticeships for poor children, and to raise taxes—"poor rates"—through annual levies on property.[37]

This basic plan—local relief of poverty with locally raised funds—continued in England and Wales through the 1700s, although Parliament and the parishes would continue tinkering with it. In 1723, Parliament passed the Workhouse Test Act of 1722/1723, also called Knatchbull's Act in honor of its sponsor, Sir Edward Knatchbull. Knatchbull's Act gave individual parishes the right to restrict poor relief to the workhouses: if you were poor but able-bodied and wanted to claim aid, you could be forced to live in a workhouse as a precondition, where you could pay for your supper with hard labor. In parishes that did not have their own workhouses, local administrators could contract their paupers to "farmers," who would agree to meet the paupers' basic needs in exchange for their labor. The farmers were free to pocket any profit from that labor. Thus, Knatchbull's Act enabled local administrators to test the depth and sincerity of an applicant's stated need for relief. If you accepted its terms (including its requirements for working and residing in a workhouse, with its attendant loss of liberty and lousy living conditions), the parish could reasonably conclude that your need for aid was serious. If you were unwilling to live in a workhouse, it meant that you believed you would be better off by making your own way and keeping your freedom.[38]

Other innovations associated with the Elizabethan Poor Laws were so progressive as to seem centuries ahead of their time. For example, as early as the 1750s, some rural parishes began testing a relief scheme that came later to be known as the Speenhamland system. Named after the tiny corner of County Berkshire where magistrates sat down in 1795 to work out the details, the Speenhamland system was an experiment in providing welfare disbursements for even the *working* poor, pegging the size of those disbursements to the size of the applicants' families and the price of bread rather than to documented need or infirmity.[39]

By the standards of the eighteenth century, these experiments with poor relief were quite expensive. By 1800, England's total expenditures for poor relief had risen to roughly 2 percent of GDP. Two percent doesn't sound like a lot today, but in 1800 it was roughly half of what the national government spent on maintaining its military, and 20 percent of the nation's total peacetime expenditures. It was also enough to fully cover the subsistence needs of 7 to 8 percent of the population, or enough to provide a 9 or 10 percent income subsidy to the poorer half of England's population, or enough to provide a 32 percent subsidy to the poorest 30 percent.[40]

Each English parish set its own poor rate. When work and food were plentiful, they could reduce the poor rates. When times were tougher, they could raise them again. Historians have assembled data sets that document these tiny year-over-year adjustments. With those data in hand, they have been able to test statistically whether the Poor Laws influenced societal well-being on a larger scale. In two independent studies, researchers found that relief under the Elizabethan Poor Laws did in fact reduce death (presumably due to famine and epidemic), quell food riots, and even stem the rate of population growth (up until the end of the eighteenth century, anyway).[41] Vives was thus (posthumously) vindicated: aggressive efforts to relieve poverty could indeed prevent not only hunger, but disease and disorder as well.

## GOING DUTCH

The English used to delight in stereotyping the Dutch as "a bunch of stolid, tightfisted, cowardly, bad-tempered drunks." English wordsmiths as far back as Shakespeare used the adjective "Dutch" as a synonym for "inferior, opposite, irregular, contrary, inadequate, cowardly, deceitful, strange, fake, awkward, outlandish, false, debased, and generally contrary to the English idea of normal," according to the historian Peter Douglas.[42] Dissing the Dutch, Douglas noted, reached its climax following the Anglo-Dutch wars of the seventeenth and eighteenth centuries, leaving us with a rich supply of "Dutch" idioms to refer to cheap stand-ins for virtuous and valuable things. Drunken bravery came to be known as *Dutch courage*. *Dutch wife* was pressed into service as a synonym for prostitute. And when you invite your sweetheart out to see a movie, but you pay only for your own ticket, you've *gone Dutch*—another passive-aggressive allusion to Dutch cheapness.

The Dutch system for poor relief during the seventeenth and eighteenth centuries, however, was a rebuke to this stereotype, because the Dutch Republic spent a bigger fraction of its wealth on poor relief than England and Wales did. At the zenith of eighteenth-century Dutch charity, the Dutch were dedicating about 3 percent of gross domestic income to formalized poor relief—50 percent more as a percentage of their income than the English were.[43] And the Dutch did it without specialized taxes for poor relief.

Instead, the Dutch financed their welfare programs through private gifts, door-to-door solicitations, and collections in churches and other public meetings. In this respect, the Dutch approach was largely voluntary, though "voluntary" did not mean "without coercion." Town officials required residents to be at home for prearranged collection times, which prevented people from dodging their civic duties. At church services and other public gatherings, alms collectors gathered donations in open plates so that donors would feel the maximum threat of social censure. And when it was time to collect poor offerings in church, ministers often began with preambles to remind parishioners that indifference to the poor was the moral equivalent of theft and murder. Voluntary giving was thus transformed into a moral ordeal.[44]

As the nineteenth century approached and poverty became more widespread, voluntary giving couldn't keep up, so the Dutch found other ways to finance their charitable activities. It's a testimony to Dutch financial ingenuity that they began to support their welfare programs with the interest earned from government bonds and other assets. And when times were really tight, the cities stepped in to provide stopgaps. Through this combination of voluntary donations, clever investment schemes, and support from city revenues, the Dutch justly acquired a reputation as the most charitable people on the continent.[45]

Why did the British and the Dutch respond so aggressively to their poverty problems? One big factor was the continued intellectual influence of reformers such as Vives, who had convinced many that poverty was a disease that rotted cities from the inside out. As the economist Peter Lindert observed, British elites had been especially worried about preventing social unrest, "probably in part because of the sound of the guillotine from across the channel and the sight of food riots at home."[46] British elites also read the war with the American colonies, no doubt, as an omen that they needed to become better at tending to their own workers' needs. By the 1700s, most of the nations of Western Europe had embraced the prevention mindset, however, so we have to look to two other factors to explain Britain's and Holland's peculiarly high levels of generosity.

The first was their unusually strong economic growth. The economists Roger Fouquet and Stephen Broadberry have shown that the total value of English goods and services per capita grew by about 30 percent between 1700 and 1800. In Holland, growth was a more modest but still respectable 15 percent. In every other European country for which we have good data (Spain, Portugal, Italy, and Sweden), per capita GDP either stagnated or declined during the eighteenth century.[47]

There's a second factor that helps explain England's and Holland's unique enthusiasm for poor relief as well: the elites discovered that it was a useful lever for preventing volatility in the labor supply. Leaders in the Dutch cities regularly tweaked poor relief schemes in order to keep workers in the city from departing in favor of rural farmwork when urban jobs became seasonally scarce. In England, the urban-rural difference in aid was reversed: the working poor received more aid in the rural counties than in the cities.[48] Lindert has interpreted this rural advantage in British poor relief as evidence that England's rural landowners were using the poor rates to prevent farmworkers from leaving the countryside in search of industrial work in the cities. Because England's parishes set their poor rates by voting on them, and because only landowners could vote, England's eighteenth-century gentlemen had the political power to manage the rates in a way that favored their interests—and their main interest was in making sure that their farms got tilled, planted, and harvested. What they most ardently wanted to prevent during the Age of Prevention was fallow fields.

By the end of the eighteenth century, the prevention mindset had thoroughly suffused British thinking about poor relief. The most surprising consequence of the prevention mindset was still to come, however, when British political theorists would, by turning Vives's reasoning on its head, reverse more than two hundred years of English law and conventional wisdom about poor relief.[49]

## CRUEL TO BE KIND

By the early 1800s, many of Britain's leading social thinkers had fully embraced the laissez-faire version of capitalism that had received its fullest expression in Adam Smith's 1776 book *The Wealth of Nations*. According to this new economic philosophy (which many people today call "classical economics"), government interference in workers' compensation was a bad idea: it distorted wages away from their so-called natural prices, which the naked laws of supply and demand would otherwise set, all by themselves, invisible-hand style. An overcompensated worker, the new laissez-faire thinking went, was an overfed worker, and an overfed worker, unmotivated by hunger, was a bad worker. And Great Britain needed good workers.

Many of the social theorists influenced by Smith also became convinced that if poor relief was too generous, it would have the same demotivating effect as overcompensation: if workers knew they could fall back on poor relief, they would be tempted to slack off at work. Several writers therefore began to call for reforms to the long-standing Elizabethan Poor Laws. A British polymath named Joseph Townsend found a memorable way to make this point in his 1786 *Dissertation on the Poor Laws*:

> Hope and fear are the springs of industry. It is the part of a good politician to strengthen these: but our laws weaken the one and destroy the other. For what encouragement have the poor to be industrious and frugal, when they know for certain . . . that if by their indolence and extravagance, by their drunkenness and vices, they should be reduced to want, they shall be abundantly supplied, not only with food and raiment, but with their accustomed luxuries, at the expence of others. The poor know little of the motives which stimulate the higher ranks to action—pride, honour, and ambition. In general it is only hunger which can spur and goad them onto labour.[50]

Political economists such as Thomas Malthus and David Ricardo proceeded to worry that the Elizabethan Poor Laws were encouraging poor people to marry earlier and have more children than they otherwise would. As the number of poor people increased with each successive generation, they reasoned, the amount of national revenue devoted to poor relief would necessarily increase as well. Ricardo catastrophized, "Whilst the present laws are in force, it is quite in the natural order of things that the fund for the maintenance of the poor should progressively increase, till it has absorbed all the neat [*sic*] revenue of the country."[51]

Although the political economists didn't have any evidence to support their concerns, their point of view gained popularity because of its consistency with the prevailing economic theory of the time and because of its resonance with long-standing stereotypes of the poor as lazy, drunken spendthrifts. In his *Essay on the Principle of Population*, Malthus wrote:

> The parish laws of England appear to have contributed to raise the price of provisions, and to lower the real price of labour. They have therefore contributed to impoverish that class of people whose only possession is their labour. It is also difficult to suppose that they have not powerfully contributed to generate that carelessness and want of frugality observable among the poor, so contrary to the disposition generally to be remarked among petty tradesmen and small farmers. The labouring poor, to use a vulgar expression, seem always to live from hand to mouth. Their present wants employ their whole attention; and they seldom think of the future. Even when they have an opportunity of saving, they seldom exercise it; but all that they earn beyond their present necessities goes, generally speaking, to the alehouse.[52]

Influenced by the rhetoric coming from Townsend, Malthus, Ricardo, and other economists like them—in combination with the character assassinations of the working poor in print and in Parliament—calls for reform of the old Poor Laws grew louder. As they did, new groups of British men were gaining access to the ballot box. And unlike the property-owning gentlemen who had monopolized the British vote until 1832, the first wave of these new middle-class voters included Britain's urban industrialists, who had little interest in maintaining a cheap supply of farm labor in England's rural areas. Their interest instead was in securing cheap labor for their factories in London, Manchester, and Liverpool. Those new workers would have to come from the countryside, so cuts to the poor rates probably sounded attractive: they would have been expected to drive Britain's rural poor to the cities in search of work.[53]

In 1834, the classical economists and the new voters of Britain's middle class got what they wanted: with that year's passage of the Poor Law Amendment Act, or Poor Law reforms, poor relief began to decline, and by 1840 it was only half what it had been before passage.[54] The able-bodied poor were particularly hard-hit: those who couldn't or wouldn't work were offered room and board in one of the nation's many workhouses, which became even more unpleasant than they had been in the previous two hundred years.[55] Mimicking Knatchbull's 1723 Workhouse Test Act, administrators worked studiously to ensure that the workhouse benefits never exceeded the wages that could be obtained from labor.[56] This policy came to be known as the principle of less eligibility (here, *eligibility* is used in its archaic sense to mean *desirability*). The 1834 act described this new view of poor relief from the point of view of an able-bodied worker: "His situation on the whole shall not be made really or apparently so eligible as the situation of the independent labourer of the lowest class."[57] The principle of less eligibility was intended to make dependency on the state as miserable as possible. No more coddling, no more handouts.

The principle of less eligibility proved to be one of the most influential memes of nineteenth-century British public life. By 1850, the historian Brian Rodgers noted, the very concept

> was slowly losing any hypothetical quality it ever had and was becoming a belief. It was an idea that *must* be right. It was the *only* explanation that could be given for the obvious fact that people were stubbornly continuing to be poor in the richest country in the world, and if it were possible to enforce "strict" administration of the Poor Laws on such terms, poverty *must* disappear.[58]

The idea of less eligibility was so infectious, in fact, that its logic was applied not only to policies about the able-bodied poor, but also to policies about the penal system (if prison became too comfortable, the reasoning went, then people would be tempted to commit the crimes that might land them there). It was also applied to policies about old age and disability (if the elderly and disabled were treated too kindly, then the young and able-bodied would fail to save for their futures), and even to policies about the treatment of orphaned children (if the orphans of this generation were treated too kindly, then the parents of this generation would fail to make proper arrangements for their children's futures). British citizens who had ever received relief under the new Poor Law were even denied the right to vote. Whatever it took to convince Britain's able-bodied poor to get to work.[59]

The Poor Law reforms helped to keep hundreds of thousands of vulnerable British families in poverty for decades. Dutch poor relief fell after

1800 as well, probably as a consequence of the French takeover and the subsequent default on public bond debt, which had been the main source of income for many Dutch charities.[60] Adding insult to injury, the Poor Law reforms had none of the beneficial effects their designers had hoped: they didn't increase wages, they didn't drive rural workers into the cities en masse, they didn't reduce fertility, and they didn't raise incomes.[61] The only beneficial effect of the Poor Law reforms, it seems, was their beneficial effect on the tax payments of the wealthy.

## LESSONS LEARNED

Despite its dreary conclusion, the Age of Prevention yielded some important advances in our understanding of how to deal with poverty. First, Vives and his successors gave us a new reason to care about poverty in the first place. They planted the seeds for the conviction, controversial at the time but almost universally accepted today, that poverty should be addressed because of its corrosive effects on the health of the larger society. Second, the Age of Prevention left us with the insight that poverty has multiple, knowable causes, each of which must be addressed in its own unique way. Third, and perhaps most consequentially, the Age of Prevention left us with the idea that the prevention of poverty—or at least the prevention of its broader societal effects—was a responsibility of the state rather than a responsibility of the church or a spiritual obligation upon the rich. This expanded view of the state's role, when coupled with the revolutions in science, medicine, and even ethics that were going to come over the next two centuries, would lead to still more changes in how we understand and address the plight of the poor. And actually, those changes had already begun.

## CHAPTER 10

# THE FIRST POVERTY ENLIGHTENMENT

As a child of the 1970s, I have never known a time when *welfare* was not one of the dirtiest words in American politics. "The current welfare system," President Richard Nixon declaimed in 1971, "has become a monstrous, consuming outrage—an outrage against the community, against the taxpayer, and particularly against the children it is supposed to help." President Jimmy Carter went on characterize welfare as "anti-work, anti-family, inequitable in its treatment of the poor and wasteful of the taxpayers' dollars."[1] On the campaign trail and from the presidential bully pulpit, Ronald Reagan stoked Americans' resentment about abuses of the social safety net by conjuring up bogeymen (and -women) who were taking advantage of the rest of us, such as the "strapping young buck" in line at the grocery store who purchased T-bone steaks with food stamps, and the "Welfare Queen" who paid for her Cadillacs with money she bilked from Social Security.[2] A decade later, President Bill Clinton fulfilled his 1992 campaign promise to "end welfare as we know it" by signing new laws abolishing federal cash assistance to poor families and slashing the federal food stamp program.

Two decades after Clinton, it was still possible to squeeze political mileage out of snotty cheap shots about the waste, fraud, and abuse surrounding America's social safety net. Here is Newt Gingrich campaigning for the Republican Party's presidential nomination in 2011: "Remember, this [president, Barack Obama] is the best food stamp president in history. . . . We have people who take their food stamp money and use it to go to Hawaii. They give food stamps now to millionaires because, after all, don't you want to be compassionate?"[3] British politicians disaffected with welfare have never been quite as outspoken as their American counterparts, but Prime Minister Margaret Thatcher's infamous (though slightly misunderstood) "There Is No Society" interview hints at the attitudes of British welfare-skeptics in the 1980s:

> I think we've been through a period where too many people have been given to understand that if they have a problem, it's the government's job to cope with it. "I have a problem, I'll get a grant." "I'm homeless, the government must house me." They're casting their problem on society. And, you know, there is no such thing as a society. There are individual men and women, and there are families. And no government can do anything except through people, and people must look to themselves first. It's our duty to look after ourselves and then, also, to look after our neighbour. People have got the entitlements too much in mind, without the obligations. There is no such thing as entitlement, unless someone has first met an obligation.[4]

The term *welfare* has become such a political albatross, in fact, that it has been largely replaced by words such as *support*, *assistance*, or just *programs*, whose meanings are far too broad to be debased by slick political messaging. *Welfare* hasn't always been a derogatory term in the United States or in Britain, however, and in much of the rest of the world it never has been. Indeed, for most countries, *welfare* in its original modern sense—the state's active involvement in the health, well-being, and education of its citizens—is recognized as one of civilization's greatest achievements. In this chapter I'll explain how that achievement came about and what ideas and institutions fostered it.

The eighteenth century and the beginning of the nineteenth century marked a difficult period for Europe's working class. In London and Antwerp, wages seemed to be frozen in place. And the British and the Dutch were the lucky ones: in other European cities, the wages of unskilled workers had fallen well below the poverty line. An unskilled worker in Krakow, Vienna, or Valencia could work sixty-five hours per week and still not earn enough to properly feed a family.[5]

Mercifully, starting around 1850, workers' wages began to keep pace with the economy.[6] Nutrition improved as more efficient production and faster transportation brought food to working-class tables in greater quantities and varieties than ever before.[7] By 1880, the typical working-class adult was living a richer, longer, and healthier life than at any other time in history. Europe's workers were finally beginning to share in the blessings of the Industrial Age.

Right about then, something weird began to happen. Just as life became more secure, European governments set out to build expansive social programs to buffer their citizens from the economic risks of illness, disability, and old age. With time, new programs also emerged to promote health, education, and economic opportunity. What followed was a century-long expansion of the state's involvement in the welfare of its citizens. Through that long expansion, humanity got swept up into an era that the development economist Martin Ravallion has winsomely called the First Poverty Enlightenment.[8]

But how did this enlightened period of expansion come about? Before examining what happened, we need to describe what existed just prior to it. Before the Industrial Age, people hedged against life's hazards by establishing guilds and voluntary societies—typically composed of merchants or craftsmen—that operated as risk pools: everyone contributed money to the common fund when times were good so they could draw from the fund when times were bad.[9] As early as the 1840s, a few European nations began to use this same notion of risk-pooling in order to provide insurance for specific groups of workers: Belgium and Austria instituted plans to provide sickness benefits for seamen and miners as early as 1844, for instance. However, the first comprehensive social insurance plans did not appear until 1883, when Germany and Poland established programs to insure workers against lost income due to illness (in Germany) and injury (in Poland). In the German scheme, employers and workers made mandatory contributions to a national fund on the worker's behalf, which then entitled the worker to thirteen weeks of income while recovering from illness. A year later, Germany added its own plan for accident insurance, and Poland returned the compliment by adding a plan for insurance against illness. And why stop with illness and accidents? An excellent question! In 1889, Poland and Germany responded to that question with the world's first universal programs to provide old-age pensions.[10]

Other European countries had been debating their obligations to their working-class citizens for a while, but they hadn't yet done anything concrete. Post-unification Germany was a big country with big aspirations, however, so when Germany took bold steps in that direction in the 1880s, other countries interpreted it as a signal that they could now safely dip a toe into the waters of nationalized social insurance.[11] In 1890 and 1891, Iceland and Denmark followed Germany in establishing old-age pensions, which they funded through general government revenues rather than through deductions from workers' paychecks. New Zealand and Australia implemented similar noncontributory schemes a few years later.[12]

The early twentieth century saw more innovation. By 1907, Denmark had passed enough social insurance legislation to bypass Germany as the world's leading welfare state, with programs in place to fund old-age pensions, workers' compensation, and voluntary insurances against illness and unemployment. A year later, Britain's liberal government passed a universal old-age pension program that was also funded in part through general revenues.

In the wake of the laissez-faire thinking that had led Britain to reform its Poor Laws only a few generations earlier, one might have expected the idea of a universal economic safety net for old age that included the "undeserving poor"—with Britain's richest taxpayers footing much of the bill—to bring howls of outrage from Parliament. But no such howling was heard. The motion to establish the pension passed in the House of Commons with 315 of 325 votes in favor.[13] The support was so overwhelming, in fact, that its major architect, a member of Parliament named David Lloyd George (who would go on to become prime minister) could admit without blushing that he wasn't even sure how to pay for it. "When the bill was pending in committee of the House," the British American journalist and political historian Maurice Low wrote,

> Mr. George was asked which of his nest eggs would be broken for the benefit of the old age pension omelette, and he somewhat startled the house by replying: "I have no nest eggs at all, and I have got to rob somebody's hen roost next year. I am on the lookout which will be the easiest to get and where I shall be least punished, and, of course, where

> I shall get the most eggs, and, beyond that, which hen roost can most easily spare the eggs.[14]

The innovations kept on coming. In 1911, Britain introduced programs for universal health insurance and for the world's first compulsory unemployment insurance, extending the safety net to cover farmworkers, white-collar workers, and others who worked outside the new industrial infrastructure.[15] In 1913, Sweden introduced the first universal program for old-age pensions.

By the end of the nineteenth century, governments began to see themselves not only as being in the business of supporting social insurance programs to help workers look after their own interests, but also as being in the bigger business of promoting health and well-being for every one of their citizens. The social *insurance* programs of the late nineteenth century paved the way for twentieth-century programs with social *security* as their goal. Initially, the shift from social insurance to social security merely extended social insurance benefits to workers' families.[16] However, other innovations were in the offing as well, including free meals and medical exams for school-aged children, minimum quality standards for urban sanitation and housing, and new programs to promote the health and welfare of mothers and young children. These family-focused efforts included publicly funded health clinics, income allowances to subsidize large families, and even maternity leave.[17]

The rise of the welfare state also involved a big expansion of the state's involvement in the education of its citizens. Until the second half of the nineteenth century, many European governments had little interest in the education of the general public. If anything, they viewed public education as a counterproductive nuisance, as this quotation from 1796 illustrates:

> Today many peasants can not only read and write, but they also begin to master arithmetic; some even start to read books. Does this make them better men? do their lives become less dissolute? have they become more obedient subjects, or better cultivators of the soil? On the contrary: is it not true that manners have visibly declined? and that lords experience far more difficulty in maintaining authority over their serfs than they did when the latter were still illiterate? . . . It is a general rule today that the most honest peasant is invariably the stupidest and the most ignorant. The officer on the drill ground has exactly the same

> experience as the landlord on his estate: the most uncouth and ignorant peasant will invariably make the best soldier. He can be treated as if he were a machine, and when he is so treated one can rely upon him absolutely.[18]

And in England, when an 1807 bill to introduce a tax-funded plan for universal primary school education came up on the floor of Parliament, a Tory lawmaker fretted about an up-ended class system, the decline of morality, and the breakdown of law and order. Not that education would do anything useful for the masses anyway:

> However specious in theory might be the project of giving education to the labouring classes of the poor, it would in effect be found to be prejudicial to their morals and happiness; it would teach them to despise their lot in life instead of making them good servants in agriculture and other laborious employments to which their rank in society had destined them instead of teaching them subordination, it would render them factious and refractory; it would enable them to read seditious pamphlets, vicious books, and publications against Christianity; it would render them insolent to their superiors; and in a few years the result would be that the legislature would find it necessary to direct the strong arm of power against them."[19]

There were some exceptions to this pattern of hostility toward mass education. Prussia and Norway, for example, along with the United States, had enrolled many of their school-aged children in primary schools by the first few decades of the 1800s. Still, as of 1820, only one in five school-aged children in even the most forward-thinking nations attended school. Schooling beyond sixth grade was extremely rare, and for most people, advanced education was completely out of the question.[20] The neglect of mass education is illustrated especially well by the neglect of women's education.[21] In 1820, 60 eligible girls were enrolled in primary school for every 100 eligible boys. The sex ratio for secondary schooling was even more skewed—44:100—and the sexual apartheid at the college and university level was almost complete, with only 2 eligible women for every 100 eligible men.[22]

Educational opportunities for the masses began to expand quite rapidly, however, as government leaders realized that an educated workforce was economically useful in the same way that a healthy and financially secure workforce was economically useful. From one decade to the next, nearly every country for which we have reliable data increased both its enrollment rates and its per-pupil investments in education. By 1900, most nations that had been early adopters of welfare innovations had also drastically increased their rates of primary school enrollment. The gender gap in education receded as well: by the time World War I broke out, advanced nations were sending eligible boys and girls to primary school in equal proportions.[23]

Several global shocks in the first half of the twentieth century led to even greater investments in social spending. Following the decimation of World War I (1914–1918), several European nations created programs to encourage married couples to have more children. In parallel, an international labor movement gained momentum, with the newly founded International Labor Organization (ILO) promoting the idea that every citizen of every nation was entitled to a minimal slate of social benefits.[24] Following the Great Depression (1929–1939), the US federal government enacted "New Deal" legislation that provided universal social security benefits to the elderly and the disabled, supplemental nutrition assistance to low-income citizens, and unemployment benefits to those who were temporarily out of work. World War II (1939–1945) ushered in further changes, galvanizing British and American resolve to provide security and create opportunity for as many people as possible. The Atlantic Charter of 1941, drawn up by US president Franklin Delano Roosevelt and British prime minister Winston Churchill, declared the two countries' desire "to bring about the fullest collaboration between all nations in the economic field with the object of securing for all, improved labour standards, economic advancement and social security."[25] In 1942, the twenty-six founding members of the United Nations endorsed the Atlantic Charter as a statement of their common purpose.[26]

After World War II, the United Kingdom and the United States expanded their welfare programs still further. Following economist William Henry Beveridge's blueprint for a postwar welfare state, the United Kingdom, under the leadership of Prime Minister Clement Attlee, led the way with a vast expansion of social security, unemployment insurance, and housing support.[27] Chief among Britain's postwar accomplishments was the establishment in 1948 of its National Health Service (NHS). The new NHS legislation entitled every citizen to cradle-to-grave health care. (In

2016, Britons were still proud enough of the accomplishment to celebrate it extravagantly in the opening ceremony for London's Olympics.) France, Belgium, Sweden, and Switzerland extended their welfare systems as well, as did the Axis powers of Italy, Japan, Austria, and Germany.

The United States also expanded its postwar welfare state, albeit at a slower tempo. Despite Roosevelt's aspirations, his only major welfare achievement was the 1946 Employment Act, which laid the responsibility for controlling inflation and unemployment at the feet of the federal government. In 1964, however, President Lyndon B. Johnson picked up the pace by declaring a War on Poverty. What followed was the largest expansion of America's welfare state since the Great Depression. During this expansion, the United States added programs to provide healthcare benefits for older adults and persons with disabilities (Medicare and Medicaid), along with a permanent food stamp program, a meal program for low-income senior citizens, consumer protection legislation, federal standards for minimum housing, rent subsidies for low-income families, federal support for public education, a school breakfast program, investments for the renovation of urban housing, and increases to the minimum wage. A decade later, Congress created a program to provide Supplementary Security Income (or SSI) to low-income Americans who were blind, disabled, or age sixty-five or older.

These expansions of the social safety net were huge endeavors with huge price tags to match. All in all, US and British government expenditures on social programs totaled about 1 percent of GDP in 1900, and expenditures for education came to another 1 percent. These expenditures increased only slightly until the eve of World War I, but after the war, government expenditures for health, welfare, and education grew rapidly for a century. The rapid growth of social spending was by no means limited to the United States and Britain, either: on average, the world's developed nations today spend about one-fifth (20 percent) of their GDPs on social expenditures and another 5 percent on education.[28] Figures 10.1 and 10.2 illustrate how these changes unfolded in five wealthy countries (Japan, Norway, the United Kingdom, the United States, and France) from the end of the nineteenth century until the end of the twentieth and beyond. Where on earth did we get the idea of increasing social and educational spending from 2 percent of GDP to nearly 25 percent of GDP in just a single century?

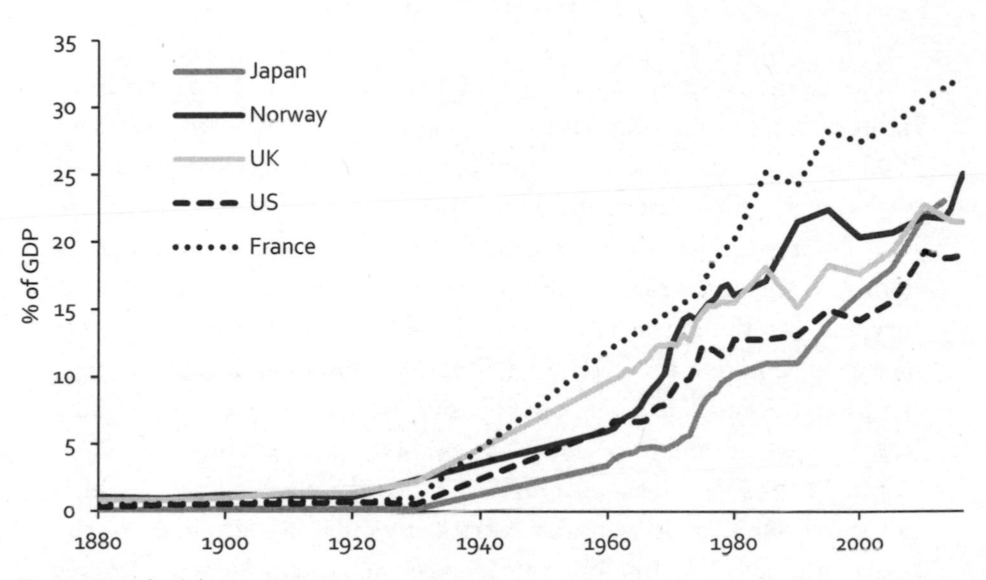


Figure 10.1. Social expenditures as a percentage of gross domestic product for five wealthy countries, 1880–2016.
*Ortiz-Ospina and Roser 2019; OECD 2019c; OECD 1985; Lindert 2004b.*

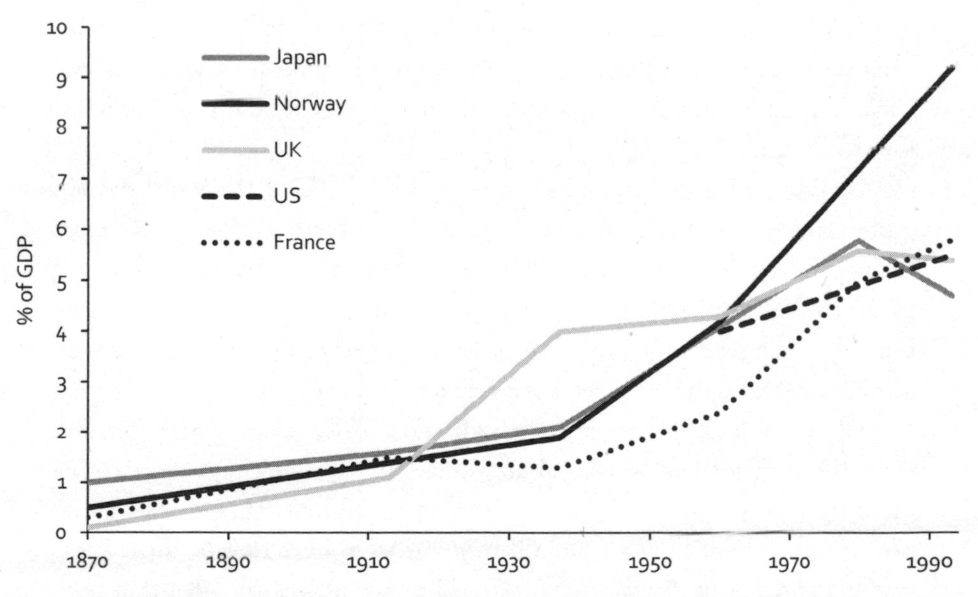


Figure 10.2. Education expenditures as a percentage of gross domestic product for five wealthy countries, 1870–1993.
*Roser and Ortiz-Ospina 2019; Tanzi and Schuknecht 2000.*

## LET THERE BE LIGHT

Those twentieth-century changes in fiscal priorities began with an eighteenth-century change in intellectual priorities: between 1700 and 1800, we began to think more about poverty and about poor people, and we began to think differently about them. This is Martin Ravallion's First Poverty Enlightenment, which we know about today because of its impact on the books that people wrote (and read) in the eighteenth century and into the nineteenth. (When I use "writing about poverty" as a proxy for "thinking about poverty," I am obviously making an assumption, but it seems like a safe one, because we have to get our concepts from somewhere, and books seem like as good a place as any.)

In 2010, Google released a product called the Ngram Viewer, which can search for a single word (or a series of words) in 5.2 million of the books in Google's online library. Because the Ngram Viewer provides a year-by-year account of how often a given word (or series of words) appears in those 5.2 million books, it becomes possible, as Ravallion discovered, to study the evolution of thinking about, in this case, poverty, by examining the frequency with which the word *poverty* was used in English-language books beginning in 1700. My research assistant Brooke Donner and I went back to the Ngram Viewer to redo some analyses that Ravallion published in 2011, and I have displayed the results of our analyses in Figure 10.3.[29]

In 1700, authors of English-language books used the word *poverty* quite rarely—about once out of every 270,000 words. It was used more and more frequently each year, however, so that by 1795 it appeared once in every 33,000 words or so—a 700 percent increase in ninety years. Once English writers became interested in poverty, they stayed interested all the way through the nineteenth century.

What were authors doing with all those extra *povertys*? In general, they were using them to contribute, in small ways and large, to two big ideas that were slowly congealing in Europe and North America about poverty and about how we should respond to it. The first of those ideas was the concept of *distributive justice*. The second was the idea that poverty and its symptoms could be understood and ameliorated—and perhaps even eradicated—by taking advantage of the ideals and tools of social science.

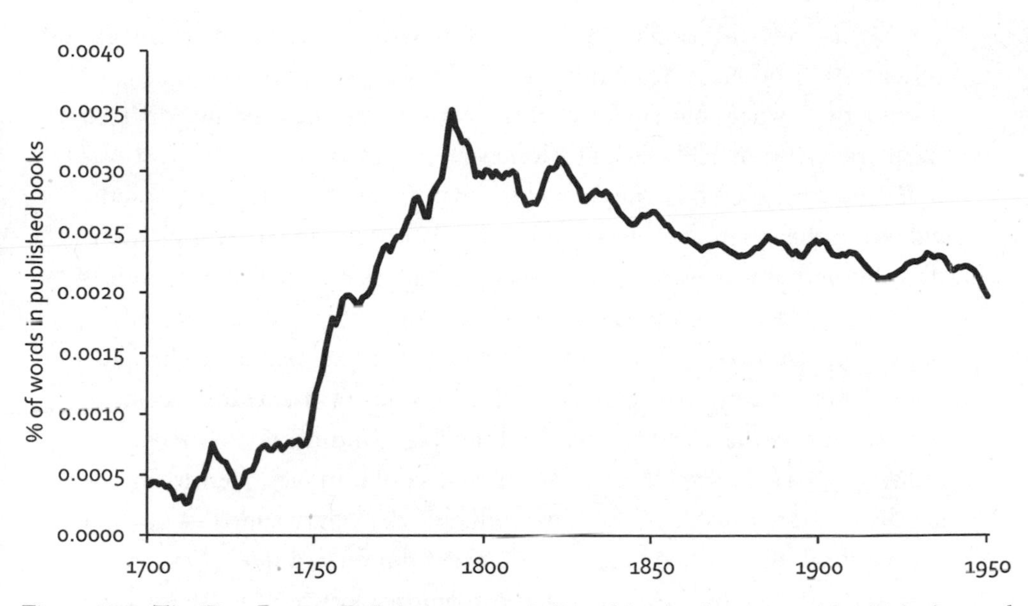


Figure 10.3. The First Poverty Enlightenment as indicated by the frequency with which the word *poverty* was used in English-language books, 1700–1950.
*Google Books Ngram Viewer 2019; Michel et al. 2011; Ravallion 2011.*

## THE FIRST IDEA: DISTRIBUTIVE JUSTICE

Distributive justice is an ethical ideal founded on the assertion that all persons are entitled to enough resources to meet their material needs. Generally, questions about distributive justice are questions about how to fairly balance two competing rights: the right to own private property and the right to stay alive. Because societies pursue distributive justice by taking property from those who possess it and distributing it to those who do not, the concept also raises questions about the prerogatives of the state and the duties of its citizens.

Political theorists concerned themselves with questions about redistribution well before the First Poverty Enlightenment, but few of them thought fairness had anything to do with it. For instance, the English philosopher Thomas Hobbes and the French philosopher Montesquieu had conceded, Vives-style, that it was wise for the state to do a little taxing in order to feed, clothe, and house the poor: the ill effects of widespread poverty on public health and government stability were simply too severe to ignore.[30]

But as the philosopher Samuel Fleischacker elucidated in his excellent 2004 book, *A Short History of Distributive Justice*, the notion of distributive justice that emerged in the early nineteenth century differed from the views of Hobbes and Montesquieu in five ways.[31] First, it was *individualistic*, rooted in the conviction that it was the well-being of individuals, rather than of society as a whole, that required our ethical attention. Second, it was *materialistic*, concerned exclusively with the resources people needed to avoid starving or freezing to death. Third, it was *rational and secular*, based on the conviction that there are reasons to help the poor that don't rely on religious justifications, just as there are reasons to oppose murder that don't rely on the Ten Commandments. Fourth, it was *practicable*, based on the conviction that we could, in fact, identify public actions that are capable of actually producing distributive justice. Fifth, it was *political*: that unlike the views that had dominated during the Axial Age and the Age of Prevention, the distributive justice of the First Poverty Enlightenment involved the assertion that government had an ethical obligation to arrange society in ways that were favorable to the poor. As Fleischacker described the process, it fell to three eighteenth-century writers—a Frenchman, a Scotsman, and a German—to provide the intellectual foundations upon which this modern concept of distributive justice could be built.

## THE FRENCHMAN

The French political theorist Jean-Jacques Rousseau contributed the idea that all citizens must be considered equal if a democratic society is to succeed. Rousseau despised how life in Europe in his day created positive feedback loops that made the rich and powerful even richer and more powerful, while the poor and vulnerable became even poorer and more vulnerable. The institutions that were in place—relating to trade, politics, legal systems, and other important human endeavors—were not a net boon to humanity, Rousseau asserted. Instead, they created an environment in which natural inequality—small individual differences in people's luck, skill, and strength—gradually evolved into a "moral or political inequality" that "depends on a kind of convention, and is established, or at least authorized by the consent of men," and "consists of the different privileges, which some men enjoy to the prejudice of others; such as that of being more rich, more honoured, more powerful or even in a position to exact obedience."[32] As political inequality grew, Rousseau argued, the rich and powerful became ever more proficient at using their wealth and influence to tilt societies' institutions even further in their favor. What Rousseau particularly loathed was the impunity with which the rich and powerful could dominate their neighbors in full public view:

> The social confederacy . . . provides a powerful protection for the immense possessions of the rich, and hardly leaves the poor man in quiet possession of the cottage he builds with his own hands. Are not all the advantages of society for the rich and powerful? Are not all lucrative posts in their hands? Are not all privileges and exemptions reserved for them alone? Is not the public authority always on their side? . . . How different is the case of the poor man! the more humanity owes him the more society denies him. Every door is shut against him, even when he has a right to its being opened: and if ever he obtains justice, it is with much more difficulty than others obtain favours. . . . I look upon any poor man as totally undone, if he has the misfortune to have an honest heart, a fine daughter, and a powerful neighbour.[33]

The solution, Rousseau insisted, was to require "government to prevent extreme inequality of fortunes; not by taking away wealth from its possessors, but by depriving all men of means to accumulate it; not by building hospitals, but by securing the citizens from becoming poor."[34] Many of Rousseau's specific policy ideas for obtaining this end, such as regulating society's expenditures on the arts, industrial goods, and luxuries, look somewhat misguided by modern standards. Others (such as progressive taxation and universal education, with an emphasis on civics) have hardly been improved upon in nearly three centuries.[35]

## THE SCOTSMAN

When the name Adam Smith comes up today, we immediately think of his idea of the "invisible hand," but in the one million words he put to paper, Smith used the term "invisible hand" exactly three times—once in a book about astronomy, once in a book about morality, and once in a book about economics.[36] Smith's most original intellectual contribution was a different idea altogether: the proposition that *common folk didn't deserve to be poor any more than anyone else did.* "If *The Wealth of Nations* was less than novel in its theories of money, trade, or value," the historian Gertrude Himmelfarb wrote, "it was genuinely revolutionary in its view of poverty and its attitude towards the poor."[37] The historian James Buchan concurred, writing, "Smith's strongest characteristic, after his hypochondria and solitude, was probably his concern for the poorest sections of society."[38] Smith's advocacy for the poor is nowhere better illustrated than in a passage in *The Wealth of Nations* in which he wrote, "No society can surely be flourishing and happy of which the far greater part of the members are poor and miserable. It is but equity, besides, that they who feed, clothe, and lodge the whole body of the people, should have such a share of the produce of their own labour as to be themselves tolerably well fed, clothed, and lodged."[39]

Smith tried to change people's attitudes toward poverty through four lines of attack. First, he argued against tax and trade policies that artificially inflated food prices. Second, he advocated for progressive taxation, universal education, and public works programs that differentially benefited the poor. And despite his latter-day reputation as a fierce advocate for free-market capitalism (which, to be sure, is a well-deserved one), he thought the government should regulate labor, land rental, and public works "because of their particular influence on common people's ability to make a living."[40] Third, he delivered a scorching critique of the eighteenth-century political writer Bernard de Mandeville's mercantilist idea that the balance of trade (the value of nation's exports minus the value of its imports) was the cardinal measure of a nation's economic health. What really mattered, Smith argued, was whether its citizens had access to the commodities that make for a good life. On Smith's view, a nation that supplies its people with food, shelter, clothing, health, and free time is a nation with a successful economy. This change in outlook, noted the economist Martin Ravallion, "opened the way to seeing progress against poverty as a goal for development, rather than a threat to it."[41]

Smith's fourth line of persuasion—and arguably his most effective—was his approach to depicting the poor and their plights. Poor people, as Smith described them, were just like everybody else. They just happened to be poor.

Consider how Smith talked about the cognitive differences between the common people and "the people of fashion." "The difference in natural talents in different men is, in reality, much less than we are aware of," Smith wrote in *The Wealth of Nations*. He wagered that the differences in the raw cognitive endowments of the typical philosopher and "the common street porter" (a laborer you hired to carry heavy things) were largely the result of nurture—habit, custom, and education—rather than nature. The intellects of two different breeds of dogs, he ventured, differ more than the intellects of philosophers and street porters. Indeed, Smith believed it was impossible to distinguish between the intellects of rich children and poor children until they reached the age of six or eight. It was only then that the richness or paucity of their early childhood environments began to produce observable differences in talent or skill. But Smith went even further in his efforts to ennoble the poor: he thought that everybody, no matter their background or intellect, had a "genius" for something.[42]

The common street porter as a *genius*? Nobody had ever written this way about the poor.

Smith expressed continuous admiration for the ingenuity, practical knowledge, and work ethic of laborers, craftsmen, and village farmers. He dispelled stereotypes that allowed the people of fashion to justify their indifference to the harshness of poor people's daily lives. No, the poor are not lazy. If anything, their sin is that their work ethic is *too strong*: with enough incentives, they are prone to work so hard that they make themselves ill. And no, it's not the poor who tend toward sexual licentiousness and substance abuse: in fact, it's just the opposite. A rich man who spends years getting drunk and going to prostitutes can rehabilitate his reputation quickly; a poor man who made such choices for even one week would be forever ruined. And so what if common people are attracted to exuberant religion? Intense religious communities give the young worker, having just arrived in the big city with all of its temptations, some social accountability and moral guardrails so he doesn't stumble into a moral pitfall that could destroy his prospects for a better life.[43]

Smith had a knack for portraying the desperate straits of the poor in a vivid and sympathetic style that was almost clinically detached at the same time—a style the historian Thomas Laqueur termed the "humanitarian narrative."[44] Indeed, Smith could bring the same humanitarian verve to his description of the big city's treacherous moral landscape in the eyes of a country boy as he did to his description of the perpetual grief that afflicts mothers who cannot feed and clothe their children:

> It is not uncommon, I have been frequently told, in the Highlands of Scotland for a mother who has borne twenty children not to have two alive. . . . This great mortality, however, will everywhere be found chiefly among the children of the common people, who cannot afford to tend them with the same care as those of better station. Though their marriages are generally more fruitful than those of people of fashion, a smaller proportion of their children arrive at maturity. . . . Every species of animals naturally multiplies in proportion to the means of their subsistence, and no species can ever multiply beyond it. But in civilised society it is only among the inferior ranks of people that the scantiness of subsistence can set limits to the further multiplication of the human species; and it can do so in no other way than by destroying a great part of the children which their fruitful marriages produce.[45]

Despite his concerns for the poor, however, Smith did not endorse anything close to a modern conception of distributive justice. Granted, he did propose that the state provide education in reading, writing, and arithmetic for all children, which was radical enough in Smith's time, but he was too much of an Enlightenment liberal to think it could ever be lawful to appropriate a citizen's property for any purpose. And besides, Smith rarely encountered a big-government antidote that was not at least as bad as the poison it was designed to counteract. Best to leave the enrichment of the poor to the market, assuming the market is operating fairly and freely.[46]

## THE GERMAN

It was Immanuel Kant—philosopher, lifelong bachelor, and workaholic—who added the final touches to the modern conception of distributive justice. These included an Enlightenment conception of morality based on universal principles, an Enlightenment conception of universal human dignity, an Enlightenment critique of traditional notions of charity, and an Enlightenment proposal for how a modern state should fulfill its obligations to its citizens.

How do people figure out how to behave morally? According to Kant, morality was not to be found through religious teachings, or tradition, or even by following one's natural sympathies for people in need. Instead, actions are morally good if the principles that motivate those actions are morally good. Fine, but which principles are morally good? Those that fulfill what he called *the categorical imperative*: "Act only upon that maxim whereby thou canst at the same time will that it should become a universal law,"[47] or, equivalently, "upon a maxim which, at the same time, involves its own universal validity for every rational being."[48] If your action is motivated by a principle that you would wish to see universalized—if you would want everyone else to live by it at all times—then, by Kant's lights, your action is morally good.

Kant's categorical imperative is such a nifty bit of philosophical apparatus because it offers an approach to discovering morality through conceptual analysis alone. By testing whether stealing obeys the categorical imperative, for example, we can discover that stealing is morally wrong: you and I might be tempted to steal on occasion, but neither of us would wish to live in a world in which taking other people's stuff was permissible. Thus, stealing is immoral. Kant also showed that theft is morally wrong by showing that if stealing *were* morally permissible, the notion of private property would make no sense, so the concept of *theft* itself would fall apart.

Kant's belief in humans' capacity for rationality led him to make another bold claim: all humans have equal and absolute moral worth. Because of our capacity for reason, he argued, we can identify stable moral truths (for example, we are capable of discovering that stealing is morally wrong) and then aspire to live by those moral truths. No other species possesses this power to find good and to do good, which on Kant's view endows humans with a basic dignity that other creatures lack. Crucially, Kant argued, this universal human dignity, which all people share in equal measure, entitles us to others' moral concern. Morality, according to Kant, requires that you and I "act as to treat humanity, whether in thine own person or in that of any other, in every case as an end withal, never as a means only."[49] Kant's morality requires us to make our choices in light of how they will affect others' dignity and welfare. From this starting point, it was not a far leap for him to argue that every person, by virtue of his or her common dignity and his or her duty to help others, "has an equal right to the good things which nature has provided."[50]

With these pieces in place—the conviction that moral actions are those that obey the categorical imperative, and the conviction that our capacity for moral reasoning endows us with equal and infinite worth—Kant was ready to demolish the traditional understanding of charity and to offer an alternative.[51] He was suspicious of the traditional concept of charity, primarily because it placed the giver and the receiver in a hierarchical relationship that made a mockery of their equal dignity. "The giving of alms flatters the giver's pride, requires no trouble and no consideration of whether the recipient is worthy or unworthy," he wrote. "Alms degrades men."[52] Instead, he asserted, "charity to one's fellows should be commended rather as a debt of honour than as an exhibition of kindness and generosity."[53] Kant also disliked how traditional charity depended on the whims and dispositions of the giver. On one day, a giver might help a stranger in need because it makes her happy or impresses her friends, but on another day, she might resentfully withhold such help because someone had recently taken advantage of her kind heart.[54] Most fundamentally, charity isn't really morally good (it's not morally bad, either), because it is not based on duty—that is, a desire to adhere to a moral principle of beneficence:

> To be beneficent when we can is a duty; and besides this, there are many minds so sympathetically constituted that, without any other motive of vanity or self-interest, they find pleasure in spreading joy around them and can take delight in the satisfaction of others so far as it is their own work. But I maintain that in such a case an action of this kind, however proper, however amiable it may be, has nevertheless no true moral worth, but is on a level with other inclinations. . . . For the maxim lacks the moral import, namely, that such actions be done *from duty*, not from inclination.[55]

To illustrate the distinction between charity and beneficence—that is, between helping from inclination and helping from duty—Kant has us imagine two men. The first is an ordinarily charitable man whose worries about a personal problem have stifled his usual tendency to sympathize with others. For the first time in his life, this man has the opportunity to base his generosity toward others on a firm moral foundation:

> While he still has the power to benefit others in distress, he is not touched by their trouble because he is absorbed with his own; and now suppose that he tears himself out of this dead insensibility, and

> performs the charitable action without any inclination to it, but simply from duty, then first has his action its genuine moral worth.[56]

Kant also encourages us to consider a second man who is temperamentally cold and indifferent to others' suffering:

> Would he not still find in himself a source from whence to give himself a far higher worth than that of a good-natured temperament could be? Unquestionably. It is just in this that the moral worth of the character is brought out which is incomparably the highest of all, namely, that he is beneficent, not from inclination, but from duty.[57]

What makes the beneficence of both men morally praiseworthy, according to Kant, is that they acted on the basis of a principle that is universalizable. We discover our moral duty not to be indifferent to others' welfare because this principle fulfills the categorical imperative: helping others, when we can do so without significantly reducing our own welfare, is a principle that all rational creatures would want to universalize, because all rational creatures can imagine scenarios in which their lives might depend on similar regard from others.[58]

Before Kant's thoughts on the ethical obligations of the state had fully crystallized, he had asserted that states should behave exactly like individuals: just as individuals must avoid dignity-debasing acts of charity (which encourage a servile relationship between donors and recipients), governments must avoid dignity-debasing handouts. In his lectures during the 1770s, for instance, Kant taught his students that he longed "to see whether the poor man could not be helped in some other way which would not entail his being degraded by accepting alms."[59] Kant's insistence on the inviolability of property rights also limited the range of public interventions he could recommend in good conscience.

By the time he wrote *The Metaphysics of Morals* in 1785, however, Kant's view of the state's duties to its citizens had changed. The state was required by duty, he concluded, to take an active role in aiding its citizens, and to fulfill that duty, the state was permitted to tax its citizens. In order to reconcile this conclusion with his beliefs about the inviolability of property rights, Kant argued that the state and its citizens were participants in a special kind of social contract. In this contract, the state's most fundamental duty is to protect our rights, including our right "to the good things nature has provided."[60] In pursuit of discharging this duty, the state acquired a right to levy taxes. And because all citizens had "bound themselves to contribute to the support of their fellow citizens," they incurred a parallel duty to help the state to fulfill its duty, "and this is the ground for the state's right to require them to do so."[61] And that, according to Kant, is that.[62]

## JOINING THE PIECES TOGETHER

Rousseau's arguments for preventing inequality, Smith's efforts to humanize the poor, and Kant's assertion that all people possess equal and infinite worth were easily fashioned into a tidy intellectual scaffolding. From this scaffolding, political theorists began to make even more audacious claims about citizens' rights and the state's redistributive duties. In his 1792 *Rights of Man*, for example, Thomas Paine proposed that states were obliged to provide each citizen with an old-age pension after a life of labor, "not [as] a matter of grace and favour, but of right."[63] Likewise, Johann Gottlieb Fichte, one of Kant's contemporaries and devotees, out-Kanted Kant himself by asserting a natural right to productive employment.[64] And toward the end of the French Revolution, a Parisian journalist and revolutionary named Gracchus Babeuf drew on Rousseau, Smith, and Kant to popularize the idea that people possessed a right to equal economic status. This was a position that not even the most radical of Babeuf's predecessors had dared to defend.

Babeuf's views gained prominence through a short but widely circulated pamphlet that outlined his radical conception of distributive justice in twelve numbered points:

1. Nature has given every man an equal right to the enjoyment of all wealth.
2. The aim of society is to defend this equality, often attacked by the strong and the wicked in the state of nature, and to increase, by the co-operation of all, this enjoyment.
3. Nature has imposed on each man the duty to work; no one can, without committing a crime, abstain from working.

4. Labour and enjoyment ought to be in common.
5. Oppression exists when one man exhausts himself working and wants for everything, while another wallows in abundance without doing anything.
6. No one can, without committing a crime, appropriate to himself alone the wealth of the earth or of industry.
7. In a true society there should be neither rich nor poor.
8. The rich who will not give up their superfluity to help the needy are enemies of the people.
9. No one should be able, by monopolizing the means, to deprive another of the education necessary for his happiness; education ought to be in common.
10. The aim of the [French] Revolution is to destroy inequality and establish the common happiness.
11. The Revolution is not finished, because the rich absorb all wealth and rule exclusively, while the poor work like veritable slaves, languishing in poverty and counting for nothing in the State.
12. The constitution of 1793 is the true law of the French nation, because the People have solemnly accepted it.[65]

This was distributive justice in its most radical eighteenth-century expression.

Beyond the blood and upheaval of the French Revolution, few social reformers were interested in adopting Babeuf's vision of distributive justice in full: it was a lot to take in. One idea from Babouvism that did have sticking power, though, as Samuel Fleischacker observed, was the notion of a political right "of all people to a certain socioeconomic status—not because poverty gets in the way of people's ability to be good citizens, but because poverty is an affront, indeed a justiciable injury, to people as human beings."[66] You have a *right* not to starve.

This was an idea that *did* catch on. British workers and their supporters began to articulate their grievances in the language of rights, which helped them coordinate their efforts to secure political influence. Joseph Townsend, Thomas Malthus, David Ricardo, and other laissez-faire political economists had gotten much of what they wanted through the Poor Law reforms of 1834, but champions of the working poor answered back with outrage, often couched in the style of the humanitarian narrative.

Social novelists in England and in France, including Charles Dickens, Anthony Trollope, and Honoré Balzac, published sympathetic accounts of the urban poor and their struggles. For his part, Dickens spent forty years pumping out social novels that dramatized poverty and humanized the poor. Many of the social novelists' works were published in serialized format, which made them accessible in small monthly portions even to people who could not afford to buy books for themselves.[67] The rising popularity of lending libraries, along with publishers' experiments with quickly converting expensive first-runs into cheaper editions, meant that the social novels became available to a wide readership of rich and poor alike. Even Queen Victoria is said to have read *Oliver Twist* and to have found it "excessively interesting."[68]

Satisfying the public's interest in the lives of the poor turned out to be a great way to sell newspapers, too. By 1835, London's *Guardian* (its full name was *The Poor Man's Guardian*) had reached a daily circulation of fifteen thousand—considerably higher than that of the more fashionable *London Times*.[69] Between 1849 and 1852, the journalist Henry Mayhew published dozens of pieces in London's *Morning Chronicle* (and later, dozens more as two-penny pamphlets) documenting how London's street folk managed to make a living. Mayhew's pieces were eventually published in four volumes titled *London Labour and the London Poor*. Readers turned to Mayhew's writings for peeks into the daily lives of London's vagabonds, dockworkers, costermongers (street vendors), mudlarks (scavengers who worked the banks of the Thames), toshers (scavengers who worked London's sewers), rag-gatherers, pure-finders ("pure"—dog feces, as it is more commonly known—was used to prepare leather for bookbinding), bone-grubbers (bones were used in plant fertilizer), and rat-killers.[70] Mayhew's accounts were in high demand well into the 1870s, inspiring a new genre of journalism on the London poor.

## THE SECOND IDEA: A SCIENTIFIC MINDSET

Ironically, literary and journalistic attention to poverty reached its zenith precisely at a time when rising wages and improved social services had already begun to improve poor people's lives.[71] Nevertheless, amid a soaring British confidence in science and technology, British philanthropists and reformers in the last few decades of the nineteenth century became convinced that the lives of the poor could be improved even further by adopting a scientific mindset.[72]

London's Charity Organisation Society (COS), established in 1869, provides a fitting example. The founders of the COS, a small group of social reformers that included Octavia Hill, John Ruskin, Lord Shaftesbury, William Gladstone, and Cardinal Henry Edward Manning, hoped to improve the efficiency of London's many private charitable institutions through careful coordination and record-keeping. In 1882, it began publishing an annual register of all of London's charities. By 1890, it was able to identify more than 1,700 of them.[73] The COS also refined the practice of casework—that is, carrying out regular visits to people's homes in order to make evidence-based distinctions between the families that could be helped out of poverty and those that were "lost causes."

Meanwhile, Octavia Hill worked to improve housing for the Victorian working class through "the application of conscious, rational, and systematic principles."[74] And in 1884, Samuel Barnett founded the first settlement house—Toynbee Hall—in London's poor East End, which set off a bona fide settlement house movement. The settlement houses were experiments in communal living for middle-class men and women who had an interest in social problems. These social reformers would stay for a few months or years to work with the residents of the poor urban neighborhoods around the houses. The settlement houses did not exist to provide food, shelter, or alms. They existed instead, as Gertrude Himmelfarb put it, "to provide a place where workers in the area and the residents of the house could come together for meetings, discussions, classes, lectures, exhibits, outings, and whatever else might be edifying and serviceable."[75] In any given year, Toynbee Hall's residents organized enough university-level coursework in the sciences and humanities to fill the course catalog of a small junior college. They also arranged classical music concerts, art exhibits, and Saturday-evening lectures by well-known academics and writers.

The fact that Toynbee Hall's educational and cultural offerings were so popular in a neighborhood of dressmakers and dockworkers is a testament to the broad societal consensus that knowledge was ennobling and empowering. In addition to their direct salutary effects on the working people they served, the settlement houses were also meeting places where reformers interested in providing social services could exchange notes with reformers who were more interested in empirical research, or in civil service, or in legal advocacy work. Many Toynbee Hall alumni went on to prominent careers in civil service and government, including William Henry Beveridge (the architect of Britain's postwar welfare state) and Clement Attlee (the Labour Party prime minister who brought Beveridge's vision to reality).[76] It has been said that the settlement house movement ended up serving as a sort of graduate school for social research and policy.[77] For an idealistic wonk, it must have been a stimulating place to spend a gap year.

Unsurprisingly, given the reformers' trust in the value of scientific thinking for addressing poverty, social science proper also came to play an important role. In 1890, with the publication of his *Principles of Economics*, the economist Alfred Marshall reinvigorated the dismal science by renewing enthusiasm for the conviction, originating with Adam Smith, that the alleviation of poverty was the chief reason to have a science called economics in the first place.[78] Meanwhile, the London businessman, philanthropist, and statistician Charles Booth, along with his wife, Mary, were busy with a series of books called *Life and Labour of the People in London*.[79]

Over the course of fourteen years (1889–1903), the Booths produced seventeen volumes of maps, data, and statistical analysis in pursuit of answers to two questions: Who in London was poor? And why? To answer them, the Booths first needed to figure out how much money was needed for a family to lead a decent, independent life (eighteen to twenty-one shillings per week, by their reckoning).[80] They also needed to determine how many families made that much money, and where they got it from. These questions led the Booths to produce a system for placing people into eight different social classes, which they labeled from A ("The lowest class of occasional labourers, loafers, and semi-criminals") through H ("Upper middle class"). The Booths' class system enabled them to produce detailed poverty maps, for which they used a color-coding system to indicate the economic status of every household in the entire city.

So how many of London's 4.3 million people were poor? According to the Booths, precisely 1,292,737 of them. And how did they get that way? The Booths attributed more than half of London's poverty to "lack of work or low pay," about 20 percent to "circumstance, sickness, or large families," and, to the surprise of many of their readers, only about 15 percent to "habit, idleness, drunkenness, or thriftlessness." By and large, the Booths argued, poor people were not poor because they were lazy, stupid, or intemperate. They were poor because they couldn't find enough work to make ends meet.

As the Booths were finishing up the final volumes of *Life and Labour*, Seebohm Rowntree—another British businessman-turned-social-scientist—was finishing his own statistical study of poverty. For his 1901 book, simply called *Poverty*, Rowntree developed the concept of a *poverty line*, the amount of income necessary to buy what one needed to maintain *physical efficiency*. (As a confectioner with an interest in the chemistry of food, Rowntree was in a particularly good position to consider the nutritional value of the working-class diet.) Although Rowntree's approach to estimating poverty differed from the Booths', he arrived at the same conclusion for the northern city of York that the Booths had arrived at for London: three out of every ten Yorkers were living in poverty, mostly due to underemployment and low wages.[81]

With *Poverty*, Rowntree showed that poverty was not a London problem, but a British one. And once Rowntree had drawn a straight line from poverty to physical health, it did not take much imagination for his readers to draw a straight line from the health of Britons to the health of Britain. People came to accept the fact that Britain could not compete in the international arena if three out of every ten Britons were not getting enough to eat. Indeed, following Rowntree's work, poor nutrition was blamed for Britain's humiliating performance during the Second Boer War (1899–1902), a misadventure in which 450,000 trained British soldiers struggled to subdue an ad hoc Boer army of 50,000 farmers. Large-scale poverty reduction came into the public eye not merely as matter of charity or duty, but also as a matter of "national efficiency."[82]

In time, American philanthropists and reformers adopted nearly every single one of Britain's innovations. After spending the summer of 1877 with London's Charity Organisation Society, Reverend Stephen Humphreys moved to Buffalo, New York, to found America's first COS. Others soon followed in over one hundred American cities, including New York, Boston, Philadelphia, Detroit, Cincinnati, Baltimore,

Chicago, Cleveland, Indianapolis, New Haven, and Washington, DC.[83] Sister societies began showing up in European cities such as Paris and Geneva, but nowhere was the concept more enthusiastically embraced than in the United States.

Americans copied the British settlement houses as well, beginning with one founded by Stanton Coit on New York's Lower East Side in 1886. Coit had lived for a time at Toynbee Hall, as did Jane Addams (the so-called mother of American social work) and Ellen Gates Starr, who came back to the States to found Chicago's famous Hull House in 1889. Hull House quickly became a model for the rest of the country. By 1913, America had more than four hundred settlement houses spread across thirty-two states.[84]

The social surveys that the Booths and Rowntree conducted anticipated many similar inquiries in the United States.[85] In 1890, one year after the Booths published the first volume of *Life and Labor*, *New York Tribune* reporter Jacob Riis published his book-length exposé of the deplorable tenement housing on New York's East Side. *How the Other Half Lives* had an immediate impact, spurring the city to pass minimal housing standards in 1894.[86] In 1895, Florence Kelley, one of America's first college-educated female social scientists, documented the lives of the families in the neighborhood around Hull House in a book called *Hull-House Maps and Papers*.[87] Robert A. Woods's 1898 book *The City Wilderness* did for Boston's South End what *Hull-House Maps and Papers* had done for Chicago's Near West Side.[88] Likewise, in his 1899 book *The Philadelphia Negro*, W. E. B. Du Bois documented the economic and social conditions of the four thousand African American residents of Philadelphia's Seventh Ward.[89] And in his 1904 *Poverty*, sociologist Robert Hunter—who at the time was a resident at New York City's University Settlement—used Rowntree's *physical efficiency* method to derive a poverty rate for the entire United States. Hunter put the US poverty rate at 12 percent, most of which was concentrated in the northern industrial states, where 20 percent of the population was poor. Like the Booths and Rowntree, Hunter attributed most American poverty to unemployment and low wages. Only four million of America's ten million poor people, Hunter noted, were receiving assistance of any kind.[90]

Next, there was the 1909 Pittsburgh survey, which resulted in six volumes and dozens of articles on the hard lives of Pittsburgh's industrial workers.[91] It was the Pittsburgh survey that really introduced the nation to industrial poverty. As the historian Robert Bremner described it,

> Supporters of nearly every good cause drew inspiration and ammunition from its pages; yet it was not a propagandistic or polemical work. It commanded respect because it was an outstanding piece of research—honest, informative, reliable. The facts it disclosed were their own best advocates. The survey findings, especially as they related to the waste of human resources, were taken to heart by a nation just awakening to the realization that conservation was as vital to its future as exploitation had been characteristic of its past.[92]

One final innovation that turned out to be uniquely important in the United States was the private charitable foundation. Industrialization had made a few Americans extraordinarily rich, and some of those industrialists viewed their good fortune as an opportunity to address America's woes. The oil magnate John D. Rockefeller, for example, used his wealth to eradicate hookworm and to promote education, particularly in the American South. The steel tycoon Andrew Carnegie supported education and libraries. And Margaret Olivia Sage, the widow of a Wall Street investor, established the Russell Sage Foundation, which devoted most of its considerable resources to accomplishing two goals. The first of those goals was to professionalize the field of social work. The second was to promote research and advocacy that would help to pinpoint—and ultimately, eradicate—the social conditions that were inimical to human flourishing.[93]

The First Poverty Enlightenment, built as it was on the concept of distributive justice and an ebullient faith in the power of science, created a new view of poverty in which it was neither a moral failing nor a necessary evil of the modern economy. Instead, it was viewed as the predictable outcome of a conspiracy of blights—some ancient, some modern—that included illness, accidents, inadequate nutrition, inadequate housing, and the vicissitudes of a complex labor market. Although the First Poverty Enlightenment provided a sturdy justification for the welfare state, it doesn't explain where cash-strapped governments found the nerve to heavily tax their citizens in order to pay for it. That political will came from two other sources.

First, it came from the spread of democracy. The economist Peter Lindert hypothesized that "the primary reason social transfers accelerated only after 1880 is that the groups that would have pushed for such transfers lacked political voice," and the historical record supports his hypothesis quite nicely.[94] Through the early 1800s, only a tiny fraction of British men (perhaps 3 percent) were allowed to vote. Not surprisingly, public policy reflected their interests almost perfectly. However, some lawmakers realized that an enlarged electorate would help to check the political influence of the upper class; some hoped it might even create enough electoral leverage to force politicians to tackle the nation's big collective challenges (such as the particularly annoying problem of urban sanitation).[95] By reducing the amount of wealth a man needed to possess in order to hold the right to vote, the Reform Act of 1832 brought the size of the electorate to 20 percent of the British male population. Later extensions of voting rights extended suffrage to all men, and then to all women, until the right to vote was universal.

In extending the right to vote so broadly, lawmakers had to become responsive to the new voters' interests, and the new voters were interested, above all, in a bit of help. As the government extended other civil liberties (including free speech and a free press) and passed legislation that brought Britain's literacy rate to 80 percent, it became easier for voters to coordinate their actions at the ballot box. By the end of the nineteenth century, the working class was an electoral force that could not be ignored by any MP who wanted to keep his job.[96]

Or by any MP who hoped to keep the peace. The late 1800s and early 1900s were a time of tremendous social unrest. Workers' organizations in Europe and America flirted with socialism, anarchism, and a variety of other *isms* that Western leaders were eager to keep at bay. As assassinations, worker strikes, labor stoppages, riots, and bombings became barometers of workers' dissatisfaction, it became clear that concessions would be required in order to minimize violence and prevent radicals from gaining hold of the levers of government power.[97] Democracy created similar pressures in other nations as well. Germany, for instance, took its first steps toward a welfare state in hopes of weakening the appeal of the Socialist Party.[98] Taking these facts into account, the democracy explanation for the rise of the welfare state looks like a compelling one.

## WAR AND THE RHETORIC OF SHARED SUFFERING

The shared suffering of two world wars provided the second source of political will. As the political scientists Kenneth Scheve and David Stasavage observed in their book *Taxing the Rich*, the monies that paid for the postwar welfare state made their way into the public coffers as the result of tax policies that had originally been put in place to fund World Wars I and II. Those wartime taxes fell mostly on society's wealthiest people, so what we really need to understand is how society became willing to raise taxes on the wealthy in order to support the war effort.

Scheve and Stasavage's hypothesis, which they support with many sources of data, is that the taxpaying public was persuaded by new arguments about what fairness and equal treatment under the law ought to mean in a democracy—particularly a democracy that needed to coordinate a massive war effort. The consensus that emerged from those debates was that it was impossible to mount a massive war effort, while also respecting the equality of all persons, without figuring out how to share the burdens of war among all citizens. "The reason wartime governments increased taxes on the rich more than the rest," Scheve and Stasavage wrote, "was because war mobilization changed beliefs about tax fairness. It created an opportunity for new and compelling compensatory arguments that increased support for taxing the rich."[99]

The argument that ended up winning the day went like this: War is a burden that is not equally shared. It's the poor who bear the brunt of it. Rich people avoid most of it. They tend to be older, for instance, and their age exempts them from military service. Their money and personal connections also enable them to seek out favors or legal loopholes that exempt their sons from service. What's even more galling is that society's rich industrialists often find ways to profit from war by producing military supplies in and hustling for fat government contracts.

War, then, isn't fair. The least disruptive remedy for this unfairness is to conscript rich people's wealth for the war effort, just as we conscript poor people's sons. Few made this compensatory argument more convincingly than the British prime minister David Lloyd George:

> Talk to a man who has returned from the horrors of the Somme, or who has been through the haunting wretchedness of a winter campaign, and

> you will know something of what those gallant men are enduring for their country. They are enduring much, they are hazarding all, whilst we are living in comfort and security at home. You cannot have absolute equality of sacrifice. In a war that is impossible, but you can have equal readiness to sacrifice from all. There are hundreds of thousands who have given their lives, there are millions who have given up comfortable homes and exchanged them for daily communion with death. Multitudes have given up those whom they love best. Let the nation as a whole place its comforts, its luxuries, its indulgences, its elegances on a national altar consecrated by such sacrifices as these men have made.[100]

Arguments like George's became commonplace during World War I. Indeed, Scheve and Stasavage found that parliamentary discussions about tax rates prior to the war were almost always based on simple "equal treatment" arguments (which favor regressive programs that tax the rich and the poor at equal rates), or "ability to pay" arguments (which call merely for taxing the rich at higher rates). Only 6 percent of their speeches featured David-Lloyd-George-style compensatory arguments. During the war, however, 62 percent of parliamentary speeches relied on compensatory arguments. The notion that the rich should pay more in order to compensate the poor for their unequal sacrifices began to win the debate. Even the venerable *Economist* magazine, ordinarily unsympathetic to the prospect of raising taxes on the wealthy, was persuaded by the compensatory argument. Compensatory arguments played a large role in other Allied nations as well, including France, Canada, and the United States.

British and American politicians also made regular use of compensatory arguments as they argued about how to pay for World War II—and, when the war was over, how to pay for the recovery. Veterans had to be compensated for their injuries, for lost earnings, and for lost educational and occupational opportunities. And they deserved not to have to worry about getting good medical care. In the United States, these compensations were realized through the 1944 GI bill and a massive expansion of the Veterans Administration hospital system. To pay for these benefits, top tax rates would stay high for decades—far longer than was strictly necessary, in fact. Once the wealthiest taxpayers got used to paying more, however, it became possible to use that additional tax revenue to extend the welfare state. And that is exactly what happened.[101]

It took a Poverty Enlightenment—along with boatloads of cash, democratic rule, and convincing rhetoric about shared wartime suffering—to bring about the modern welfare state. Even then, it came to fruition only because the ideas undergirding the enlightenment were hammered into the heads of politicians and everyday people alike. Why did these ideas need such forceful hammering? Because they were in direct competition with some other very strong ideas about poverty, and those ideas had already been hammered in by natural selection. The ancient ideas counsel us to lend a helping hand to the unlucky, but they also whisper conspiratorially that we should withhold help from the lazy. The First Poverty Enlightenment asked people to ignore those murmurings and to focus instead on first principles: Inequality breeds more inequality. Every person should be treated as an end withal, never as a means only. Poor people are just like rich people; they just don't have any money. You have a right not to starve. Science is the reformer's friend.

Because our evolved intuitions about when and whom to help are so powerful, the principles upon which the First Poverty Enlightenment was built are easily ignored or forgotten. This is how small-government propagandists have succeeded in convincing several generations of Americans that the typical recipient of government aid is a lazy, parasitic welfare queen rather than a poor kid, a chronically ill adult, or a disabled veteran. Welfare is still one of the dirtiest words in politics, but it's also one of the greatest innovations in the history of generosity.

## CHAPTER 11

# THE HUMANITARIAN BIG BANG

At daybreak on November 1, 1755—All Saints' Day—Lisbon was Europe's fourth-largest city and one of its most important commercial centers. By the afternoon, three natural disasters had reduced the city to rubble and ash. The first wave of devastation came in the form of an 8.5-magnitude earthquake, the largest to have ever hit a major European city. In the second wave, a great fire—caused by all the overturned candles, which had been lit in observance of the religious festival—devoured thousands of homes and other buildings. Finally, the smoldering city was inundated by a trio of tidal waves more than sixty feet high that the earthquake had pushed up from the ocean floor. All told, forty thousand of Lisbon's residents were killed and more than 80 percent of its buildings were damaged or destroyed. By sunset, Portugal was paupered, and one of the most important centers of the international economy was a rock quarry. In today's dollars, the total loss would amount to the damages incurred by a hundred Hurricane Katrinas.[1]

Modern Europe had never before experienced a natural disaster on the scale of Lisbon, so its effects went beyond the death toll and the immediate economic impacts: the Great Lisbon Disaster had deep conceptual effects. "The eighteenth century used the word *Lisbon* much as we use the word *Auschwitz* today," wrote the philosopher Susan Neiman.[2] For decades, people throughout Europe and the New World walked around with the "Lisbon Question" on their lips: *If horrors like Lisbon can happen, what kind of world are we living in? And how should we live in it?* Lisbon drastically altered how nineteenth-century people thought about their relationship to nature, and about how to brace themselves against nature's enormity and caprice.[3] It also revolutionized how they thought about their ethical obligations to suffering people in distant lands. We are the heirs of this conceptual revolution, which the political scientist Michael Barnett has named the Humanitarian Big Bang.[4]

## A PROTOTYPE FOR INTERNATIONAL ASSISTANCE

We remember the horrors of Lisbon much better than we remember the outpouring of concern from Portugal's neighbors, but pour out their concern they did. As soon as Spain's King Ferdinand and Queen María Bárbara (the brother-in-law and sister of Portugal's King José I) learned of the disaster, they dispatched orders to send supplies, along with the equivalent of about £37,000 in cash ($10.2 million in today's dollars). Three weeks after the disaster, when the news reached France (at the time, news traveled by horseback at less than four miles per hour), King Louis XV offered £16,600 ($4.6 million in today's dollars) in aid (though Portugal's king politely declined it, probably from fear of provoking his English allies).

Speaking of England, as soon as he received the news, King George II directed Britain's prime minister to come up with a plan for sending relief as well. Only days later, a British man-of-war set sail for Lisbon with £50,000 in gold and silver and nearly £50,000 in food, tools, and clothes, along with three warships to protect Lisbon from pirates. Another £100,000 would follow ten days after that. In total, Britain's contribution was worth $55 million in today's dollars. News of Lisbon's decimation reached Hamburg, Portugal's second most important trading partner after Britain, exactly four weeks after the catastrophe. Within two weeks, a supply ship left Hamburg laden with relief supplies. Three more were sent soon thereafter. Collectively, Hamburg's four ships delivered £160,000 ($44 million in today's dollars) in money, food, and goods.[5]

The supportive responses from Spain, France, Britain, and Hamburg were a great surprise to the Portuguese. On one hand, certainly, it's easy to see the varieties of self-interest—love of family, concern for one's citizens and assets in foreign lands, and efforts to buy friendship and favor—that may have motivated Portugal's friends to get involved. "Humanitarian concerns existed," the historian Mark Molesky acknowledged in *This Gulf of Fire*,

> but they were intermingled with political and economic interests. The city of Hamburg, for example, was motivated to send aid to alleviate the suffering of both the Portuguese and its own merchants as well as to secure its position as Lisbon's second most important trading partner. Spain . . . was moved by the close familial ties between the two royal couples and, perhaps, to gain some political leverage over the Portuguese. France's offer of aid most likely reflected both the personal influence of Madame de Pompadour and a clever, albeit unsuccessful, attempt to encourage Portugal to remain neutral (i.e., not help Britain) during the imminent conflict.[6]

Even so, the international response was unprecedented. In the zero-sum view of international relations that reigned at that time, most states would have seen Portugal's loss as their gain. Not that most sovereigns would have wanted to humiliate themselves by accepting foreign assistance anyway. "For reasons of national pride and a jealous sense of sovereignty," wrote historian Nicholas Shrady in *The Last Day*,

> the notion of foreign relief would previously have been unthinkable. By 1755, however, military and political alliances, mutual commercial interests, and improvements in travel and communications had made the states of Europe interdependent, and the Lisbon earthquake struck a chord of collective, if not universal, compassion.[7]

The international response to the Great Lisbon Disaster was a case study in what could happen when nations took an interest in other nations' welfare, inspiring some political theorists to propose that nations might have ethical responsibilities to each other. This notion was a major theme in Emmerich de Vattel's 1758 masterpiece, *The Law of Nations*, published only a few years after the disaster. Vattel argued that natural law confers the same rights on states, and imposes the same obligations on states, as it confers and imposes on individuals. Just as natural law affords equal dignity to all *individual* human beings, makes *individual* people so interdependent that they must live in societies in order to flourish, and obligates *individual* people to concern themselves with other *individuals'* welfare, it also affords equal dignity to all *states*, makes *states* interdependent in order to flourish, and obligates *states* to concern themselves with other *states'* welfare. Vattel called these obligations the Offices of Humanity: "One state owes to another state whatever it owes to itself, so far as that other stands in real need of its assistance, and the former can grant it without neglecting the duties it owes to itself."[8] In illustrating how to execute those offices, Vattel wrote admiringly of Spain's and England's assistance after the Lisbon disaster.

According to Vattel, the Offices of Humanity extended far beyond disaster relief, however: natural law also obligated nations to assist each other in developing their political, economic, and human capabilities:

> A state is more or less perfect, as it is more or less adapted to attain the end of civil society, which consists in procuring for its members every thing of which they stand in need, for the necessities, the conveniences and enjoyments of life, and for their happiness in general—in providing for the peaceful enjoyment of property, and the safe and easy administration of justice—and, finally, in defending itself against all foreign violence. Every nation therefore should occasionally, and according to its power, contribute, not only to put another nation in possession of these advantages, but likewise to render it capable of procuring them itself. Accordingly, a learned nation, if applied to for masters and teachers in the sciences, by another nation desirous of shaking off its native barbarism, ought not to refuse such a request. A nation whose happiness it is to live under wise laws, should, on occasion, make it a point of duty to communicate them.[9]

## A PROTOTYPE FOR DISASTER RECOVERY

Lisbon also set a precedent for two other ideas: that the state was responsible for disaster recovery, and that the recovery effort should be organized in accordance with scientific principles. Every detail of Lisbon's recovery and reconstruction was planned and orchestrated by an extraordinary

Portuguese prime minister named Sebastião José de Carvalho e Melo, 1st Marquis of Pombal. Straightaway, Pombal had the courts grant emergency powers to a dozen district leaders who then supervised the recovery and cleanup operations. But Pombal micromanaged the recovery effort himself, tirelessly crisscrossing the city by carriage in order to monitor the effort, advise his subordinates, and strengthen morale.

As soon as the waters had subsided, Pombal launched plans to haul out the corpses and the rubble, haul in food and supplies, build temporary hospitals, establish marshal law to prevent looting, freeze the prices of goods to prevent price gouging, set up a plan to repopulate the city, and even reestablish religious services at the local churches.[10] Although he used military force to keep citizens out of the disaster area, he also refused to let them wander so far that they resettled elsewhere: Pombal was going to rebuild Lisbon, period. He even made sure Lisbon's newspaper continued running without interruption.

Later, after the crisis had passed, Pombal created a uniform building code for the city. With the blessing of the king, he was given full rein to design a new plan based on Enlightenment principles of architecture and construction. To achieve this goal, Pombal worked for two and a half years with a group of highly experienced military architects and engineers. The new Lisbon would be safer, cleaner, more earthquake-resistant, more efficient, and more economically productive than the Old Lisbon had been. They built scale models of the new constructions and tested their earthquake resistance by marching soldiers around them to see whether the vibrations from the stampede caused the models to collapse.[11] They even succeeded in forcing the city's religious leaders to reconstruct their churches at lower heights in order to bring them into conformity with the new building code.[12] Every government's effort toward a systematic post-disaster response in the past three hundred years is an homage to Pombal's ambitious project.[13]

## THEISTIC INTERPRETATIONS GIVE WAY TO NATURALISTIC ONES

Lisbon also marks a turning point in how we explain natural disasters. Before Lisbon, large natural disasters were customarily viewed as God's rebukes for humanity's hubris and sin. Following the Great Fire of London, for instance, which in 1666 laid waste to most of the old city of London, preachers of every denomination exhorted Londoners to repent for the many varieties of iniquity that had provoked God into torching the city.[14] After Lisbon, however, people increasingly came to see natural disasters as the outcomes of long chains of interaction between matter and energy—that is, as purely physical events that were uninterested in human morality and indifferent to human welfare. The transition from supernatural explanations to naturalistic ones was not easy to make because it required people to turn their backs on strong intuitions and firmly held religious convictions.

For most people throughout most of history, supernatural explanations for earthquakes were the only ones that made any sense.[15] Even today, earthquakes, tsunamis, and other natural disasters can lead to outbreaks of religious fervor as people try to find some meaning in them and figure out how they ought to respond.[16] Psychologists have discovered two features of human cognition that explain why we so readily turn to religious explanations.

First, there is our tendency to explain natural phenomena in terms of preferences, goals, and desires—that is, in terms of the same causal forces we use to explain the behavior of humans and other animals. Under the right laboratory conditions, children, college students, and even scientists with PhDs can be made to reveal their preferences for goal-based explanations for natural occurrences. For example, when we don't have enough time for reflection, we're more likely to endorse the idea that hurricanes circulate sea water *in order to gather heat energy for themselves*, or that the Earth has an ozone layer *in order to protect the Earth from the damaging effects of ultraviolet rays*. Because of our affinity for goal-based explanations, the psychologist Deborah Kelemen has proposed that we are *intuitive teleologists* (from the Greek word *telos*, which means "end, goal, or purpose").[17] If you ask why Lisbon was destroyed by an earthquake, a fire, and a tsunami, an intuitive teleologist concludes that something or someone must have wanted it destroyed.

Second, there is our penchant for *just-world thinking*, which involves the intuitive belief that some sort of karma-like force brings us happiness and misery in proportion to the good things and the bad things we have done. When bad things happen to others, our just-world ruminations lead us to wonder what they might have done to deserve them.[18] When we conclude that they are in fact not responsible for their plights, we feel that an injustice has been done, and we try to help them. Conversely, when we hold them responsible for their plights—perhaps because we think they took unnecessary risks, failed to take reasonable precautions, or offended God—we feel callous or angry toward them, and we are content to let them suffer.[19]

In light of our tendencies toward intuitive teleology and just-world thinking, it is unsurprising that Europe's religious leaders interpreted the Lisbon disaster as God's punishment for human wickedness. It didn't help, of course, that the earthquake struck on All Saint's Day, one of the most important holidays in the church calendar: surely *that* was no coincidence. Religious writers in non-Catholic Europe blamed the disaster on the Catholics' inquisitions against suspected heretics.[20] Comically, some Catholics concluded that God had used the disaster to chastise Lisbon's religious leaders for being too *lenient* with heretics.[21] Where the Protestants and Catholics concurred was in their conviction that God used Lisbon to chastise Europe's intellectuals for their brash confidence that science and reason could explain the natural world.[22] The outspoken Jesuit priest Gabriel Malagrida used his pulpit week after week to try to bring Lisbon back to its senses, which is to say, back to a supernatural explanation: "It is scandalous to pretend the earthquake was just a natural event," he preached in an exuberant display of question-begging, "for if that be true, there is no need to repent and try to avert the wrath of God, and not even the Devil himself could invent a false idea more likely to lead us all to irreparable ruin."[23]

To prevent further ruin, many leaders directed their people to make amends. King George II, intuitive teleologist that he was, led Britain in a national day of prayer and fasting. Pope Benedict XIV called the priests, monks, and laypeople of Rome to three days of supplication.[24] At least one parish priest in Portugal called for nine days of religious ceremonies in hopes of staying God's hand.[25] Because most people had a religious understanding of the disaster, a religious response was the one that seemed most fitting.

Not everyone was content to embrace the traditional theistic responses, however. For his part, Pombal encouraged scientific reflection and the application of reason—cognitive activities that would not only *not* breed passivity and resignation, but would also help Lisboetas rebuild their city and prevent future misfortunes. In a research project that some scholars have called the birth of seismology, Pombal asked every parish priest to complete a questionnaire about the disaster. No mere damage assessment, Pombal's questionnaire was a serious effort to gather scientifically meaningful data about the earthquake's precursors, its origin, its force, its direction of propagation, its duration, its relationship to the tidal waves that succeeded it, and its enduring effects on the natural landscape.[26] No religious preamble, no questions about local rates of wickedness. "Not only had God been left out of the picture," wrote Nicholas Shrady, "but an enlightened state had stepped to the fore."[27]

The post-Lisbon retreat of supernatural explanations also put wind in the sails of scientists who wanted to understand earthquakes in naturalistic terms. In the post-Lisbon years, the official journals of the major European scientific societies published scores of scientific papers on the causes of earthquakes. Even a young Immanuel Kant joined the burgeoning field of seismology, writing three scientific papers about earthquakes in 1756.[28] Although many of the natural philosophers who were studying earthquakes continued to pay lip service to God's involvement, many of them also adopted a Pombaline indifference to theistic speculations. What causes earthquakes? Kant and many of his contemporaries made it clear that only one kind of explanation really mattered to them: "We have the causes under our feet."[29]

The retreat of supernatural explanations had one other important downstream consequence: optimism—about humanity's place in the world, about humanity's capacity for understanding the world, and about humanity's capacity for improving humanity's lot in the world. Post-Lisbon optimism was different from the earlier optimism of the German philosopher Gottfried Wilhelm Leibniz, who had reasoned that our world must be the best of all possible worlds because it was created by a perfect and loving God. It also differed from the earlier optimism of the deist poet Alexander Pope, who had written in *An Essay on Man* that "whatever is, is right."[30] Instead, post-Lisbon optimism was utterly incurious about whether world events were good or just or part of God's grand design. Instead, the new optimism was grounded in the conviction that science, reason, and large-scale international efforts could be combined to reduce human misery, including misery in distant lands.

## THE EXPANDING HUMANITARIAN UNIVERSE

Over the next decades, an optimistic humanitarian ethos would begin to spread throughout Europe, spurred on by the lessons of Lisbon and the same Enlightenment ideals that had set the stage for the First Poverty Enlightenment. International humanitarian undertakings became more common, and were generally organized not by governments, but instead by nongovernmental actors operating with the blessing of their sovereigns. (Governments did not yet recognize an individual right to free association, so if you wanted to organize a group of people around a common cause, you needed to get official permission.) The new humanitarian organizations were mostly secular, highly specialized, and self-consciously international.[31]

The first of these new organizations, 1767's Society for the Recovery of the Drowned, was devoted to spreading lifesaving techniques that Dutch mariners had picked up in China (where they had been practiced since the Middle Ages). By the 1790s, new chapters had sprung up in London, Lisbon, Vienna, Copenhagen, Algiers, and throughout the British Empire and America. Next, in 1775, Philadelphia became home to the world's first antislavery society. It was soon emulated elsewhere in the United States, in France, and in Britain, with British Quakers and evangelicals such as William Wilberforce setting the pace.[32]

Over the next few decades, international societies cropped up to promote prison reform, the eradication of smallpox, the welfare of workers, world peace, the abolition of war, and the prevention of shipwrecks, along with others promoting temperance and women's suffrage. Basically, any issue concerning human welfare that transcended political borders, and that could be better addressed by pooling effort across those borders, became ripe for internationalization.[33]

The Humanitarian Big Bang even prompted evangelical Christians in Britain and America to begin considering a deeper connection between the material needs and the spiritual needs of the people whom they sought to convert. In one paragraph of his pathbreaking *Enquiry into the Obligations of Christians to Use Means for the Conversion of the Heathens*, the pioneer missionary evangelist William Carey described the world's heathens as "poor, barbarous, naked pagans, as destitute of civilization as they are of true religion," while in the next passage lauding them with an admiration echoing Adam Smith's praise for the genius of "the common street porter": "Barbarous as these poor heathens are, they appear to be as capable of knowledge as we are; and in many places, at least, have discovered uncommon genius and tractableness."[34] Carey went on to reason that good missionary work *was* good humanitarian work, and vice versa:

> Can we as men, or as christians, hear that a great part of our fellow creatures, whose souls are as immortal as ours, and who are as capable as ourselves, of adorning the gospel, and contributing by their preaching, writings, or practices to the glory of our Redeemer's name, and the good of his church, are inveloped in ignorance and barbarism? Can we hear that they are without the gospel, without government, without laws, and without arts, and sciences; and not exert ourselves to introduce amongst them the sentiments of men, and of Christians? Would not the spread of the gospel be the most effectual mean of their civilization? Would not that make them useful members of society? . . . [M]ight we not expect to see able Divines, or read well-conducted treatises in defence of the truth, even amongst those who at present seem to be scarcely human?[35]

As the humanitarian ethos propagated outward, it began to disrupt conventional thinking about the entire European colonial enterprise. France and England, for instance, increasingly came to talk about their colonies in Asia, Africa, and the Americas not as assets to be exploited, but as trusts—fiduciary arrangements that demanded responsibility and respect for the rights and well-being of indigenous peoples.[36] Toward the end of a parliamentary debate in 1783 about the East India Company's governance of India (where the company's mistreatment of the locals had become a source of considerable public outcry), the British parliamentarian Edmund Burke (hailed today as one of the champions of political conservatism) argued that the Indian people's natural rights were as central to the debate as Britain's commercial interests were, and as such, must be respected:

> The rights of *men*, that is to say, the natural rights of mankind, are indeed, sacred things; and if any public measure is proved mischievously

> to affect them, the objection ought to be fatal to that measure. . . . [A]ll political power which is set over men, and all privilege claimed or exercised in exclusion of them, being wholly artificial, and for so much a derogation from the natural equality of mankind at large, ought to be some way or other exercised for their benefit. If this is true with regard to every species of political dominion, and every description of political privilege . . . then such rights or privileges, or whatever else you choose to call them, are all in the strictest sense a trust; and it is of the very essence of every trust to be rendered accountable; and even totally to cease, when it substantially varies from the purpose for which it alone could have lawful existence.[37]

Burke's humanitarian talk, it must be said, didn't lead to much humanitarian action. However, it stands as a marker of how the humanitarian impulse was seeping into many important conversations about each nation's ethical obligations to other nations and other peoples.

War is hell, but it was also the mother of several humanitarian inventions in the 1800s. During the Napoleonic Wars (1803–1815), for example, the French battlefield surgeon Dominique Jean Larrey developed the concept of triage (from French, meaning "separating out"). By distinguishing between the patients who would likely benefit from immediate medical care and those likely to die even if they received immediate care—and, conversely, those likely to survive regardless of whether they received immediate care—triage sought to do the most humanitarian good possible in a milieu of scarce resources. Likewise, the American Civil War (1861–1865) led to improvements in how wounded soldiers were carried to safety and cared for once out of harm's way.[38] And through her work as a nurse during the Crimean War (1854–1865), Florence Nightingale found her lifelong calling: improving the nursing profession and disseminating basic nursing skills to the working classes.[39]

One of the most important humanitarian advances in the century following Lisbon was the founding of the International Committee of the Red Cross, or ICRC. Established in 1863, the ICRC was the brainchild of Henry Dunant, a Swiss banker and member of Genevan high society. In 1859, Dunant had been eyewitness to a battle between French and Austrian troops in the Italian town of Solferino. Horrified by the lack of medical attention given to the wounded soldiers, Dunant led the townspeople of Solferino in setting up an ad hoc battlefield hospital in a local church. For three days and nights they cleaned wounds, changed dressings, provided food and water, wrote letters on behalf of the wounded, and took down last words. Inspired by the townspeople's enthusiasm, and their commitment to caring for every wounded soldier regardless of nationality ("*Tutti fratelli*," or "All are brothers," he recalled the Italians saying[40]), Dunant returned to Geneva with the idea of building an international civilian corps of medical personnel who would follow Europe's armies into battle. "Would it not be possible," he wrote, "in time of peace and quiet to form relief societies for the purpose of having care given to the wounded in wartime by zealous, devoted and thoroughly qualified volunteers?"[41]

By 1863, Dunant and his friends in Geneva had persuaded twelve European states to form the ICRC. In 1864, those twelve founding members also signed and ratified the Geneva Conventions, which declared their commitment to providing timely medical care to wounded soldiers. To the surprise and delight of those twelve original signatories, who assumed that their appreciation of the ICRC's mission and the Geneva Conventions came from their Christian faith and their cultivated Western moral sense, the sultan of the Ottoman Empire wrote in 1865 to indicate his intentions to adopt the Geneva Conventions as well and to put Red Cross principles into practice (though he insisted, not unreasonably, on using a red crescent instead of a red cross as an insignia). After Japan joined the ICRC in 1877, the member states accurately prophesied that the ICRC and the Geneva Conventions would become the most influential humanitarian institutions in the world. Today, 191 states are members of the ICRC and 196 are party to the Geneva Conventions.[42]

## EXPANSIONS IN COMMUNICATION AND TRANSPORTATION WERE CRUCIAL

The Humanitarian Big Bang was a *conceptual* revolution. However enlightened or noble they might be, though, naked concepts don't build army hospitals or get food into the mouths of starving people. The humanitarian ethos found a lasting foothold around the world only because of two changes to the global matrix of human interaction: for the first time in history, it became possible to obtain accurate and timely information about the world's humanitarian crises, and it became possible to send timely assistance in response. Innovations in communication and transportation were critical to the expansion of the humanitarian universe.

## Communication

Obviously, you can't help people if you don't know they're in trouble. To a greater extent than is often appreciated, the Humanitarian Big Bang, like the First Poverty Enlightenment, therefore owes a tremendous debt to the growth of newspapers in the nineteenth century. Before newspapers were widely available, the only news that people received about tragedies in distant lands was weeks out of date and fuzzy on details. Newspapers solved this problem by publishing news about world events that was timely, factual, and free of excessive editorial commentary. The earliest newspapers tended simply to reprint the news verbatim as it arrived by post, in fact—that is, by messengers on horseback: as the telegraph was not invented until the 1830s.[43]

The first newspapers had begun popping up in Europe 150 years before the Great Lisbon Disaster (though not so much in England, where newspaper publication was prohibited until the late 1600s).[44] But it wasn't until 1702 that newspapers began to publish the news every day, beginning with London's *Daily Courant*. As the number of newspapers increased, and as the weeklies became dailies, more and more Europeans were able to keep up to date on current events in the wider world. European newspapers published hundreds of items about the Lisbon earthquake, for instance.[45]

Judging from what we know about the historical trends in newspaper circulation, the public's hunger for information about world events was hard to satisfy. At the beginning of the eighteenth century, the average Briton could get his or her hands on a newspaper about four times per year (usually second-, third-, or tenth-hand, actually: scholars estimate that the typical newspaper copy was passed on between five and forty times, or read aloud in the pub or coffeehouse for the benefit of those who couldn't read by those who could.[46] By 1755, however, the year of the Lisbon earthquake, British newspaper consumption had more than doubled, to ten views per capita per year. By 1800, British news exposure was up to fifteen views per capita per year—a 275 percent increase in news exposure from only a century earlier.

The humanitarian universe expanded further when newspapers added war reporters in the late eighteenth century. As early war correspondents (including John Bell of London's *Oracle*, Henry Crabb Robinson of the *London Times*, and Charles Lewis Gruneison of London's *Morning Post*) began filing articles from the front lines of one conflict after another, readers came to understand the bloody awfulness of war more vividly than ever before. War reporting also opened readers' eyes to the failures of their governments to provide proper medical care for wounded soldiers. A common observation of the time was that armies went into battle with more veterinarians than doctors.

Warmongering governments and their neglect of their wounded soldiers created a public relations problem: in earlier generations, sovereigns had fought their wars with foreign mercenary armies, but after the French Revolution governments began to heed Machiavelli's advice to build their armies by conscripting their own young men. Thanks to war reporting from the front lines of the Napoleonic Wars, the Spanish Revolution, the Mexican-American War, and the American Civil War, as the political scientist Michael Barnett observed, "parents could now imagine that at the very moment they were reading battlefield accounts in the newspaper, their loved ones were being allowed to suffer and die."[47] As a result, the public began to question not only conscription but also war itself, stirring pacifist sentiments in some corners. To maintain the public's confidence in their war efforts, governments needed some sort of device to address the public's humanitarian concerns. The Red Cross served this goal perfectly, as did innovations in military medicine such as triage and ambulances.

## Transportation

Humanitarian ideals and timely information about humanitarian crises are indispensable—you can't help people you don't care about or know about—but you also need to be able to get the help where it's needed when it's needed. Despite England's best intentions, for example, the relief ships they dispatched to Lisbon in November 1755 didn't arrive until February 1756. England's ships were among the strongest and fastest in the world at the time, but they were still no match for the rough winter sailing conditions.[48] England's assistance did arrive, and no doubt it was received gratefully despite the delay, but that assistance would have been much more useful if it hadn't taken three months to get to Lisbon's shores.

Fortunately, the eighteenth century brought big improvements in transportation, which reduced the time and expense involved in getting humanitarian resources to the places they needed to be. International shipping was becoming faster and cheaper even before 1750, but sailing speeds continued to increase well after 1750 as a result of technological improvements in hulls, sails, and rigging.[49] By the 1830s and 1840s, steam-powered ships had slashed travel times even further. In 1842, the *Hindustan*—the fastest steamship in Britain's Peninsular and Oriental Steam Navigation Company's fleet—had a maximum speed of ten knots. By 1912, the company's *Majola* went nearly twice as fast, with a maximum speed of nineteen knots. Improvements in steam-engine technology also created a virtuous cycle that reduced shipping costs: rising fuel efficiency in the late nineteenth century meant less coal per pound of cargo, which meant that less coal had to be stored in the ship's hold for each voyage, which meant more room for cargo per voyage, which meant more cargo per voyage, which meant lower shipping prices per pound of cargo. Additional progress brought even more efficiency. The 1859 Suez Canal, for example, cut the sailing distance from London to Mumbai by 41 percent, and the 1914 Panama Canal cut the distance from London to Shanghai by 32 percent.[50] Transportation prices continued to plummet as well throughout the twentieth century. Today, travel by sea and by air costs only 15 to 20 percent of what it did in 1930.[51] Thanks to these and other advances, we now get more humanitarian bang out of each humanitarian buck than ever before in humanitarian history.

## HUMANITARIAN MISSION CREEP

The humanitarian ethic is an insidious thing. Once you make up your mind that one group of needy people is entitled to humanitarian assistance as a natural right, it becomes hard not to apply the same humanitarian reasoning to the next group of needy people. So once people began to accept the humanitarian conviction that wounded soldiers were entitled to medical care and protection from torture (thanks to the Red Cross and the Geneva Conventions), it was hard to resist applying the same humanitarian reasoning to the other horrors of war. Shouldn't the children left starving after World War I be fed, for instance—even the German ones? The British philanthropist Eglantyne Jebb certainly thought so, and in 1919 she founded an international humanitarian organization called Save the Children, which had one mission: getting food to Europe's starving children, even the German ones.[52]

And if humanitarian ideals applied to children, shouldn't they apply to all noncombatants? Well, why wouldn't they? In 1914, as World War I began, food aid flowed into Belgium through the American Committee for the Relief of Belgium. And in 1921, well after the war had ended, President Woodrow Wilson warned that Europe was going to need a "second American intervention" to prevent a massive humanitarian crisis. The American Relief Administration (ARA) was thus born. During its lifetime, the ARA cooperated with the Red Cross, Save the Children, and other organizations to distribute millions of tons of food and supplies throughout Europe, most notably in Russia, where it helped to ease the brutal Russian famine of 1921–1922. Europe operated on a parallel humanitarian track. The League of Nations (a sort of European dress rehearsal for the United Nations) established the High Commissioner for Refugees and the International Relief Union in hopes of better managing the millions of displaced Europeans who were unable to return home. The High Commissioner even succeeded in crafting the first set of internationally recognized rights of refugees.[53]

And why should a humanitarian's interest in war's victims apply only to wars between states? Why indeed! As the political scientist David Forsythe noted in *The Humanitarians*, one of the questions that confronted an earnest humanitarian was this: "If prisoners of war morally mattered in international armed conflict, why not detained combatants in internal wars; and why not other 'political enemies' when detained? Were not all of these detainees in potential danger and thus in need of a humanitarian intermediary when in the hands of an adversary?"[54] The Red Cross responded to this very good question by adding political detainees and prisoners captured in civil wars to its clientele.

And why stop with Europe? Or with war? More excellent questions, to which Save the Children responded by expanding its portfolio of causes to include child labor, child marriage, and childhood education. And not just in Europe, but in Iran, China, and Africa as well. Save the Children also set out to create a "Children's Charter" that would lay out the fundamental and inalienable rights of children. With time, the Children's Charter morphed into the Declaration of Geneva, which the League of Nations endorsed in 1923. With this enthusiastic endorsement, Save the Children pressed on to fight for children's rights worldwide. By that time, the Red Cross had also expanded its ambit to include relief and recovery from natural disasters such as hurricanes, floods, and earthquakes.[55]

By 1930, humanitarianism had grown from an exciting innovation to a permanent feature of international civil society. A global network of humanitarian agencies was now staffed by specially trained, highly experienced professionals. The humanitarians had learned how to use laws, charters, and declarations of rights to press their causes. And just like the reformers of the First Poverty Enlightenment, the humanitarians had also learned to use statistical data and empirical analysis to identify problems and assess the effectiveness of solutions. And they did so, as Michael Barnett observed, for the same reason they revered the humanitarian ethos itself: "Both were neutral and universal in scope and application."[56] All in all, it was a brawny and optimistic time for the humanitarian movement. World War II was about to test its limits.

## SAVING A SAVAGE CONTINENT

Through the gauzy lens of history, we recall the end of World War II as a time of rejoicing—of ticker-tape parades; of throngs of celebrants in the streets of London, New York, and Paris; of an American sailor embracing an American nurse on the cover of *Life*. However, most of Europe remained a savage continent for years after the war's official end.

Numbers help to put the postwar horror into comprehensible form. Between 35 and 40 million people died as a result of the war—roughly as many as live in California or in Poland today. Twenty-seven million Soviets alone died, mostly civilians, along with 6 million Poles and 6 million Germans. Of the 6 million European Jews captured by the Nazis, only a remnant of 300,000 survived until Allied forces could liberate the death camps in 1945. Another 600,000 Serbs, Muslims, and Jews were slaughtered in a Croatian genocide.

Of the Europeans who survived the war, tens of millions were displaced from their homes by Allied bombs, Nazi slavers, and ethnic cleansings. Germany alone held 17 million displaced persons in 1945, including 12 million that the Third Reich had captured and pressed into forced labor. Following the war, Europe experienced the greatest refugee crisis in history as 11 million people clogged Europe's roads trying to return home. Exhausted and malnourished, these rivers of humanity became sluices for communicable diseases such as typhus and malaria.[57] The war also made millions of orphans and widows who had no choice but to turn to begging, stealing, and prostitution. Europe's women and children were raped millions of times, mostly by departing Russian troops.[58]

The loss of housing and infrastructure was enormous. Bombing had reduced many European cities and towns to moonscapes. The German air raids that leveled the English city of Coventry occasioned the coining of a new German verb. *Coventriren*: to obliterate into nothingness. In Germany, the Allies bombed and shelled 3.5 million German dwellings into brick piles, leaving some 20 million Germans homeless. Some 1,700 Soviet towns and cities were similarly devastated. Across Europe, thousands of factories, hospitals, and public buildings were gone. Anything of value that survived the aerial bombings was ruined by looting, pillaging, or scorched-earth maneuvers.[59]

Although no one could have envisioned the hell that Europe would become by 1946, Britain and the United States began early on to plan for the postwar reconstruction of Europe. After the fall of France to an advancing German army in 1940, Britain made a pledge to stockpile food and supplies around the globe for eventual distribution in war-torn European nations. To help bring this plan to fruition, British prime minister Winston Churchill arranged for the establishment of an Inter-Allied Committee on Post-War Requirements (IACPR), charged with predicting and preparing to meet each nation's postwar relief and reconstruction needs.[60] For its part, the United States worked in 1942 to establish the International Wheat Council (IWC), whose other members included Argentina, Australia, Britain, and Canada. The IWC aimed to create a global network of stockpiled wheat that could be delivered to war-stricken countries. As the historian Grace Fox noted, the IWC was the world's first official international effort to help secure the "Freedom from

Want" that Britain and the United States had endorsed in their Atlantic Charter just a year earlier.[61]

By 1943, after the war in Europe had spilled into Asia and North Africa, it became clear that postwar relief and reconstruction were going to require work in many nations, and not just in the European ones. President Roosevelt therefore appointed New York's governor, Herbert H. Lehman, to direct the State Department's new Office of Foreign Relief and Rehabilitation Operations (OFRRO). He gave Lehman two tasks: to make a plan for getting aid to any nation liberated from Axis control, and to lay the groundwork for a new relief organization within the United Nations. The United Nations Relief and Rehabilitation Administration (UNRRA) was thus born.

UNRRA, as Grace Fox noted, has been called "the first blueprint of the post-war order," and with good reason. During its short existence, it pulled off the largest international humanitarian effort in history. It quickly grew into a massive global enterprise that involved support from 47 member nations.[62] By 1945, UNRRA had more than 10,000 trained employees in Europe and was coordinating the activity of scores of private relief organizations.[63]

It was a display of humanitarian action on a scale the world had never before seen. Between 1945 and 1947, UNRRA delivered more than $4 billion ($56 billion in today's dollars) in food, medicine, clothing, supplies, equipment, utilities, education, and technical assistance to people in 17 different countries in Europe and Asia. China, Poland, Italy, Yugoslavia, and Greece were the biggest beneficiaries.[64] By 1947, UNRRA was also running 800 resettlement camps to look after more than 7 million displaced persons, mainly in Germany, Austria, and Italy.[65] By 1951, all but 177,000 of those displaced persons had been returned to their homes or resettled.[66]

## THE HUMANITARIAN MISSION KEEPS CREEPING

Hundreds of private organizations in Britain and the United States pitched in as well. As early as 1940, Save the Children was running a child sponsorship program to get food to European children, and in

1942 a collective of British charities called the Oxford Committee for Famine Relief (OXFAM) began raising money for the starving people in Axis-occupied Greece.[67] The Red Cross was omnipresent throughout the war. Out of a sense of ethnic or religious solidarity, dozens of other organizations also worked to get food to select groups in Europe: Greek American organizations wanted to help Greeks, Polish American organizations wanted to help Poles, Italian American organizations wanted to help Italians.

On one hand, these organizations contributed immensely to the humanitarian effort. Their money and expertise made a big difference in many European lives. They also gave ordinary Britons and Americans a sense of personal responsibility for the reconstruction project, which strengthened their resolve to keep paying for it. Moreover, these organizations added international prestige to the recovery cause and combated the perception that the Allies were spending so much time in postwar Europe simply to advance their geopolitical interests.[68]

On the other hand, it was a little too much activity to manage efficiently, so UNRRA (and, back in the States, the War Relief Control Board) worked to consolidate the private organizations down to a more manageable number. UNRRA also wanted to limit some of these organizations' activities to raising funds and shipping supplies, rather than sending workers to provide direct services. They didn't want Europe to become so crowded with humanitarians that they began tripping over each other.

In late 1945, the need for consolidation led 22 private voluntary organizations to create the Cooperative for American Remittances to Europe, or CARE. What CARE did, and learned to do very well, was send CARE Packages. The original CARE Packages, unlike the care packages that have long brought comfort to summer campers and dorm rats, were the famous government-issue "10-in-1" ration parcels. Each 10-in-1 contained enough food to feed ten soldiers for one day or one soldier for ten days. After the army canceled the invasion of Japan, 2.8 million 10-in-1s became army surplus. CARE bought them all and shipped them back to Philadelphia. There, they were purchased by American citizens in local churches, post offices, train stations, and department stores. CARE then shipped them out to Europe. By 1947, every single one of the 10-in-1s had been delivered, so CARE started producing its own

CARE Packages, which often contained clothing, medicine, and books for children.

After its success in Greece during the war, OXFAM wanted to get food aid to the rest of Europe after the war, particularly to Germany, where the humanitarian needs were almost beyond comprehension. An understandably bitter British public, still shell-shocked and cash-strapped, was not exactly delighted to send relief aid to the nation that had tried to *Coventry* all of Britain just a few years earlier, even if it was full of starving orphans and refugees. So OXFAM took a page from Save the Children's post–World War I playbook and started a "Save Europe Now" campaign to raise money for food relief everywhere in Europe, with a winking acknowledgment that Germany was, as a matter of fact, part of Europe. In 1949, OXFAM's impartiality, and its reliance on need alone as the humanitarian litmus test for deciding whom to help, would prompt OXFAM's central committee to change its official mandate to "the relief of suffering arising as a result of wars or of other causes in any part of the world."[69]

Humanitarian mission creep hit CARE as well, as American donors came to better appreciate the concept of humanitarian impartiality. By 1948, CARE began accepting undesignated gifts, which allowed private citizens to send food and supplies to wherever they were needed—addressed "For a hungry person in Europe"—and not just to second cousins in Poland or ex-girlfriends from their GI days in Italy.[70] Eventually, CARE would deliver 100 million European CARE Packages—1 billion meals in all, with 400 million of those in Germany alone. An orphaned Holocaust survivor named Margot Meurtens recalled receiving a CARE Package from an anonymous donor while she was recovering in a US Army hospital. "When the day-and-night shooting and bombing were over," she wrote,

> when we found out that the doors were not locked any longer; that we could go home, I experienced very slowly the worst hours. I had no strength to hold to anything, to stand, to put one foot in front of the other one. I had no shoes, no food, no home, no friends, no relatives. No one looked at me or called my name. And then a miracle happened. A CARE Package came for me. With a blanket, with some food, some warm socks. I never found out who paid for it. It was a beginning.[71]

In the early years of CARE's work in Europe, the Germans didn't know quite what to think about those anonymous packages of American food. Many assumed they were poisoned or booby-trapped. After staring skeptically for days at a CARE package, one young German came to this conclusion: "I think I have the answer. These Americans are just different. They want to help those in need."[72] "What also touched me, maybe not until later, maybe not when I was a child," wrote another German who had received a package in 1946, "was that the people that sent us the packages had been our enemies just a few years earlier, and the war was a terrible thing, and they had every right to feel hostile toward us, and they didn't. They helped us."[73]

## THE END OF THE BEGINNING

Americans lost their enthusiasm for the United Nations Relief and Rehabilitation Administration in 1946 when they learned that Yugoslavia was trying to shoot American airplanes out of the sky at the same time that its new communist government was enthusiastically receiving UNRRA support.[74] Without American backing, there was little point in keeping UNRRA going. Besides, UNRRA was never meant to be permanent anyway. So in 1947, UNRRA transferred its responsibilities to other agencies in the fledgling United Nations. Then it closed down for good.

The postwar world had hardly recovered, and the impending Cold War promised to make the recovery even messier. So when some observers took stock of the work of UNRRA, CARE, Save the Children, the Red Cross, OXFAM, and the dozens of other organizations that had lent a hand, they saw only a qualified success. An economically self-sufficient Europe was still a long way off. It would come only after another $13 billion in American investment via the Marshall Plan.

Even so, the postwar blueprint for humanitarian action was set. The United Nations geared up for much deeper investments in humanitarian causes as the 1940s gave way to the 1950s. The missions of CARE, OXFAM, the Red Cross, and other humanitarian organizations kept on creeping as well. By 1950, enough of Europe had been restored to a reasonable degree of self-sufficiency that CARE moved on to Japan, China, the Philippines, Korea, Vietnam, Latin America, and the Middle East, which led to a rebranding: the Cooperative for American Remittances to Europe became the Cooperative for American Remittances to Everywhere. By 1959, CARE's deliveries of emergency food and supplies began to take a back seat to a broader focus on development, disease prevention, and health promotion. By 1961, CARE had developed such a good reputation in international development that President John F. Kennedy tapped it to train volunteers for the new Peace Corps. As CARE's mission changed yet again, the name changed again, too. With typical humanitarian pluck, CARE became the Cooperative for American Relief Everywhere.[75]

By the middle of the twentieth century, the world's richest democratic nations had acknowledged the wisdom in the conviction, which Churchill and Roosevelt had championed, that freedom from war requires freedom from want. The humanitarian successes of the previous two centuries had created an optimism that global humanitarian crises could be addressed—and should be addressed—through multilateral action, democratic decision-making, and contributions from the private voluntary sector. The civilized world wasn't going to treat earthquakes, tsunamis, fires, floods, volcanoes, droughts, crop failures, famines, and the horrors of war as punishments from God anymore. Thanks to the Humanitarian Big Bang, these crises came to be understood instead as tests of donors' ingenuity and moral character, as challenges to the world order and world peace, and as opportunities to acknowledge the dignity of every person everywhere.

## CHAPTER 12

# THE SECOND POVERTY ENLIGHTENMENT

On a chilly January 20, 1949, Harry Truman was sworn into office for his second term as president of the United States. From the inaugural podium, he surveyed the achievements of his first term: the fascists had been defeated, most of the postwar mess had been mopped up, and Congress had approved the Marshall Plan, through which the United States would eventually provide $13 billion in economic aid to a Europe in recovery. The United Nations was growing stronger as well. All in all, not a bad four years.

But Truman saw danger ahead. Humanity's enemies were lying in wait. First, there was a political enemy: communism was on the move—"a false philosophy which purports to offer freedom, security, and greater opportunity to mankind," and for which "many peoples have sacrificed their liberties only to learn to their sorrow that deceit and mockery, poverty and tyranny, are their reward," as Truman put it. Communism, he asserted, was an enemy to the Enlightenment values upon which the world's democratic nations intended to build a stable and peaceful global order.

There were other enemies to worry about as well. Truman called them humanity's "ancient enemies—hunger, misery, and despair." Without defeating these ancient enemies, Truman predicted, the world's poorest nations would remain vulnerable to the blandishments of a strengthening Soviet Union. If the world's democracies could defeat humanity's ancient enemies, maybe they could bring humanity's newest political enemy to heel as well. Truman therefore portrayed the United States as the protagonist in one of the most critical chapters in humanity's ongoing saga. "It may be our lot to experience, and in a large measure bring about, a major turning point in the long history of the human race," he declared, "to help create the conditions that will lead eventually to personal freedom and happiness for all mankind."[1]

To achieve these lofty ambitions, Truman outlined a four-point plan that emphasized America's deepening engagements with the rest of the world. Points One through Three—nurture the fledgling United Nations, support the European economic recovery, and complete the establishment of the North Atlantic Treaty Organization (NATO)—came as a surprise to no one: these initiatives were well under way. It was Truman's Point Four that was truly novel. The same generosity that America and the Allies had shown to a postwar Europe through UNRRA, CARE, OXFAM, Save the Children, and all of the others, Truman argued, had to be extended to the rest of the world:

> We must embark on a bold new program for making the benefits of our scientific advances and industrial progress available for the improvement and growth of underdeveloped areas. More than half the people of the world are living in conditions approaching misery. Their food is inadequate. They are victims of disease. Their economic life is primitive and stagnant. Their poverty is a handicap and a threat both to them and to more prosperous areas. For the first time in history, humanity possesses the knowledge and skill to relieve the suffering of these people. . . . Our aim should be to help the free peoples of the world, through their own efforts, to produce more food, more clothing, more materials for housing, and more mechanical power to lighten their burdens. . . . This should be a cooperative enterprise in which all nations work together through the United Nations and its specialized agencies whenever practicable. It must be a worldwide effort for the achievement of peace, plenty, and freedom.[2]

In 1950, Congress approved what came to be known as the Point Four Program, a plan to reduce global poverty by sharing technical knowledge and skills with the world's developing countries. Over the next decade, the United States would provide millions of dollars of Point Four aid to many developing nations, mostly in Southeast Asia, where the communist threat seemed greatest. The African and South American nations most vulnerable to communism received aid as well.

The Point Four effort yielded some grand successes. The Green Revolution in agriculture stands as one of the most impressive. Scientists developed strains of wheat, maize, and rice that produced much higher yields than the traditional varieties did. Mexico, India, Pakistan, and other countries in Asia and Latin America took advantage of these new hybrids, along with innovations in fertilizers, pesticides, and irrigation, to pull themselves up by their agricultural bootstraps until they were capable of meeting 100 percent of their food requirements. No more dependency on other countries' generosity to feed themselves.[3] Thanks to the widespread adoption of these new techniques, global grain production quadrupled between 1950 and 1992, even though the amount of land being farmed increased by only 1 percent.[4]

Meanwhile, the United Nations, along with the United States and Canada, built its institutional capacity to deliver humanitarian interventions around the world, particularly during famine crises. The voluntary agencies that had been so instrumental in the European postwar recovery had already moved on from Europe and had begun to work in Asia and sub-Saharan Africa, but their work would be limited by a lack of cash. The international community eventually conceded that the effort to help the developing world was going to require more intensive (and expensive) involvement, so the UN launched its Freedom from Hunger campaign to put additional pressure on donor countries. Even so, by the end of the 1960s—a decade of brisk economic growth for the world's rich nations—aggregate aid from rich countries to poor ones had actually *fallen* as a proportion of gross national income. The developed world had been getting richer (indeed, it had been getting richer ever since the start of the postwar recovery), but its aid to developing nations wasn't getting any more generous.[5]

That's when the international aid community decided to change tack: the ancient enemies were not going to be defeated, they concluded, merely through technical assistance, economic interventions, and humanitarian relief. Global poverty would need to be tackled directly and forcefully.

It would be a tall order: in the early 1970s, global poverty was so poorly understood that experts couldn't really say how much of it there was—or, for that matter, even where it was. Nevertheless, enthusiasm for reducing it was building. The UN proclaimed the 1970s "the Second UN Decade for Development" and called on the world's richest nations to devote seven-tenths of 1 percent of their gross domestic incomes to international development assistance. Why seven-tenths of a percent rather than a nice round number such as, say, half a percent, or 1 percent? Because economists agreed that seven-tenths of a percent was exactly what was required to get the job done.[6]

The world's rich nations—particularly the twenty-four countries of the Organization for Economic Cooperation and Development (OECD) known as the Development Assistance Committee (DAC)—responded to this call to action by upping their commitments. Over the thirty years between 1970 and 2000, as Figure 12.1 shows, official development aid from the DAC countries increased from $40 billion to $73 billion per year (expressed in 2015 dollars)—which is an 83 percent increase, even after adjusting for inflation.

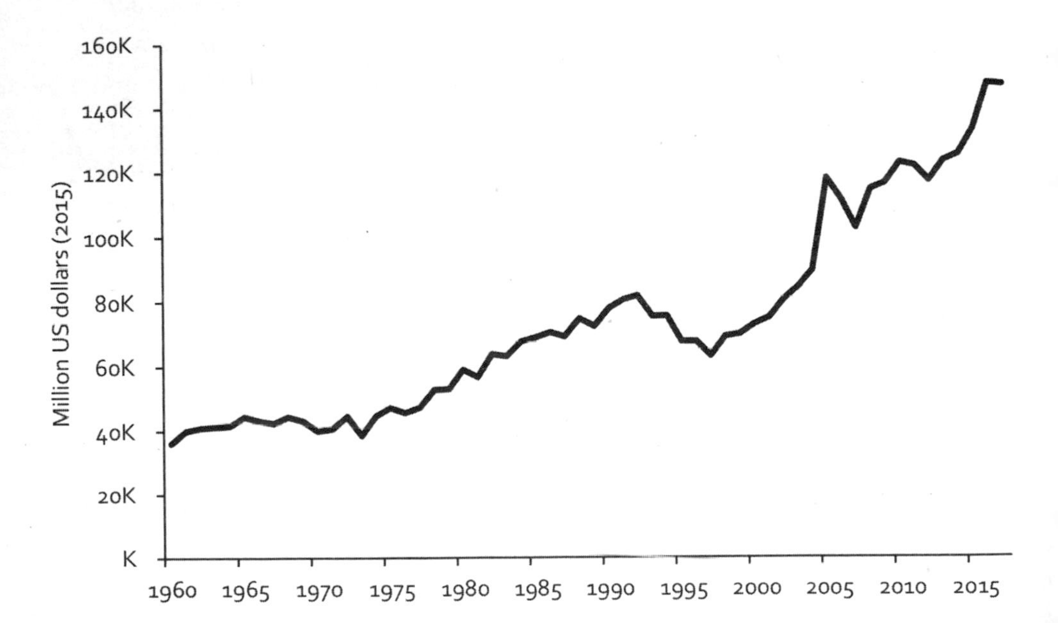


Figure 12.1. Official development assistance from the twenty-four countries on the OECD's Development Assistance Committee, 1960–2016.
OECD 2019b.

During the last few decades of the twentieth century, private donors also got more involved. Figure 12.2 displays the value of development assistance grants by private agencies and nongovernmental organizations (NGOs) from 1970 to 2016 (the most recent year for which data are available). From 1970 to the mid-1980s, grant aid to poor nations stayed mostly flat, but beginning around 1983, private aid began to grow. As the end of the second millennium approached, private aid began to grow even faster. Indeed, between 1970 and 2000, the total value of private grant aid to the poorest nations doubled, even after adjusting for inflation. Between 2000 and 2016—in half as much time—it quadrupled. As the twentieth century dissolved into the twenty-first, we were not only becoming more generous with our private resources, but becoming more generous more quickly.

The story of these twentieth-century changes in our concern for the welfare of strangers in developing nations is a story that involves professional philosophers, newscasters, rock stars, well-dressed diplomats, policy wonks, and pencil-pushing economists. But it begins with another enlightenment.

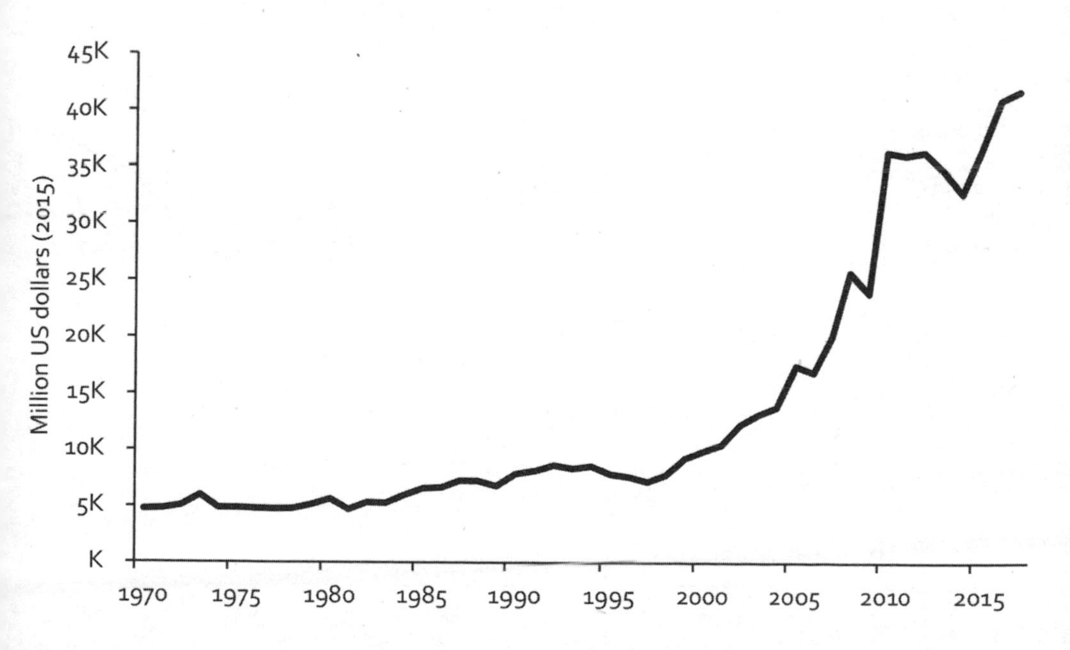


Figure 12.2. Development assistance grants by private agencies and nongovernmental organizations, 1970–2017.
OECD 2019a.

## THE SECOND POVERTY ENLIGHTENMENT

In the 1960s, the decade before the world's developed nations became committed enough to ending global poverty to start putting some real money behind it, they became interested in simply trying to understand poverty better. The smoking gun that reveals this fact of history is the same smoking gun that I entered into evidence in Chapter 10 to introduce the eighteenth century's First Poverty Enlightenment: an unmistakable uptick in the frequency with which English-language book authors used the word *poverty*, along with other poverty-related phrases, in their writings. Using Google's Ngram Viewer, the economist Martin Ravallion found that English-language authors began writing more frequently about poverty in 1960 than they had at any point in the previous six decades.

To confirm Ravallion's findings, Brooke Donner and I also used the Ngram Viewer to plot the frequency with which the word *poverty* was used in English-language books from 1900 to 2008 (the last year for which data are available). As Figure 12.3 shows, *poverty* was not used any more frequently in 1960 than it had been in 1900. If anything, writers

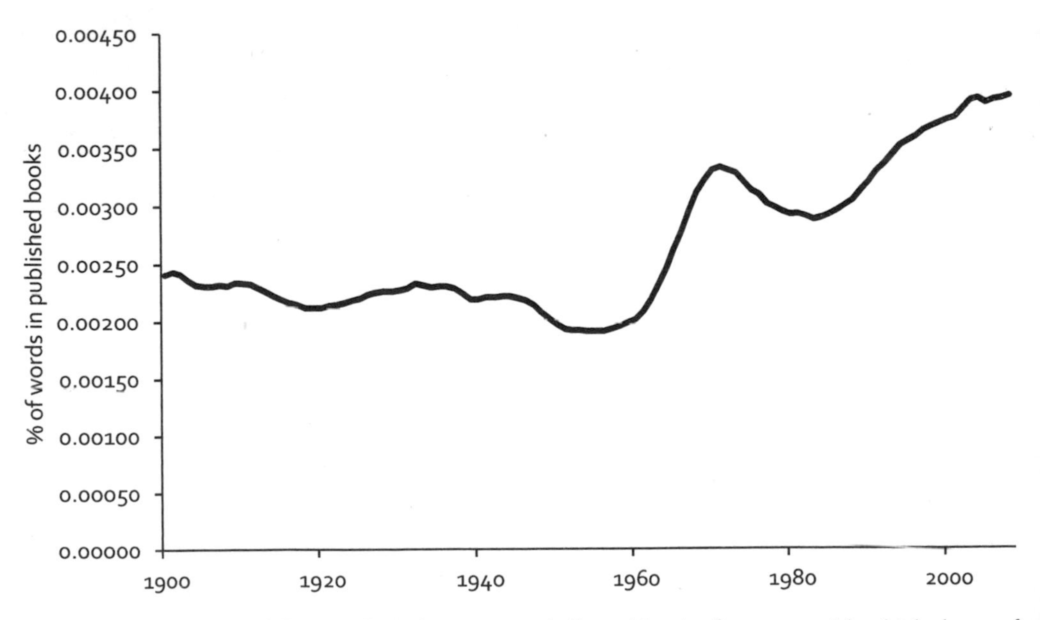


Figure 12.3. The Second Poverty Enlightenment as indicated by the frequency with which the word *poverty* is used in English-language books, 1900–2008.
Google Books Ngram Viewer 2019; Michel et al. 2011; Ravallion 2011.

were writing about poverty slightly less than they had been in the sixty previous years. This situation began to change almost as soon as we entered the 1960s, however, and between 1960 and 1970 authors increased their use of *poverty* by about two-thirds. Our growing concern for poverty in the developing world showed up in our books before it showed up in our checkbooks.[7] It is therefore fitting that Ravallion refers to the 1960s as the beginning of the Second Poverty Enlightenment.[8]

During the Second Poverty Enlightenment, people didn't simply think about poverty *more*. They also thought about poverty with more precision and nuance. During the 1960s, hair-splitters such as *social inequality*, *absolute poverty*, *relative poverty*, and *relative deprivation* escaped the ivory tower and insinuated themselves into lay readers' vocabularies (see Figure 12.4). And if you've ever used poverty-related terms such as *culture of poverty*, *working poor*, *underclass*, or even *third world*, then you've got the Second Poverty Enlightenment to thank: as Figure 12.5 shows, few people outside academia encountered those words in print before 1960, but after 1960, plenty of people did.[9] It seems people really did become more enlightened about poverty during the Second Poverty Enlightenment.

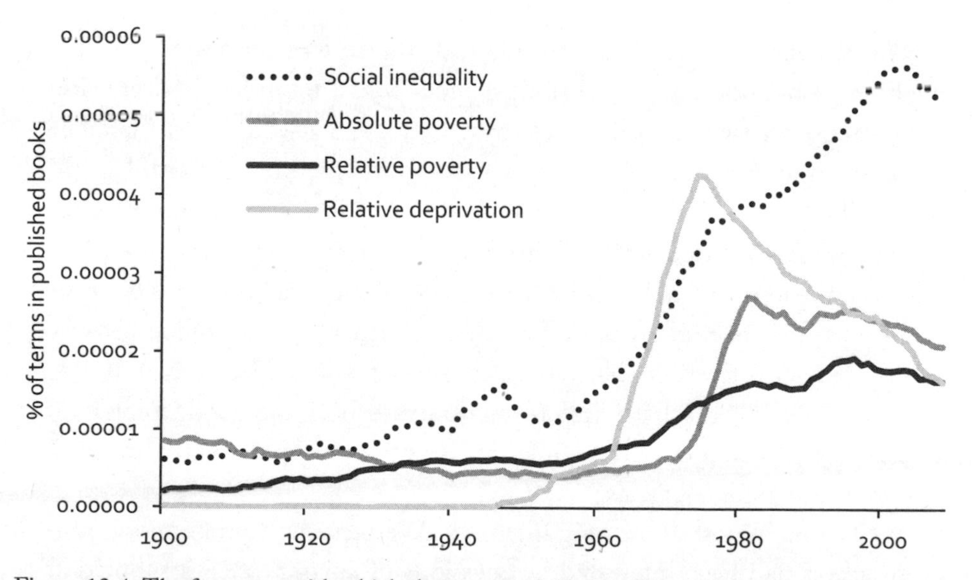


Figure 12.4. The frequency with which the terms *social inequality*, *absolute poverty*, *relative poverty*, and *relative deprivation* are used in English-language books, 1900–2008.
Google Books Ngram Viewer 2019; Michel et al. 2011; Ravallion 2011.

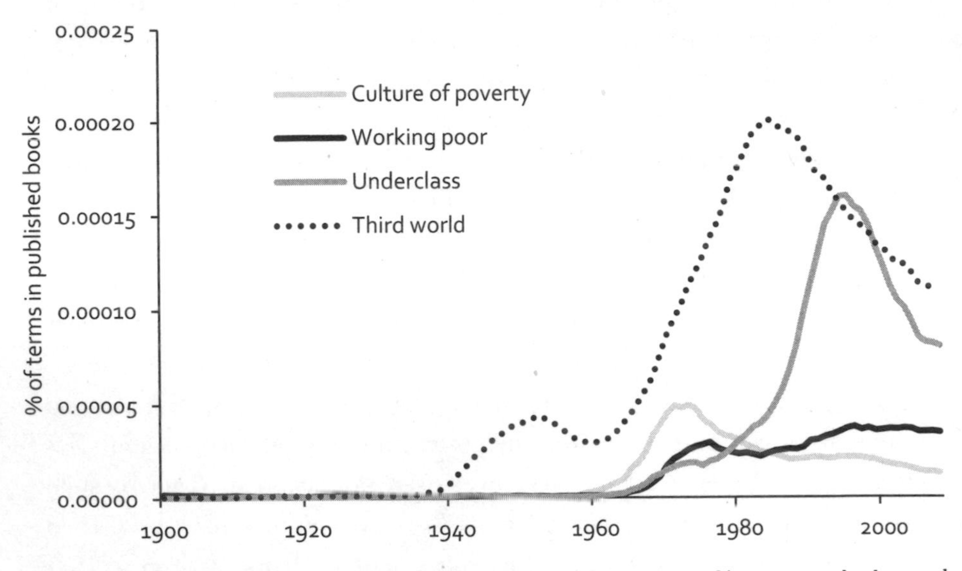


Figure 12.5. The frequency with which the terms *culture of poverty*, *working poor*, *underclass*, and *third world* are used in English-language books, 1900–2008.
Google Books Ngram Viewer 2019; Michel et al. 2011; Ravallion 2011.

## I SHOULD CARE

The Second Poverty Enlightenment also gave us new reasons to care about global poverty. Moral philosophers began to argue that relieving global poverty was not just a nice thing to do, or a good strategy for keeping communism at bay, or a good strategy for preventing a third world war. They argued instead that we needed to address global poverty because we had a moral obligation to do so. Their arguments were not novel through and through, but neither were they mere recapitulations of arguments from generations past. The new arguments were unique because they led with the fact that our actions—not only collectively, but also as individuals—can have direct, tangible effects (some good, some bad) on the well-being of poor people in distant lands—particularly those in chronically poor countries.

Through most of the long history of Western civilization, moral philosophers had been interested in questions of *applied ethics* (a subfield of moral philosophy that inquires into the morality of specific forms of conduct). In the early decades of the twentieth century, however, they mostly abandoned the whole applied enterprise. Philosophers' esteem for their own ability to shed light on ethical matters had become so low by 1940 that the English philosopher C. D. Broad was able to write that

> moral philosophers, as such, have no special information, not available to the general public, about what is right and what is wrong; nor have they any call to undertake those hortatory functions which are so adequately performed by clergymen, politicians, leader-writers, and wireless [radio] loudspeakers.[10]

Around 1970, some moral philosophers began to push back against Broad's broad denunciation of applied ethics. The Australian-then-American ethicist Peter Singer, for one, argued in a journal article called "Moral Experts" that it would be surprising if philosophers did not in fact have a special kind of expertise for addressing questions of applied ethics: after all, philosophers' training prepares them well for distinguishing between the facts that are relevant to moral arguments and those that are irrelevant. It also gives them expertise in the theoretical systems that might be brought to bear on moral problems. Finally, their training teaches them how to prevent the biases and errors of reasoning that might lead us astray as we think about moral matters. Shouldn't all of this expertise qualify moral philosophers to help the rest of us figure out how to live morally better lives? Singer surely thought so.[11]

Along with Singer's spirited defense of applied ethics, the profound moral issues raised by the civil rights movement, nuclear weapons, and the Vietnam War convinced many professional philosophers that their moral expertise was more relevant to society's concerns than it had been for quite some time. Applied ethics found its moxie again.[12]

Wasting no time, they began applying their moral expertise to global poverty and hunger. In 1972, Singer published an essay called "Famine, Affluence and Morality." In this essay, he used a utilitarian approach to moral philosophy (which holds, roughly, that the moral goodness or moral badness of a policy should be evaluated on the basis of whether it advances or diminishes the welfare of sentient beings) to make the following argument: if it is morally wrong to ignore a child who is drowning in a shallow pond, even when there are personal costs involved in helping (for example, the cost of a ruined pair of shoes), then it is also morally wrong to refuse to spend one's money to help starving people in distant lands. Singer went on to state the moral principle that should govern our behavior in such situations, which has come to be known, unsurprisingly, as Singer's Principle: "If it is in our power to prevent something very bad from happening, without thereby sacrificing anything morally significant, we ought, morally, to do it."[13]

To most people who think the moral goodness of an action has something to do with how it affects human flourishing, Singer's Principle is easy to accept: we are morally obligated to help others, even when it creates some hardship for us, if by so doing we prevent much worse things from happening to the people we help. But, as Singer observed, this principle has some weighty implications:

> If it were acted upon, even in its qualified form, our lives, our society, and our world would be fundamentally changed. For the principle takes, firstly, no account of proximity or distance. It makes no moral difference whether the person I can help is a neighbor's child ten yards from me or a Bengali whose name I shall never know, ten thousand miles away. Secondly, the principle makes no distinction between cases in which I am the only person who could possibly do anything and cases in which I am just one among millions in the same position.[14]

Other ethicists also discovered ethical obligations to the distant poor. In a 1975 essay titled "Lifeboat Earth," Onora O'Neill started from the premise that people possess a natural right not to be killed unjustifiably by other persons, which is perhaps the single most unexceptionable moral premise there is. She asserted further that your right not to be killed places ethical obligations upon others—chiefly, an obligation not to kill you without justification, and an obligation to try and prevent other people from killing you without justification. From these premises, O'Neill argued that anytime our economic activities play a causal role, however small and however remote, in hastening the deaths of vulnerable people in poor nations, we have played a hand in killing them unjustifiably.[15] Through the products they buy (or don't buy) and sell (or don't sell), O'Neill said, people going about their daily business can end up guilty of involuntary manslaughter. It is morally wrong, she concluded, to make economic decisions that harm people in distant lands, no matter how trivially or indirectly they do so.

A third contribution to our understanding of the ethics of global poverty came from the political theorist Charles Beitz, who used a 1971 book called *A Theory of Justice*, written by the political philosopher John Rawls, as his starting point.[16] Beitz was taken by Rawls's conception of justice as fairness, which he believed had implications for how we should work to raise the welfare of people in the world's poorest countries. (Rawls, it should be said, wasn't so sure that his theory of justice as fairness had any implications at all for how people from one nation should seek to address poverty in other nations.[17]) In Beitz's view, the main reason some countries are rich and others are poor is that they start out with different degrees of access to resources. Some societies are blessed with resources and others are not, and this inequality ends up making economic winners out of some nations and economic losers out of others. It's simply not fair, Beitz contended, that some societies are so much healthier and wealthier than others just because they have the good fortune to be situated in places with plentiful timber, valuable minerals, predictable climates, and access to the ocean. A conception of justice as fairness, Beitz concluded, requires us to combat Lady Luck's caprice by taking active steps to level the global economic playing field.[18]

These works by Singer, O'Neill, and Beitz were serious philosophical contributions: they were original, they were rigorously derived, and they were highly relevant to world events. Singer's "Famine, Affluence and Morality" makes it onto every philosopher's list of the most important philosophical works of the twentieth century. These works and others like them are still widely taught in universities today—some fifty years after their publication—no doubt in part because their relevance to world affairs makes them easy introductions to philosophical thinking. As a result, thousands upon thousands of college students have now confronted their rigorous logic and challenging conclusions. Some of those college graduates now run governments, nongovernmental organizations, and corporations, where they make decisions about how to allocate millions or even billions of dollars, pounds, euros, or other forms of currency in official or private assistance each year. I like to think that their encounters with the likes of Singer, O'Neill, Rawls, and Beitz had an enduring influence on their ideas about helping the global poor, even if they've long since forgotten that their convictions were shaped in late-night cram sessions in library study carrels.

## MOVING PICTURES

At the very moment that moral philosophy began to bring global poverty to America's university campuses, television began to bring it into America's living rooms. America's involvement in the Vietnam War (which foreign policy expert Michael Mandelbaum dubbed "the Television War"[19]) was tapering off, and as the monthly body counts declined from numbers in the thousands to numbers in the hundreds, and then to numbers in the dozens, the TV time devoted to the war declined as well.[20] Foreign disasters and humanitarian crises fight for media attention with other important matters of national interest (and nothing was more interesting to Americans than Vietnam, with those nightly body counts), so America's exit from Vietnam left room for coverage of other international events, including famines and similar humanitarian crises.[21] In this way, television came to serve the same function that the social novels and newspapers of the eighteenth and nineteenth centuries had: they made people more concerned about the welfare of suffering strangers, and they made them ask whether there was some way they might help.

Research suggests that news coverage of this sort was indeed one of the factors responsible for the increase in aid to poor nations during the 1970s, at least in the United States: when disasters are covered on the nightly news, the United States increases its official aid (primarily through the Office of Foreign Disaster Assistance), and ordinary citizens increase their private donations to humanitarian organizations such as the Red Cross.[22] That said, the nightly news was hardly the ideal medium for holding people's interest in the slow-moving humanitarian crises of the 1970s and 1980s: even the deadliest humanitarian disasters make for fairly boring television. To wit, a major study of American aid in the wake of more than five thousand disasters revealed that nightly news broadcasts give far less attention to epidemics, droughts, and food shortages than to faster-moving (and more spectacular) disasters such as volcanoes, earthquakes, fires, and storms. How much less attention? To get the same amount of news coverage as a volcano that killed a single person, an epidemic would need to kill 1,696 people, a drought would need to kill 3,395 people, and a food shortage would need to kill 38,920 people.[23]

The development agencies and NGOs found a way to overcome this limitation. They realized they could raise funds for international development with television *commercials*—particularly commercials focusing on the deprivations of individual children. Humanitarian organizations such as Save the Children had been taking out ads in magazines since at least the 1920s, and many of those ads had featured photos of suffering kids. But television allowed the humanitarian and development organizations to portray poverty more heartrendingly than a magazine ad could ever do. One noteworthy TV commercial from the 1970s began with footage of a young girl morosely tossing pebbles into a muddy puddle. The backing track is "Somewhere over the Rainbow" from the soundtrack to *The Wizard of Oz*. The mud puddle dissolves into—what else?—an elegant ice skating routine by the Olympic gold medalist Peggy Fleming. Next comes Fleming's voiceover:

> This is Peggy Fleming. Millions of children in over one hundred countries dream that somewhere, they'll find the help they need to walk again or to keep from going blind, or just to stay alive. And for all those children, that somewhere over the rainbow is UNICEF. The United Nations Children's Fund provides vitamins, vaccines, and school supplies for the children of the world. This is Peggy Fleming asking you to write UNICEF to find out what you can do to help.

As the commercial ends, the New York address for UNICEF appears at the bottom of the screen.

Other organizations also began to run commercials that featured children—children covered in flies, children in hospital beds, children scavenging in garbage dumps. And they worked. A nonprofit organization called World Vision (which focuses on childhood poverty in Southeast Asia and Africa), for instance, was able to raise its annual donations from $4.5 million in 1969 to $94 million by 1982.[24] World Vision's pioneering experiments with television fundraising deserve the credit for most of this growth. Children make such good targets for aid campaigns because their dependence on adults means that their misery cannot be blamed on their own fecklessness, laziness, or poor choices. They don't trigger the alarm bells that evolved to prevent us from wasting our resources on people who are unlikely to repay us. The poverty of children seems unfair in a way that the poverty of able-bodied adults does not.

On the success of such advertising campaigns, many private organizations adopted a child-sponsorship fundraising model in which they encouraged their viewers to sponsor a single, identifiable child through regular monthly or annual subscriptions (a typical sponsor might pay $25 or $40 per month to help a single child). The child-sponsorship model continues to be wildly successful even today.[25] According to a recent reckoning, approximately nine million children around the world have sponsors, mostly through the ten leading international programs. The aid that flows to the developing world through child sponsorships amounts to roughly $3 billion annually, which is a substantial fraction of the total aid from both official and private sources that the United States sends to the developing world each year.[26]

## THE TELEVISION FAMINE

The media had still other roles to play. In the last week of October 1984, the world was introduced to a massive Ethiopian famine that had been worsening for months. Until then, most media outlets had overlooked the famine completely, but that changed when extended coverage by the British Broadcasting Company (BBC) finally made it onto British airwaves. It was the first time most people had an opportunity to learn of the awful realities on the ground in Ethiopia, even though many of the world's leading humanitarian organizations (including Doctors Without Borders, Save the Children, OXFAM, World Vision, and the Red Cross) had been working in Ethiopia for some time. Even by today's standards, when the Internet has dulled our sensibilities, the scenes of the dying, the dead, and the mourning from that first BBC report are difficult to watch.

After the broadcast, news outlets around the world picked up the story, and in the weeks that followed, the BBC footage was rebroadcast by 425 news organizations to a global audience of perhaps 470 million people. The world's newspapers increased their coverage of the Ethiopian famine by almost 1,000 percent. Donations from concerned citizens began to flood into the coffers of the major humanitarian organizations.

Western governments also vowed to take a closer look at what they could do to help.

The public's response to the October 1984 reporting closely resembled their response to a similar television broadcast a decade earlier, in October 1973, which also drew attention to a disastrous and largely unheard-of Ethiopian famine.[27] There was one crucial difference between the television coverage in October 1973 and the coverage in October 1984, however: on October 23, 1984, the British musician Bob Geldof was at home watching the BBC's coverage on *The Nine O'Clock News*. Heartsick and appalled, he resolved to do more than just send money.

## COOL AID

Geldof, along with another musician named Midge Ure, decided to raise money and awareness by assembling a pop-music supergroup. Their plan was to record a catchy charity single, put it up for sale, and devote the proceeds to Ethiopian famine relief. On November 25, 1984, Geldof and Ure mustered a volunteer musical army comprising more than forty of the most popular musicians of the time (including Sting, Phil Collins, Bono, George Michael, and Boy George). They recorded the song "Do They Know It's Christmas" under the name Band Aid. Days later, they released the tune, and it immediately went to number one on the UK singles chart. They had hoped the single might raise £70,000. Instead, it sold more than three million copies, raising more than £8 million (more than $10 million) within the twelve months following its release.

When the American entertainer and social activist Harry Belafonte heard the song, he decided to organize a similar project in the United States. Belafonte recruited so many recognizable voices from the American pop pantheon (including Lionel Richie, Stevie Wonder, Paul Simon, Kenny Rogers, James Ingram, Michael Jackson, Tina Turner, Diana Ross, Dionne Warwick, Willie Nelson, Ray Charles, Bob Dylan, Cyndi Lauper, Kim Carnes, Bruce Springsteen, Steve Perry, Huey Lewis, and Al Jarreau) that it makes my ears ring to imagine them all in the studio together. The band was called USA for Africa. Their song "We Are the World," written by Michael Jackson and Lionel Richie, sold twenty million copies and raised more than $50 million for African relief.

Musical artists from Canada, France, and the world of Latin pop would record their own charity records later in 1985, and a collective of the biggest heavy-metal and hard-rock bands of the time produced a nine-song compilation called *Hear 'n Aid*, raising another $1 million. The Mormon Church sponsored two days of fasting that raised an impressive $11 million.[28] And for computer enthusiasts who wanted to buy something to support famine relief, there was Soft Aid. This late-1985 compilation of video games, which could be played on Sinclair's ZX Spectrum and the Commodore 64, raised about £350,000 (about $450,000) for the cause.

There were live events as well. In the summer of 1985, Geldof and Ure pulled off an ambitious one-day music festival and fundraising event that they named Live Aid. Taking place on both sides of the Atlantic at once, half of the Live Aid acts played in London's Wembley Stadium while the other half played in Philadelphia's John F. Kennedy Stadium. It was a day of Anglophone rock-and-roll extravagance, sent over the top by rapt coverage of Phil Collins's transatlantic journey from London to Philadelphia. After finishing a Wembley set with Sting, Collins ran to a waiting helicopter that whisked him off to the tarmac at London's Heathrow Airport, where he then boarded a Concorde for New York's John F. Kennedy Airport. From New York, Collins took a second helicopter to JFK Stadium in Philadelphia, where he arrived in time to perform some of his own tunes. At the end of his set, and after a moment to catch his breath, he introduced his friends Robert Plant, Jimmy Page, and John Paul Jones, whom he proceeded to back on drums (along with the funk drummer Tony Thompson—whose idea was it to use two drummers?) for what is widely regarded as the worst Led Zeppelin performance in the band's history.

Other star-powered fundraising events followed later in the year. Sport Aid was an all-star athletic exhibition that culminated in a ten-kilometer run called Race Against Time. Globally, nearly twenty million people ran, jogged, or walked to raise money for African famine relief. Fashion Aid was a high-profile fashion show at London's Royal Albert Hall. Over five thousand people attended. Finally, there was Comic Relief, a comedy-based charity event that debuted on Christmas Day 1985 with live coverage from a refugee camp in Somalia.

These star-studded efforts had more impact than anyone could have predicted. For starters, the public outcry pressured Western governments into increasing their involvement in African famine relief. In the 1983 fiscal year just before the first BBC broadcast, official aid from all sources to Ethiopia totaled about $361 million. By the end of 1985, that figure had more than doubled, reaching $784 million. The charity records and live events also generated impressive sums. By the time Band Aid called it quits in 1987, it had raised $228 million, and USA for Africa had raised another $51 million.[29] Comic Relief and Sport Aid raised respectable hauls of their own (Comic Relief, by the way, continues to raise money to fight global poverty to this day). Private donations poured into the NGOs as well, doubling their annual incomes and enabling them to double or even triple their staffs in just a few years' time.[30]

Another effect of 1984–1985 that can't be ignored was its influence on young people. Before Band Aid, humanitarian agencies had had little success in reaching people under the age of thirty-five. But the *annus mirabilis* between the Christmases of 1984 and 1985 changed all that. It's young people who buy records, of course, and it was young people who watched Live Aid on TV.[31] When they saw that their favorite rock stars were concerned about global poverty, they became concerned about it as well, brashly optimistic in their new conviction that anyone with enough money to buy a record or enough time to write a letter could make the world a better place. Live Aid was their Woodstock. Asked what he thought Live Aid could realistically hope to accomplish, Bob Geldof posed the worst-case scenario like this: "What USA for Africa did, and what Band Aid did in England, is to make compassion hip. . . . If all that comes out of this is the perceived attitude that it's fashionable to care, then it's worth it, time and time again."[32] In previous generations, aid organizations such as CARE had used celebrities to generate support for their missions, but never before in history had a relief cause received so much celebrity attention. No wonder it became the most successful humanitarian campaign in history.[33]

## TRAGICALLY HIP

Compassion certainly did become hip: the big-name efforts of 1984 and 1985 opened the hearts and wallets of the under-thirty-five crowd in a way no other aid organization had managed to do before. However, the realities on the ground in Ethiopia were far more complex than most people had been led to believe. The Ethiopian famine was not caused, as it happened, by the drought of biblical proportions that popular depictions had implied. Instead, the Ethiopian famine was largely a human-made disaster. As the economist Amartya Sen has pointed out, modern famines are rarely caused by food shortages per se. More often, they're caused by bureaucratic failures to get the food that is available to where it is needed.[34] The Ethiopian famine bore all of these features: Yes, rainfall had been down, but hardly enough to starve hundreds of thousands. Bureaucratic paralysis killed many more people than the drought did. Even after food aid began to arrive in Ethiopia, hundreds of thousands of tons of it rotted on Ethiopian docks along the Red Sea.[35]

But it wasn't only bureaucratic failure, either: the Ethiopian government was also starving its own people as a tactic in the ongoing civil war.[36] Of the food that did make it into the country, about 90 percent of it ended up in the hands of the government, and only a trickle of it got distributed to the starving people. Most of it was used instead to feed and pay the government's armies (which came to be known as "wheat militias"). OXFAM and the other agencies operating in Ethiopia were trying mightily to make a difference on their own, independently of the government, but they couldn't get nearly enough food into the areas that were occupied by rebel forces—which, of course, was where all of the starvation was happening.

"The terrible truth," as the journalist Robert Keating put it, is that the billion-dollar outpouring of concern for a starving Ethiopia may have ended up exacerbating the misery and extending the war. Yes, the interventions by Live Aid and all of the other agencies at work in Ethiopia plausibly prevented many deaths, but that's not an assertion that is well supported by empirical research.[37] It's also not implausible that any good work they did accomplish was canceled out by the deaths to which they might have contributed by inadvertently feeding government armies and assisting the Ethiopian government with its disastrous resettlement project: in this Stalinist-style misadventure, the government sought to move three hundred thousand people out of parts of the country where they were most likely to become sympathetic to the rebels. As many as one hundred thousand Ethiopians may have died in the resettlement process.[38]

How did such good intentions go so horribly wrong? Quite simply, our evolved instincts for generosity got out ahead of good judgment. The news broadcasts, Band Aid, and USA for Africa—along with the consciousness-raising efforts of the official agencies and the NGOs—presented the world with a narrative of a blameless Ethiopian people that had been immiserated by an act of God. They had promised that each of us was in a position to help: all we needed to do was pick up the phone or write a check.

## THE DOLDRUMS

Live Aid became a byword for the power of good intentions, when combined with bad information, to lead the heirs of the Second Poverty Enlightenment into well-meaning folly. For other reasons as well, development experts began to worry out loud about things that they had, until then, discussed only behind closed doors: Perhaps aid projects feed donors' vanity and NGOs' balance sheets more effectively than they feed starving people. Perhaps development assistance breeds government corruption, fuels civil wars, and thwarts democracy. Perhaps NGOs are so addicted to short-term publicity victories that they avoid less glamorous commitments that could actually improve people's lives over the long term. Perhaps aid discourages the global poor from taking responsibility for their own lives. Perhaps aid doesn't work when humanitarian crises are complicated by complex political factors—which they almost always are. As the international development community entertained these worries during the 1980s and early 1990s, many began to wonder whether traditional aid worked at all.[39]

In light of such concerns, the major donor nations began to experience compassion fatigue. And with the end of the Cold War and the disintegration of the Soviet Union, worries about communism also subsided. As a result, many donors also lost a strategic motivation as well for funding international development. Without assurances that aid was working, and without a political enemy to fight, the West lost some of its will for defeating humanity's ancient enemies. Development aid hit the skids.

People who wanted to improve the lot of the world's poorest didn't stop trying, but they did try different things. They organized a global effort to remove millions of land mines from war-torn nations. They launched a campaign to get AIDS drugs to poor countries. And as the second millennium rolled over into the third, they set out to convince the world's rich nations, along with the international banks that make development loans, to forgive billions of dollars in debt that the world's poor nations had racked up over many decades. Geldof, along with U2's Bono and a bevy of other celebrities—and even Pope John Paul II—worked to raise awareness of the issue. The push for debt relief, which they named the "Jubilee 2000 Campaign," was an easy sell, especially in the United States, where the allusion to the biblical concept of the Jubilee year (the one year out of fifty in which the debts of the poor were canceled) resonated with religious sensibilities. By 2005, the members of the G8 (a consortium of the eight largest industrialized nations) had agreed to put pressure on the International Monetary Fund (IMF) and the World Bank to start writing off some of that debt. Fast-forward to today, and $99 billion in debt has been canceled, with more still to come for countries that can meet the eligibility requirements.[40]

## DOUBLING DOWN ON DEVELOPMENT

In the run-up to the third millennium, the UN saw an opportunity to renew enthusiasm for development aid. The 9/11 terrorist attacks on the United States in 2001 made the task seem even more pressing. Cold-war worries about the creep of communism were replaced by worries about the allure of terrorist ideologies in so-called failed states, where governance was poor, human rights were regularly trampled, and a life of poverty and ignorance was most people's lot. The need to assist those countries came to feel more urgent than it had for many years. The time had come to double down on development.[41]

Following a gathering in 2000 called the Millennium Summit, the UN adopted the Millennium Declaration, a reassertion of its historic values and a declaration of its aims for realizing them. Not soon after came the Millennium Development Goals (MDGs), a list of eight Herculean labors for promoting global development:

1. Eradicate extreme poverty and hunger;
2. Achieve universal primary education;
3. Promote gender equality and empower women;
4. Reduce child mortality;
5. Improve maternal health;
6. Combat HIV/AIDS, malaria, and other diseases;
7. Ensure environmental sustainability; and
8. Develop a global partnership for development.

These eight targets looked extremely ambitious. The Millennium Development Goals called for halving the percentage of people living on less than one dollar a day, halving the percentage of those suffering from hunger, increasing access to schooling (particularly among girls), and reducing the infant mortality rate by two-thirds. And they wanted to get it all done by 2015.

The Millennium Development Goals were a big bet, with a big price tag to match. Development experts had grown accustomed to referring to massive, coordinated aid inflows, somewhat condescendingly, as "Big Pushes," and the Millennium Development Goals were going to require one of the Biggest Pushes Ever. Fortunately, the world's developed nations had a lot of money to push, and their commitments to aid had already begun to recover from the doldrums of the 1990s. The most developed countries therefore agreed in 2002 to recommit to the old goal, set back in 1970, of devoting 0.7 percent of their gross national incomes to development. In his book *The End of Poverty*, the development economist Jeffrey Sachs proposed that the UN could achieve the Millennium Development Goals by doubling its aid commitments over a decade until it reached $195 billion in 2015.[42]

Not everyone shared Sachs's optimism. The authors of books with titles such as *Despite Good Intentions*, *The White Man's Burden*, and *Dead Aid* were extremely skeptical of Big Pushes.[43] These experts argued, not untruthfully, that there was little convincing empirical evidence to advocate for the Big Push model of development. And they had almost no faith that the bureaucrats who design massive aid packages could possibly know what they were doing: too much money was needed, there were too many complexities, and any conceivable plan would have too many unforeseen consequences. The following warning about the futility (and even harmfulness) of aid is a good example of just how low many economists' expectations for MDG-sized Big Pushes had sunk:

> Aid has become a cultural commodity. Millions march for it. Governments are judged by it. But has more than US $1 trillion in development assistance over the last several decades made African people better off? No. In fact, across the globe the recipients of this aid are worse off; much worse off. Aid has helped make the poor poorer, and growth slower. Yet aid remains a centrepiece of today's development policy and one of the biggest ideas of our time. The notion that aid can

> alleviate systemic poverty, and has done so, is a myth. Millions in Africa are poorer today because of aid; misery and poverty have not ended but have increased. Aid has been, and continues to be, an unmitigated political, economic, and humanitarian disaster for most parts of the developing world.[44]

The wonderful thing about predictions is that if they are clear enough, and if you live long enough, you eventually get to figure out who was right and who was wrong. The Millennium Development Challenge is over and we're all still alive, so let's see what came of it. Did the Big Push become the Big Success that some had hoped, or the Big Waste that others had feared?

First of all, it is important to acknowledge that the UN failed to raise all the money it was hoping to raise. Only seven countries (Sweden, Norway, Luxembourg, Denmark, the Netherlands, Great Britain, and the United Arab Emirates) met the challenge of raising their aid transfers to 0.7 percent of gross national income.[45] Nevertheless, aid did increase—up to $131 billion in 2015—even though that's only 67 percent of the way to the original $195 billion goal. In my world, 67 percent achievement earns you a grade of D+. How much did the UN manage to accomplish with a D+ level of funding for the MDGs?

In its official evaluation, the UN declared the MDGs a stunning success. By comparing the state of the world in 1990 to the state of the world in 2015 (to create a sort of before-and-after comparison), the UN credited the MDGs with huge reductions in global poverty, child mortality, maternal mortality, and deaths due to infectious diseases, as well as with improvements in education, vaccination, and sanitation.[46] But isn't there something fishy about using 1990 as the comparison year for evaluating development efforts that didn't begin until 2000? The developing world was already changing for the better during the 1990s—well before the MDGs kicked in. Why should the MDGs get credit for all of that progress? To really understand how the MDGs affected global welfare, we have to find a way to statistically control for all of the progress that was under way even before the MDGs were implemented in 2000.

Fortunately, scientists were well aware of this problem and took steps to assess the MDGs more realistically. The economists Isaac Ahimbisibwe and Rati Ram began by estimating the development progress that was already in motion between 1990 and 2000. Then they estimated where those late twentieth-century trends should have led by 2015 in a parallel universe in which the MDGs were never implemented. With those estimates from the no-MDG parallel universe in hand, they were then able to derive a cleaner estimate of how much credit the MDGs actually deserved. Ahimbisibwe and Ram concluded that the UN's official estimates gave the MDGs too much credit. Which is too bad, because even without those inflated estimates of the MDGs' effectiveness, the MDGs still appear to have done an awful lot of good.

By Ahimbisibwe and Ram's reckoning, the MDGs were responsible for as much as 40 percent of all global progress in reducing poverty and maternal mortality between 1990 and 2015. They also credit the MDGs with one-third of all reductions in child mortality and 86 percent of all increases in primary school completion rates.[47] And, in a rebuke to development pessimists everywhere, who have long assumed that the world's poorest nations are too poor to benefit from the rich nations' help, Ahimbisibwe and Ram found that the MDGs actually had their largest effects in the world's poorest nations and regions (including, notably, in sub-Saharan Africa). Poor countries with especially long histories of statehood (which may be a reasonably good measure of their governments' ability to make sure MDG resources were used wisely) enjoyed the biggest benefits.[48] The economists John MacArthur and Krista Rasmussen independently came to similar conclusions. By their numbers, the MDGs saved somewhere between 21 million and 30 million lives, mostly in sub-Saharan Africa, China, and India. Most of those lives were saved through reduced child mortality, improved maternal care, and improved access to treatments for HIV/AIDS and tuberculosis. MacArthur and Rasmussen also concluded that the MDGs educated 111 million children and brought nearly half a billion people's daily consumption over the "extreme poverty" line of $1.90 per day.[49]

None of which is to say that the MDGs were an across-the-board success. They failed to halve the global poverty rate, for instance, and they made only small dents in undernourishment, sanitation, and access to clean water.[50] Clearly, a great deal of work was left to be done. But buoyed by the successes of the previous fifteen years and the Big Push that was the United Nations' MDG effort, the rich world was ready to keep pushing.

## GETTING BETTER AT DOING GOOD

In the years between the beginning of the Point Four Program in 1950 and the close of the UN's Millennium Development Goals project in 2015, the developed world became more enlightened about global poverty, more concerned about the world's poor, and more active in trying to assist them than ever before. What began in 1949 as a worry about the creep of communism gradually gave way to the audacious idea that global poverty could be stopped in its tracks. This aspiration was underwritten by serious intellectual inquiry. Why do some people stay poor while others get rich? Can we do anything about it? What should we do? And how do we know whether what we did ended up doing any good? During the Second Poverty Enlightenment, we found answers to many of these important questions. We also became fluent with a variety of novel concepts—absolute poverty, relative poverty, the third world, and others—that enabled us to think and talk about poverty with clarity and subtlety.

We also cooked up an alphabet soup of new institutions—UNICEF, the OECD, the DAC, the IMF, the G8, and dozens of others—that were devoted to development and humanitarian assistance. Bureaucrats and activists discovered the power of sad faces and emaciated bodies to rouse our evolved instincts for caring about others. They also discovered that they could use people's love of celebrity to draw attention to the harsh realities of poor people's lives. We spent a lot of money. Some of it was catastrophically wasted, but some of it boosted human welfare to a degree that would have been unfathomable just a century earlier. All in all, not a bad sixty-five years.

With our spirits buoyed by some real successes during the Second Poverty Enlightenment, we resolved to get even better at doing good. Indeed, "getting better at doing good" could easily be the motto of the final big revolution in humanity's concern for the well-being of strangers.

## CHAPTER 13

# THE AGE OF IMPACT

Which volunteer activities, charities, humanitarian organizations, or other fine causes should you support? In the United States many people answer this question by helping organizations that fit with their religious beliefs or advance their political agendas. Others use their money to buy naming rights for the new wing of the downtown library or Tweed University's new School of Business. About one-half of all voluntary giving in the United States, in fact, goes to religion, the arts, and education (in the United Kingdom, the percentage is a much smaller 28 percent or so).[1] Giving, to be sure, but not necessarily what one would call "charitable."

Most of the rest goes to entities that are more directly in the business of helping needy human beings (or other living things). We support institutions devoted to helping children, old people, disabled people, drug-addicted people, homeless people, homeless animals, victims of foreign disasters, and the environment. But how should we decide which of those good causes are truly worthy of our support? Often, we tie our purse strings tightly to our heartstrings, supporting institutions that help us or move us. Grateful heart patients donate to the hospitals that gave them back their lives; rags-to-riches millionaires donate to the community organizations that kept them out of trouble as kids; nieces and nephews walk, run, bike, and swim in memory of beloved aunts and uncles to raise money for medical research; animal lovers give to organizations that prevent animal cruelty. Good causes each and every one, but should we choose our causes based on gratitude, grief, or a love of dogs?

We live in a world of finite resources. If you give a dollar to one cause, you cannot give it to another. If you think some causes could possibly be worthier than others—and especially if you think we should measure the worthiness of all causes in terms of their effectiveness at alleviating suffering and promoting flourishing—then you have drunk the Kool-Aid of the Age of Impact.

Unlike the six previous eras in the history of human generosity, the Age of Impact did not emerge from any specific historical confrontation with mass suffering. It's also not a response to any particular type of suffering—it's not about economic inequality, homelessness, disease, disability, unemployment, infant mortality, natural disasters, the victims of war, poverty in the developing world, or factory farming. Instead, the Age of Impact is a response to every form of suffering. In a world that involves trade-offs and limited resources—and in which you have torrents of information at your disposal for learning about every possible flavor of want and woe, and about the swarms of organizations that exist to alleviate them—how should you choose where to make your mark? The Age of Impact encourages us to answer this question by giving our undivided attention to two sets of concerns.

First, altruists who are caught up in the Age of Impact are obsessed with science and research, with data and facts. They believe that if you want to help people—as many people as possible—facts must be front and center. You cannot help people in need if you don't know what their real needs are, or how bad those needs are, or how those needs measure up against other people's needs, or whether it's actually possible to help them, or what tools are available for helping them, and whether those tools actually help. In virtually every age in the history of human generosity, questions about impact have been in the back of most people's minds, but for Impact Age helpers, the desire for facts is an obsession.

Second, altruists who are caught up in the Age of Impact are obsessed with consequences. To them, effectiveness at alleviating suffering is the sole criterion that we should use to figure out whether any given cause deserves our support, not whether it helps people that we love, or whether it jibes well with our religious or political beliefs, or whether it makes us happy. Impact Age altruists care first and foremost about how much suffering they can relieve.

These twin obsessions—a preoccupation with science and research, and a commitment to doing as much good as possible with each dollar or second or calorie we devote to helping others—work hand in hand. We use science to learn how best to help because "helping as much as we can" is the ethical gold standard for judging the worth of our efforts. Together, the Impact Age's obsessions with truth and with consequences have created a technological and ethical ecosystem that has led to the evolution of five new kinds of Impact Age helper: the Effective Altruist, the Philanthrocapitalist, the Poverty Scientist, the Efficiency Expert, and the Bathrobe Humanitarian. This chapter describes the beliefs and behaviors of these five creatures in their natural habitats.

## THE EFFECTIVE ALTRUIST

The first creature we encounter as we survey the Age of Impact is the Effective Altruist. The roots of effective altruism reach back to the eighteenth century, when an approach to ethics called *utilitarianism* was founded. The early formulators of utilitarianism, including Cesare Beccaria, Jeremy Bentham, John Stuart Mill, and Henry Sidgwick, all taught in one way or another that the heart of ethics was to be found in striving to fulfill the Maximum Happiness Principle—that is, in doing "the greatest amount of good for the greatest number." "Actions are right," Mill wrote in *Utilitarianism*, "in proportion as they tend to promote happiness, wrong as they tend to produce the reverse of happiness. By happiness is intended pleasure, and the absence of pain; by unhappiness, pain, and the privation of pleasure."[2] Effective altruism also takes inspiration from the contemporary utilitarian philosopher Peter Singer, from whom we get Singer's Principle: "If it is in our power to prevent something bad from happening, without thereby sacrificing anything of comparable moral importance, then we ought, morally, to do it."[3]

To Singer's Principle, the effective altruists have added what we might call Wesley's Corollary, after the eighteenth-century Methodist preacher John Wesley, who wrote "Having, First, gained all you can, and,

Secondly saved all you can, Then give all you can."[4] But give all you can to what? The effective altruist's answer to that question is pretty simple: give all you can to reduce the suffering of sentient beings.

The doctrine of effective altruism begins with suffering. Utilitarians, and the effective altruists who have taken up the utilitarian mantle during the Age of Impact, believe that the only things in the universe that are entitled to our moral concern are things with nervous systems: things with nervous systems, after all, are the only things in the universe that can suffer. Effective Altruists seek to obey this utilitarian mandate by applying Singer's Principle and Wesley's Corollary. To an effective altruist, the reduction of as much suffering as possible—the greatest good for the greatest number—is the metric by which the moral worth of our actions should be judged.

The modern effective altruism movement grew out of the efforts of a small handful of visionaries. In 1998, there was a wealthy pharmaceutical executive named Pat Dugan who was interested in locating reliable information about the effectiveness of the philanthropic organizations he wanted to support. In the end, Dugan had to throw up his hands in frustration because he couldn't find any reliable information at all. Based on this eye-opening experience, he founded Charity Navigator in 2001, a nonprofit organization that rates the financial health, accountability, and transparency of the 9,000-plus nonprofits in the United States with annual budgets of at least $1 million.

In 2006, two other visionaries—young hedge-fund analysts named Elie Hassenfeld and Holden Karnofsky—were also looking to invest in organizations that were effective at promoting human welfare. Like Dugan, Hassenfeld and Karnofsky were surprised by how hard it was to find objective data on the effectiveness of the world's nonprofit organizations. In an effort to meet this need, they founded GiveWell, a nonprofit devoted to promoting charitable organizations that can actually provide credible evidence about their impact. In 2018, GiveWell was responsible for the movement of $130 million from private donors to its highly recommended charities.[5]

Around 2009, Toby Ord, Bernadette Young, and Will McAskill—two ethicists and a physician from the University of Oxford—started an organization called Giving What We Can, devoted to inspiring people to set aside 10 percent of their lifetime incomes to the reduction of poverty in the developing world. As of this writing, nearly 4,300 people have taken GiveWell's pledge to do just that. They have already donated roughly $127 million to highly effective charities. Like GiveWell, Giving What We Can recommends charities that scientifically evaluate their effectiveness.[6]

Effective altruists, with their firm utilitarian convictions, espouse beliefs about charity and philanthropy that have surprising consequences for action. They sharply distinguish between doing good, which is good, and doing as much good as possible, which is better, for example. Consider two causes you might support to help people who are blind. Effective altruists are fond of pointing out that for the cost of training one seeing-eye dog for one blind person ($40,000 by some estimates), you could pay for hundreds of eye surgeries in the developing world. Some fraction of those surgeries would halt the progress of a blinding eye disease called trachoma, thereby preserving the patients' eyesight.[7] Now, if it is morally good to help a single blind person, an effective altruist asks, isn't it ethically better—doesn't it do more good—to use that $40,000 to prevent several dozen people from becoming blind in the first place?

Effective altruists also find an imperative in Kant's insight that all people have equal value. Once I embrace the idea that I have no more worth than you do, and you embrace the idea that you have no more worth than I do, and we both embrace the idea that neither of us has more worth than a stranger on the other side of the planet, the metric for evaluating the moral worth of our actions becomes the total amount of suffering those actions reduce: because we all have equal value, it doesn't matter whether that suffering belongs to you, or me, or Veronica down the street, or anyone anywhere else.

The Kantian principle of equal moral worth, particularly when combined with the economic principle of diminishing marginal utility, has a startling implication. Once you have eaten your first meal of the day, the principle of diminishing marginal utility tells us, the second one will have less value to you than to someone who hasn't eaten anything yet today. Likewise, the $3 I might use to buy myself a coffee will benefit me less than it would benefit someone who subsists on $3 a day. To me, that $3 coffee is a time-killing afternoon distraction. For a poor mother in the developing world, it's a doubling of her daily income.[8] If everyone's suffering matters the same as everyone else's, and if I get less benefit from a $3 coffee than the poor mother would get from $3 worth of additional food for her family, the simplest reason tells me to skip the coffee and give that money to her instead.

If you follow this Utilo-Kantonomic argument to its end, you reach a potentially uncomfortable conclusion: we should all be giving our wealth away until we are so close to desperation that any additional suffering we might reduce out in the world is offset by an equal increase in our own suffering. For most people, that's a conclusion that's difficult to accept, but it certainly opens one's mind to the fact that spending $600 a year on coffee is not an ethically neutral choice.

It can be discouraging to think about how far short most of us fall below the high ethical bar that effective altruism sets, but there is a hopeful way to look at things: whether we are willing to commit fully to the effective altruists' worldview or not, nearly every one of us is rich enough to profoundly improve the welfare of one or more persons in the developing world, and at little or no meaningful cost to our own well-being. Peter Singer gives several estimates of how much it would cost to save a life through activities such as providing bed nets to prevent malaria, delivering rehydration therapies to prevent deaths from diarrhea (which kills hundreds of thousands of children each year), preventing HIV/AIDS, and immunizing children. By helping to fund efforts such as these, it is possible to save a life for something in the neighborhood of $3,500.[9] In a rich nation like the United States, where the typical household income is $57,000, most families could save the life of at least one very poor person each year.

Effective altruism has other interesting implications for what we should be doing with our money and time. If what you really want to do is reduce as much suffering as possible, then you ought to choose your career based not on whether you think you would enjoy it, or even whether it is one of the traditional "helping professions." Instead, effective altruists say, you should pick the career that enables you to give as much as possible to effective charities. Singer offers this thought experiment:

> Suppose you could have worked for an effective charity but instead you accept a job with an investment bank that pays you $200,000 a year. There is usually no shortage of applicants for jobs with charities, so the charity will appoint someone else who will probably do almost as

> good a job as you would have done. . . . Working in finance, however, you earn much more than you need and give half of your earnings to the charity, which can use that money to employ two extra workers it would not, without your donation, have been able to employ at all. . . . Whereas you would have been replaceable as a charity worker, you are not replaceable as a donor.[10]

Effective altruists also don't worry overmuch about the overhead costs of charities, and they don't make invidious remarks about the high salaries of nonprofit CEOs. Just because a charity costs a lot to administer doesn't mean it's not cost-effective. And so what if the CEO is paid lavishly? If the organization is particularly large, or its ambit is particularly complex, it might need an extremely competent (and well-compensated) chief executive. Those indirect costs can be easily offset if the cost-effectiveness of what remains is extremely high.[11] (People value low overhead rates because they are relatively easy to evaluate; cost-effectiveness is much more difficult to evaluate. Research shows that when cost-effectiveness information does become available, however, people weigh it more heavily in their charitable decisions, as they should.[12])

Effective altruism is not without its critics. Many of their criticisms apply to utilitarianism in general.[13] Here's one: Effective altruists claim that I have a moral duty to help the poor until the suffering I relieve with the next dollar I invest is offset by the amount of suffering I would incur from the loss of that dollar myself. "That sounds lovely and all," I might say, "but I have a family that's depending on my salary: Should I impoverish them in the interests of doing the most good I can do? And what do I tell my niece when she wants me to support her Girl Scout troop by buying cookies? Should I tell her that I'm using that money to save a life instead?" And where does personal integrity fit in? Should I take a job because its high salary enables me to give more than if I were to take a less well-paid job that I would really enjoy? Don't I give up something important by forsaking work I love in order to become a money pump?[14]

There are more serious ethical issues lurking as well. What about other values, such as justice? The philosopher Iason Gabriel illustrates how effective altruism can undermine justice with a thought experiment he calls the "two villages" dilemma:

> There are two villages, each in a different country. Both stand in need of assistance but they are unaware of each other and never interact. As a donor, you must choose between financing one of two programs. The first allocates an equal amount of money to each community and achieves substantial overall benefit. The second allocates all of the money to one village and none to the other. By concentrating resources it achieves a marginally greater gain in overall welfare than the first project.[15]

Or consider Gabriel's "ultra-poverty" dilemma:

> There are a large number of people living in extreme poverty. Within this group, some are worse off than others. As a donor, you must choose between financing one of two development interventions. The first program focuses on those who will benefit the most. It targets literate men in urban areas and has considerable success in sustainably lifting them out of poverty. The second program focuses on those who are most in need. It works primarily with illiterate widows and disabled people in rural areas. It also has some success in lifting these people out of poverty but is less successful at raising overall welfare.[16]

For a diehard effective altruist, the answer to both dilemmas is clear: use your money where it can do the most good. If that means concentrating all of it in a single village, so be it. If you get additional advantage by focusing on literate men rather than on illiterate widows and people with disabilities, then that's what you should do. Yet there seems to be something morally objectionable in concentrating on one village and neglecting others, or in focusing on able-bodied men instead of women or disabled people, simply out of a commitment to cost-effectiveness. These unseemly conclusions seem to fly in the face of the humanitarian conviction that we should intervene anywhere we find needs.

There are still other concerns. What about people's rights, for instance? Consider a sweatshop where shirts are made for export to a developed nation. The sweatshop provides jobs that lift many people out of poverty, but the employees have to work with dodgy equipment that will, on rare occasions, kill one of them. Improving the safety of the factory would cost a lot of money—so much that hundreds of jobs would have to be cut to pay for it. What's to be done? A calculation based simply on utility might suggest that the factory owner should skip the equipment upgrade because the workers would gladly assume a small risk of workplace death in order to continue feeding their families. This course of action should please an effective altruist—it's the course of action that maximizes overall welfare—but don't workers have an intrinsic right not to be killed on the job?[17]

Gabriel offered the effective altruists some guidance through these ethical quandaries. For example, he recommends they consider justice and rights when the empirical evidence is too scanty to inspire real confidence that one intervention is more cost-effective than another. He also recommends that effective altruists conduct "rights audits" to evaluate how their interventions might adversely impact people's fundamental human rights. Moreover, they might simply steer clear altogether of interventions that create morally objectionable side effects. The effective altruists are a conscientious species, and a quickly evolving one, so it will be interesting to see how they deal with these objections in the years ahead.

## THE PHILANTHROCAPITALIST

Most of us can imagine what it would feel like to give up two hundred coffees a year in order to save a life (or part of one, at least), but the Age of Impact gave rise to a second species of helper that has to consider trade-offs of a different scale entirely. These helpers must decide whether to withhold a few extra billion from their kids, or their local communities—or other philanthropic causes they care about—in order, potentially, to alter the global geography of suffering. The Impact Age helpers who are willing to confront trade-offs of this magnitude are known as philanthrocapitalists.

As the effective altruism movement was growing, a malaise had been spreading through the world of private philanthropy because nobody in that business really knew whether their projects were having the impacts they were hoping for. Most philanthropies didn't do much serious strategic planning. They took a scattershot approach to setting their priorities, they obsessed about low overhead rates rather than cost-effectiveness, and they seemed blasé about measuring outcomes. Inattention to these questions struck some observers as especially odd because the ultrarich businesswomen and businessmen with the strongest interest in philanthropy were experts in creating value. It's what had made them rich. Why weren't they bringing the same entrepreneurial, dollars-and-cents outlook to their charitable giving?

As the irony began to sink in, some philanthropists started paying closer attention to their grantees' ability to create value.[18] A movement called "philanthrocapitalism" or "venture philanthropy" was thus born.[19] Teaming up with the super-investor Warren Buffett, the Bill and Melinda Gates Foundation led the way in making philanthrocapitalism mainstream by taking a more outcome-oriented approach to their own aid projects, including efforts to eradicate infectious disease, promote family planning, and stimulate international development. In 2010, Buffett and the Gateses began encouraging the world's billionaires to sign the Giving Pledge—a public commitment to give at least half of their wealth to philanthropy. Within the first year, forty billionaires signed the pledge. A decade on, the number of pledgers had increased to more than two hundred, including some of the biggest names in investment, technology, and entertainment.[20]

For the philanthrocapitalists, as for the effective altruists—not to mention the official agencies and nongovernmental organizations that seek to alleviate suffering around the globe—a major stumbling block to doing as much good as possible has been their lack of access to good information about which interventions work—and where, and for whom. Fortunately, a third species of Impact Age helper—the Poverty Scientist—devotes its energies to filling those information gaps.

## THE POVERTY SCIENTIST

When they resolved to eradicate global poverty following World War II, the world's rich democratic nations were hoping to avert World War III. Their expectation that a richer world would be a more peaceful world was really just a hunch, however. Indeed, it would not have been unreasonable to argue just the opposite: Couldn't transfers of food and money to poor nations actually encourage war, perhaps by providing new resources for warlords to fight over? (The unintended consequences of efforts to relieve the Ethiopian famine of 1983–1985 serve as a warning that such outcomes are real possibilities.) People who make big decisions about how to fight poverty face these sorts of on-one-hand / on-the-other-hand uncertainties all the time. Does development help poor people become self-sufficient, or does it lead to reckless spending? Do cash transfers to foreign governments promote economic growth, or do they stifle national ambition? If we provide technical assistance to small farmers so their fields become more productive, do we help them out of poverty or do we actually reduce their incomes by flooding markets with cheap coffee and cotton? Do microloans help people start small businesses, or do they leave people more deeply in debt?

For most of the twentieth century, nobody could answer questions like these with a lot of confidence. The official statistics for tracking the effectiveness of development efforts were simply too coarse to be of much use. It wasn't even easy to tally up how much aid was actually floating around out there. In a typical twentieth-century analysis, an economist would look for a statistical correlation between the amounts of aid that different countries had received and the changes in those countries' rates of economic growth in the years that followed. However, the statistics upon which they relied for those analyses often bundled up many forms of aid (including technical assistance, clinics and hospitals, schools, road-building programs, and cash payments to governments, to name a few) into a single number, which made it impossible to say anything about which of those forms of aid actually made a difference.[21] The only other information available for assessing the impact of development interventions was contained in the aid agencies' own reports of their impact. Those glossy summaries almost surely painted a rosier picture of their projects' success than rigorous science would have warranted.[22]

Fortunately, the Age of Impact has dramatically improved the quality of the data at scientists' disposal for evaluating the effectiveness of aid. The UN and the World Bank now generate high-resolution data about where aid is spent, how much is spent, and what it is spent on. With these data, the new Poverty Scientists can now plot the precise geographic locations of most of the major clinics, hospitals, schools, and infrastructure projects in the developing world. They are also able to collect high-resolution data on the health, wealth, and educational outcomes of the people who are directly touched by those aid projects. When these two streams of data are brought together—geo-coded information about specific aid projects, and information about the well-being of the people who are served by those projects—it is possible to get a much better picture of the effectiveness of aid.

Recently, a team of poverty scientists geo-coded the precise locations of every aid project in Nigeria. With those data in hand, they then superimposed official health information from approximately 300,000 Nigerians born between 1953 and 2003. They discovered that children who had been born close to aid projects were less likely to die in infancy and childhood.[23] A similar study found that health projects such as clinics and medical centers really do reduce the burden of illness among the people they serve.[24] Another project, which involved geo-coding the locations of 852 Chinese-funded aid projects in 13 different African nations, revealed that the aid projects boosted the educational attainment and household wealth of the nearby populations.[25] Still other studies have shown, reassuringly, that projects for controlling parasitic diseases reduce the prevalence of malaria and water-related illnesses.[26] One study even showed that Ugandan small farmers can pull themselves out of poverty by switching to coffee growing.[27]

Poverty scientists are also using the geo-coding approach to figure out where aid isn't working. The political scientist Ryan Briggs gridded the entire continent of Africa into 10,500 individual cells, each of which was a square of 50 by 50 kilometers (about 30 by 30 miles). By calculating the rates of poverty within each of those cells, and then superimposing that poverty data onto a map of all of Africa's geo-tagged development projects, Briggs was able to figure out whether enough aid was getting to the poorest regions. He found just the opposite: throughout Africa, people in poor, remote areas—where infant mortality was highest and life expectancy lowest—received less aid per capita than people in large urban centers did.[28] Other researchers have discovered similar effects: city people get more development projects per capita than do country people, even though city people tend to have better services even before that aid arrives.[29] And when poor rural communities do get aid projects, they are less effective than projects that go to richer urban centers, no doubt in part because poor rural areas lack other kinds of infrastructure (reliable roads, energy, and water supplies, for example) that make aid projects more efficient.

The geo-coding approach is dramatically improving our understanding of what works and doesn't work in the crusade to end global poverty, but the scientists of poverty have also found that they can learn other important lessons by thinking less like geographers and more like drug companies.

Suppose a drug company wants to know whether its new experimental drug prevents heart attacks. The best scientific tool for asking this question is a special kind of experiment called a randomized controlled trial (RCT). To conduct an RCT for this heart-attack prevention drug, the researchers working for the drug company would first identify a group of people who were at risk for heart attacks. They would then use the flip of a coin to determine which patients received the experimental drug and which received a comparison treatment (a sugar-pill placebo, for instance, or another medication whose effectiveness had already been proved). Next, they would follow the patients for months or years to see which ones eventually had heart attacks. If the patients who received the experimental drug had fewer heart attacks than those who didn't, then the researchers would be justified in inferring that the new drug was responsible for this difference. Because of their ability to deliver unambiguous conclusions about cause and effect, RCTs are widely (though not unanimously[30]) regarded as the epistemic gold standard in medicine (and other fields as well).

Rather than simply rolling out new projects and then trying to evaluate them after the fact, aid agencies have now begun partnering with researchers to take advantage of the power of the randomized controlled trial.[31] First, they measure a large group of individuals (or families, schools, clinics, or communities) on the outcomes of interest. Then they randomly assign one-half of the individuals (or families, schools, clinics, or communities) to the new intervention. The other half get some sort of comparison intervention, or perhaps no intervention at all. Later, after the intervention has had time to take effect, the researchers take a second set of measurements, which they then use to figure out whether the intervention made people better off.[32] Poverty scientists have now conducted hundreds of high-quality RCTs in more than eighty countries, and their efforts are producing straightforward yes-or-no answers to many questions about what works in development.[33]

Take school attendance as an example. Most experts believe that education lifts people out of poverty, so they want to boost school attendance. But how do you boost school attendance? Should you pay the children if they show up for school? Should you pay the parents? Should you raise teachers' salaries? Install cameras in the classrooms to ensure that teachers are actually teaching? Provide free uniforms? Provide feminine hygiene products for girls? In a world of finite resources, you can't do everything, and a dollar spent on a less cost-effective approach is a dollar that's not available to spend on a more cost-effective approach.

Randomized controlled trials take the guesswork out of questions like these. One experiment, for instance, showed that providing Nepalese schoolgirls with modern sanitary products, which no doubt is a kindness, does not boost school attendance (which may not be so surprising: the girls in this study were in fact missing less than half a day per year because of their menstrual periods).[34] Deworming those children turns out to be a much more promising way to make sure they get to school. Intestinal parasites are one of the scourges of the developing world, sapping hundreds of millions of children of vital nutrients. Severe infections can lead to anemia, low energy, stunted growth, and impaired intellect. In the Primary School Deworming Project, Kenyan schoolchildren in seventy-five randomly assigned schools received oral treatments for pinworms. At a cost of only pennies per dose, the intervention provided several health benefits. In the treated schools, moderate to heavy worm infections fell by 25 percent. Anemia and illness fell as well, and the kids stopped missing school. And because the dewormed children could no longer pass worm infections to others in their communities, the rates of worm-related problems fell even among kids who did not get the school-based treatment. Years later, the children who had received the treatment were still better off. As young adults, they even made more money.[35]

Deworming schoolchildren isn't just effective: it's also cost-effective. Spend as much money as you like on menstrual products for Nepalese schoolgirls and you still won't raise their school attendance by even one day: the effectiveness of that intervention is zero, at least in Nepal, so its cost-effectiveness is necessarily zero as well. Not so for deworming interventions. One randomized trial revealed that every $35 spent on a school-based deworming program in Kenya bought an additional ten years of schooling for one kid, or, equivalently, one year of schooling for ten kids.[36] The evidence for the efficacy of deworming so impressed the authorities in India's Bihar state (one of the poorest states in India) that in 2011 they agreed to partner with an aid agency called Deworming the World with the goal of deworming seventeen million children. By 2016, the program had actually reached over thirty-five million children in Bihar. Trials in India's Rajasthan state as well as in Nigeria and Vietnam were soon to follow.[37] Organizations such as Evidence Action and Innovations for Poverty Action seek to convert the evidence from trials like these into public policies and plug-and-play interventions that are ready to be implemented anywhere.

## THE EFFICIENCY EXPERTS

Unfortunately, the cost-effectiveness mindset has not yet caught on in development circles as well as it should have by now. When the United Nations committed in 2015 to the Sustainable Development Goals (SDGs)—a sort of sequel to their Millennium Development Goals from 2000—its primary architects got carried away, replacing the 8 Millennium Development Goals with 17 Sustainable Development Goals and *169* targets for tracking their progress. The SDGs became a Frankenstein's monster, one that the *Economist* magazine condemned in 2015 as "so sprawling and misconceived that the entire enterprise is set up to fail."[38] It's easy to understand why: a list of targets that includes reducing bribery and government corruption, increasing access to medications for high blood pressure, improving policing, bringing broadband access to the entire planet, boosting life satisfaction, promoting urbanization, supporting research to increase crop yields, increasing the number of UN peacekeepers, and raising birth rates in rich countries is nothing if not "sprawling."

To accomplish all of the Sustainable Development Goals, in fact, the rich world would need to invest between $5 trillion and $7 trillion annually, which amounts to between 7 percent and 9 percent of the gross domestic income of the entire planet.[39] If we can't get the world's rich nations to regularly contribute even 1 percent of their gross domestic incomes, then how in the name of Jeremy Bentham are we going to get them to contribute 7 percent or 9 percent? Goals this pricey are barely conceivable and not even remotely attainable. In a world of finite resources, in which a dollar committed to one goal is a dollar that is not available for another goal, it is important to be efficient—that is, if we want to do the most good we can do.

That is where the efficiency experts come in. Their argument about the Sustainable Development Goals is simple: we can reduce global poverty more efficiently, and faster, by focusing only on those SDG targets that *create* value, where one dollar spent creates more than one dollar of benefit. After all, it is easy to generate a 1-to-1 ratio of benefits to costs: we could pull every person in the world above the poverty line, for instance, simply by transferring enough money to ensure that each has at least $1.90 per day to live on. But unless those cash transfers change poor people's opportunities in more fundamental ways, then 1-to-1 is the best you can do with a cash transfer. If we know of interventions that create even more value for poor people, as the efficiency experts correctly point out, then we should favor those interventions instead.

After the UN announced its Sustainable Development Goals, the Copenhagen Consensus Center, under the direction of the economist Bjørn Lomborg, convened a group of economists to conduct cost-benefit analyses for over 80 of the SDG targets. They found that some interventions could yield $20, $50, or even $100 in benefits per $1 spent. Others created *less* than $1 for each $1 spent. If our only goal is to improve the lot of the world's poorest people as quickly and efficiently as possible, then we shouldn't blow our money on ending child marriage, guaranteeing universal employment, increasing life satisfaction, or even keeping the mean global temperature from going more than 2°C over preindustrial levels, as admirable as those objectives might be. If we focus instead on just 19 "Phenomenal Development Targets," we could achieve $20 to $40 in benefits, on average, for every $1 spent. Spreading that same amount of money across 168 targets, in contrast, would generate less than $10 per $1 spent. If a 10-to-1 ratio of benefits to costs sounds nice, then a 20-to-1 ratio or a 40-to-1 ratio should sound twice or four times as nice. I have placed some of the Copenhagen Consensus's Phenomenal Development Targets in Figure 13.1, along with estimates of their projected costs and benefits.

There's something in the Phenomenal Development Targets for people of every political persuasion. For pro-family social conservatives, there's the reduction of newborn and maternal death. For social liberals,

| Target | Cost/Year ($ billion) | Benefits/Year |
|---|---|---|
| Lower chronic malnutrition by 40% | 11 | Prevent malnourishment in 68 million children |
| Reduce malaria infections by 50% | 0.6 | Prevent 100 million cases of malaria; save 440,000 lives |
| Reduce tuberculosis deaths by 90% | 8 | Save 1.3 million lives |
| Cut early death from chronic disease by 33% | 9 | Save 5 million lives |
| Cut HIV infections through circumcision | 0.035 | Avert 1.1 million HIV infections, with increasing benefits over time |
| Reduce newborn mortality by 70% | 14 | Save 2 million newborns' lives |
| Increase childhood immunization | 1 | Save 1 million children's lives |
| Make family planning available to everyone | 3.6 | Save 150,000 mothers' lives, plus a demographic dividend |
| Eliminate violence against women and girls | Likely low | Likely substantial (305 million women are domestically abused annually at a cost of $4.4 trillion) |
| Phase out fossil fuel subsidies | 37 | Free up $548 billion in government revenue |
| Tax pollution damage from energy | Likely low | Prevent as many as 7 million deaths |
| Cut indoor air pollution by 20% | 11 | Prevent 1.3 million deaths |
| Reduce trade restrictions | 20 | Make every person in the developing world $1000 richer by 2030; lift 160 million people out of poverty |
| Boost agricultural yield growth by 40% | 2.5 | Prevent 80 million people from going hungry; produce additional food worth $84 billion |
| Achieve universal primary education in sub-Saharan Africa | 9 | Increase number of children attending primary school by 30 million |

Figure 13.1. Cost-benefit estimates for fifteen Phenomenal Development Targets. Lomborg 2018.

there's contraception and HIV/AIDS prevention. For classic liberals and libertarians, there's free trade, pollution taxes, and the elimination of fuel subsidies. Unfortunately, the UN hedged its development bets by throwing money at all 168 of the SDG targets. They will do some good, which is good, but it will not be as much good as they could have done.

## THE BATHROBE HUMANITARIAN

While the Age of Impact was revolutionizing development aid and other traditional forms of assistance, the Internet and other technologies were creating previously unimaginable new opportunities for people to improve others' lives. To take advantage of these opportunities, most people wouldn't even need to leave the house. This brings us to our fifth species of Impact Age helper: the Bathrobe Humanitarian.

The Internet has made it easier than ever to access reliable facts about the plights of the unfortunate. News of the Great Lisbon Disaster of November 1755 had to travel for weeks by land and sea before it arrived in London. Thanks to social media, which enable people to communicate with the rest of the world through Facebook, Twitter, Instagram, and YouTube, news of the Indonesian earthquake of September 2018 traveled thousands of times faster, reaching the entire globe before the tsunami had even receded from the shores of Sulawesi. And we learned about the tsunami in enough time to actually do something to help its victims.

The Internet hasn't just made the news faster; it has also made the news more informative and more vivid. Gone are the days of learning about the misfortunes of faraway people from three column-inches in the newspaper or a ninety-second spot on the evening news. Now you can swipe on a smartphone news alert about a humanitarian crisis in sub-Saharan Africa or the latest Caribbean hurricane, and then watch a high-resolution video of conditions on the ground. Through innovations like these, twenty-first-century communication technologies have drastically cut the psychological distance between people in need and people in a position to help them.

The Internet has also made our knowledge of people's hardships more actionable. Websites that go by names such as GoFundMe.com, Kiva.org, HandUp.org, and NetSquared.org enable people to advertise their needs, recruit others to join their causes, seek donations, and organize volunteers. If you are moved by a charity you've learned about online, you can lend a hand by simply entering your credit card number and pressing "Donate." Or if raising public pressure on elected officials is your favored approach to social change, you can use Care2.org to advertise your gripe and gather signatures. Thanks to web-based services like these, we can now blog, boycott, protest, petition, share, enlighten, and crowdfund with a speed and efficiency that would have stunned the reformers, philanthropists, and humanitarians of yesteryear.

The Internet has also made generosity more participatory, counteracting the increased professionalization of compassion over the past two centuries.[40] Instead of using their screen time to make memes or share photos of their dinners, some people are now using the Internet as a vehicle for channeling some of their cognitive surplus—their small shares of the trillion hours of free time that the world's Internet users have at their disposal for thinking and problem-solving—into altruistic pursuits. "The wiring of humanity," as Clay Shirky wrote in his book *Cognitive Surplus*, "lets us treat free time as a shared global resource, and lets us design new kinds of participation and sharing that take advantage of that resource."[41]

A movement called Digital Humanitarianism is an excellent example of how people are turning their cognitive surpluses into forces for good. Some digital humanitarians, operating in crowdsourcing networks with names such as Humanity Road and the Digital Humanitarian Network, sort through the terabytes of photos, videos, and messages that people post to social media as natural disasters and other large-scale crises unfold around them. Once the digital humanitarians have used ordinary human ingenuity to turn those posts into information, they are able to do all sorts of useful things with it.[42] The Standby Volunteer Task Force, for instance, sifted through social media posts following the 2010 fires in Haiti and Russia, the 2011 outbreak of civil war in Syria, the 2014 Ebola epidemic in West Africa, and Hurricanes Harvey and Michael, which devastated Texas and Florida in 2017 and 2018, respectively. From those tidal waves of online data, the Task Force volunteers were able to determine where the humanitarian needs were the greatest, along with the specific types of assistance people most needed. They were then able to integrate their social media data with official sources of data, such as imagery from satellites and drones, to create high-resolution maps that helped official agencies focus their relief efforts.[43] Journalists also are turning to the digital humanitarians for information they can use in their efforts to inform the public about ongoing humanitarian crises.[44]

The Internet has revolutionized philanthropy and humanitarianism in one last way: the Information Superhighway became a compassion superhighway in part because it is a reputation superhighway. On social media, people can blog, post, tweet, and selfie their way to glory by advertising their participation in fund drives, ice-bucket challenges, and pledges to give 10 percent of their lifetime incomes to effective charities.[45] Thanks to the Internet, even noncelebrities can now grab their fifteen minutes of philanthropic fame. This is not a cause for cynicism: a desire for others' esteem has motivated human generosity ever since we began sharing meat with each other on the African savanna hundreds of millennia ago. If advertising your virtue helps you to do the most good you can do, as any card-carrying utilitarian would tell you, then advertise away.

## PRETTY SIMPLE

Gregg Popovich is one of the greatest coaches in the history of basketball. He is also one of the most difficult-to-interview coaches in the history of basketball, frequently responding to sportscasters' questions in a manner that oscillates between unhelpful concreteness and hammy condescension. Popovich is also a highly engaged philanthropist, supporting a wide variety of organizations that serve disadvantaged people in the United States and elsewhere. A journalist once asked Popovich, I suspect with some dread, why he did so much charitable work. His response was classic Popovich: "Because we're rich as hell, and we don't need it all, and other people need it. Then, you're an ass if you don't give it. Pretty simple."

What I find most remarkable about Popovich's response is his complete confidence that any listener would find his reasoning "pretty simple." For most of our time on this planet, Popovich's rationale for his generosity would not have seemed simple at all. It would have seemed ridiculous. What kind of idiot would divert his or her wealth away from kith, kin, and countrymen in order to share it with complete strangers who will never be in a position to repay it? That Popovich's account of his own generosity is compelling to anybody is a testimony to the power of the forces that have transformed human compassion over the past ten thousand years.

As we reach the end of this chapter, we also reach the end of history. We'll find no more help in our efforts to understand humanity's concern for humanity by picking through the historical details that our forebears left behind. To do the heavy explanatory lifting that the task still requires, we will need to fashion a threefold cord of knowledge. We will braid that cord out of three different strands: first, what we have learned about human instincts; second, what we have learned about human history; and third, what we have learned about human progress.

## CHAPTER 14

# GOOD REASONS

That twenty-first-century humans take such an abiding interest in the welfare of strangers is a real curiosity. Although we owe them no debt of love or loyalty, we and our ancestors from generations past have often taken considerable time and trouble to help suffering strangers—particularly when the suffering confronting us came in vast quantities, or challenged our worldviews, or pulled at the fabric of society. Over the past ten thousand years, as one era of history gave way to the next, the circle of human concern expanded ever further. Today, it would seem negligent not to at least consider what we might do to help—even if we were separated from those in need by race, religion, language, or thousands of miles. Twenty thousand years ago, people would have barely understood the question.

Because everyone has the same basic needs and vulnerabilities, the large-scale crises that have widened our concern for others over the past ten millennia have looked much the same everywhere: orphaned children, widowed mothers, old people, blind people, disabled people, hungry foreigners at the gates, city-street scavengers, jobless breadwinners, the victims of disaster, the collateral damage of war. These confrontations with suffering seriously taxed our forebears' imagination and ingenuity. Their evolved instincts for generosity didn't give them a lot of guidance about how to respond to such afflictions, but some of the lucky bystanders who were spared from them—rulers, elites, and scholars, mostly—started casting about for solutions. Eventually, they settled on solutions that compelled them to cancel their appointments, roll up their sleeves, take out their checkbooks, and take action. Their efforts were helped along by progress in technology, science, and trade.

To weave this book's many observations into a coherent explanation for the kindness of strangers, we begin with a quick glance over our shoulders at the past. In the beginning, our stone-age ancestors didn't care very much at all about the well-being of true strangers. When they greeted strangers, it was more often with spears and arrows than with open arms. The initial innovations came during the agricultural revolution and in the first cities. The spectacular inequalities in wealth during this era provided opportunities for powerful god-kings to generate loyalty and gratitude among their subjects by cultivating reputations as defenders of the orphan and the widow. Several millennia later, the mushy historical epoch known as the Axial Age left us with a more egalitarian and universally applicable reason for helping strangers: the Golden Rule. It was more than greeting-card treacle—it was a rule of thumb for compassion that managed to be both deeply spiritual and surprisingly practical.

More than a thousand years after the close of the Axial Age, compassion become more secular and expedient. The political theorists of the sixteenth century argued that the relief of poverty made good sense for any government that wanted to keep disease, crime, and civil unrest under control. A couple of centuries after that, the intellectual gifts of the Enlightenment—new convictions about human equality, human dignity, and natural rights—led to a new way of thinking about the state's concern for the welfare of its citizens as well as the victims of humanitarian crises in other lands. Conditions that had previously been viewed as inescapable tragedies or necessary evils became eighteenth-century outrages. By the nineteenth century, those same Enlightenment values set the stage for the modern welfare state. And only decades after that, in the years following World War II, the world's most fortunate people took up an abiding interest in improving the lot of the world's least fortunate people.

We can arrange these milestones in the history of humanity's regard for humanity on a time line, but time lines don't explain anything. A real explanation requires that we understand the interaction of human events with a small set of human instincts. But which instincts?

## THE INSTINCTS THAT MATTER

Our evolved affinity for helping kin, which William Hamilton brought to scientists' attention in 1964, undoubtedly explains much of the altruism we see in the biological world.[1] Bacteria assist their bacterial clones, mothers and fathers take care of their offspring, eye-rolling big sisters look out for their dumb little brothers. The psychological adaptations that govern these manifestations of familial love, however, simply cannot explain why we help strangers. We humans and many other animals are very good at distinguishing between kin and non-kin, and with good reason: helping others is costly, so you're better at the game of natural selection if you avoid wasting your help on individuals who are not your genetic relatives. Even when we use the terminology of kinship, as when we refer to our friends as our brothers and sisters, or when we refer to our churches or communities as "families," we're using it figuratively rather than literally. Natural selection is way too smart for cheap linguistic sleights of hand to have been a major force in our evolution. Kin selection is not going to help us understand where the kindness of strangers come from.

Relying on group selection won't help either. Darwin's thoughts on the matter reveal just how seductive group selection has been as an explanation for our regard for others. "At all times throughout the world," he wrote, "tribes have supplanted other tribes; and as morality is one important element in their success, the standard of morality and the number of well-endowed men will thus everywhere tend to rise and increase."[2] For the purpose of understanding why we care about strangers, however, group selection, like kin selection, is an explanatory dead end. The group-selection ideas from the 1960s turned out to be full of holes, and later reformulations turned out simply to be restatements of Hamilton's idea that we evolve to help our genetic relatives. And even if it were the case that natural selection endowed us with traits that lowered our fitness relative to the fitness of our group-mates, while simultaneously improving our groups' success in battle against other groups, there is no reason to believe that those traits made us kind: just as likely, they made us cruel.

There are, however, two evolved instincts for helping others that probably have played a role in the expansion of compassion. The first is our instinct for reciprocity. Natural selection has endowed us with an appetite for helping nonrelatives, even at a personal cost, by virtue of the repayments we're likely to receive at a later date from those grateful beneficiaries. As Darwin conjectured in *The Descent of Man*, it's likely that in the social lives of our stone-age ancestors, in which favors granted often led to favors repaid, "each man would eventually learn that if he aided his own fellow-man, he would commonly receive aid in return."[3]

After fifty years of scientific inquiry, Darwin's conjecture has held up pretty well: humans really do help others when they anticipate that it will produce return favors.[4] Our appetite for the rewards of reciprocity has left its mark throughout the history of compassion. Since the dawn of civilization, rulers have used the power of the purse to buy the loyalty of their subjects and citizens. Likewise, the Axial Age, the Age of Prevention, the two Poverty Enlightenments, and the Humanitarian Big Bang have revealed that human societies become more sensitive toward strangers' concerns as the cost of helping declines. This is exactly as it should be if our willingness to help is regulated by our instincts for reciprocity: as the marginal cost of helping others goes down, helping others on the prospect of return favors comes to look more attractive.

The second evolved instinct for altruism that has driven the history of human compassion is our appetite for helping others in hopes of appearing virtuous. Big shots throughout history have tried to buy good names for themselves and their descendants by paying for alms, funding feasts for the poor, and building hospitals, orphanages, and soup kitchens. The religiously devout have used charity and good works to please God and avoid his wrath. To this very day, people commit themselves to worthy causes in order to call attention to their civic-mindedness, to escape opprobrium for seeming callous, and to demonstrate their intelligence and reasonableness. If you can't get return favors from the people you help—and even if you can, actually—you can improve your fortune by helping others liberally and ostentatiously.

Our love of reciprocity and reputation are not the only instincts that have played an instrumental role in the history of compassion, however. In fact, they're not even the most important ones. After all, we have possessed those instincts for as long as we have been human, and the history of our concern for strangers began only when we started living together in large numbers. Before then, strangers were treated mostly with suspicion and hostility. We need additional explanatory equipment if we are to get to the bottom of the expansion of our concern for strangers. In particular, we need to understand our instinct for reasoning, and how that instinct gets activated by mass suffering.

Philosophers and psychologists have become quite skeptical of reason in recent decades. Adherents to philosophical schools of thought called *intuitionism* and *sentimentalism* argue that our moral behavior is barely yoked to reason at all. The intuitionists think our moral judgments are little more than after-the-fact rationalizations of decisions we have already made, and without much in the way of deliberative thought: an intuitionist might argue, for instance, that we care about others not because we have reasoned it out as a desirable course of action, but instead because our evolutionary histories of group living have made us sensitive to considerations of reciprocity and fairness.[5] Similarly, a sentimentalist might think it is our emotions that are the font of our moral judgments: we judge actions as immoral because we first experience them as disgusting, for instance, and we experience a desire to care for others because we first feel empathy for them.[6] For the intuitionists and the sentimentalists alike, we cough up reasons for why we should do this thing or that thing only after other voices have had their say.

Psychologists of the past fifty years have been positively exuberant about the shortcomings of human reason. They've proffered long lists of systematic biases and untrustworthy rules of thumb that lead our judgments astray.[7] The correspondence bias, for example, is a failure of reasoning that inclines us to assume that other people's actions are caused by their personalities rather than by situational forces acting upon them. The fundamental attribution error extends the correspondence bias by inclining us to do just the opposite when making sense of our own behavior: we attribute it to situational forces rather than to our personalities. The hot-hand fallacy leads us to believe in self-perpetuating streaks of good or bad performance. We confuse correlation with causation, we neglect base rates, we deny antecedents, we affirm consequents, we hate losing five dollars more intensely than we love gaining five dollars: if you took the typical psychology textbook as your guide, you would conclude that our minds are positively brimming over with fallacies, biases, and errors.

Whether people actually commit these sins of reasoning or not (the much-celebrated fundamental attribution error may not exist at all, for example[8]), and whether they are actually errors in the first place or not (some good research suggests that basketball players really do get "hot hands"[9]), it's certainly a mistake to think that these inclinations are the totality of human reasoning. In fact, there is a species of reasoning, too often taken for granted, that is devoted solely to helping us to choose our actions wisely. It is called practical reasoning, which the *Stanford Encyclopedia of Philosophy* defines as "the general human capacity for resolving, through reflection, the question of what one is to do":

> [Practical reasoning] typically asks, of a set of alternatives for action none of which has yet been performed, what one ought to do, or what it would be best to do. It is thus concerned not with matters of fact and their explanation, but with matters of value, of what it would be desirable to do. In practical reasoning agents attempt to assess and weigh their reasons for action, the considerations that speak for and against alternative courses of action that are open to them. Moreover they do this from a distinctively first-personal point of view, one that is defined in terms of a practical predicament in which they find themselves (either individually or collectively—people sometimes reason jointly about what they should do together).[10]

Practical reasoning, then—looking for the most desirable course of action—involves looking for reasons and pondering the reasons that others offer. We might engage in practical reasoning about whether to extend a helping hand to a particular group of people by first asking ourselves whether we should help at all, and then asking what sorts of help we could or should provide. Practical reasoning is surely the form of reasoning Darwin had in mind when he asserted the role of reason in our expanding regard for the welfare of strangers. "As man advances in civilization," he wrote,

> and small tribes are united into larger communities, the simplest reason would tell each individual that he ought to extend his social instincts and sympathies to all the members of the same nation, though personally unknown to him. This point being once reached, there is only an artificial barrier to prevent his sympathies extending to men of all nations and races.[11]

Practical reasoning—the search for reasons and the consideration of others' reasons—has left an indelible mark on the history of human benevolence. Consider Uruinimgina, the king of the archaic city-state of Lagash. When he recognized that easing the burdens of inequality for widows and orphans would help him win their admiration and loyalty, he discovered a *reason* for compassion. Likewise, the Axial Age belief that God blesses the generous and punishes the hard-hearted provides a reason for generosity. The Axial Age discovery of the Golden Rule—do unto others as you would have them do unto you—was another reason to stretch our generous impulses beyond the limits of kith and kin.

Other reasons were not far off. During the Age of Prevention, reformers such as Juan Luis Vives sought to persuade Europe's rulers that it would be cheaper to alleviate poverty than to cope with its social side effects. Additional arguments in Britain for the liberalization of poverty relief arose from the worry that ignoring the welfare of the masses could bring to Britain the same sort of working-class outrage that sparked the French Revolution.

The First Poverty Enlightenment brought additional reasons for compassion. The writings of Jean Jacques Rousseau, Adam Smith, and Immanuel Kant provided the raw materials for an argument that all people, no matter their social standing, share a dignity that entitles them to certain rights—most notably, the right to possess enough wealth to meet their basic needs. And in repudiation of the old mercantilist assertion that poverty strengthens the economy by incentivizing workers to keep working, the First Poverty Enlightenment saw the use of economic theory to argue that the mercantilists had it exactly backward: poverty stunts economic growth.

The Humanitarian Big Bang brought still other reasons. As international trade made the economic health of each nation more dependent on the economic health of the others, Europe's elites discovered a reason to safeguard the welfare of their most important trading partners. They also found a reason for compassion in Emmerich de Vattel's argument that nations have the same sorts of natural rights as individuals do. As a consequence of those rights, nations owe each other the Offices of Humanity, which include development assistance and emergency aid during times of crisis. The reformers of the Humanitarian Big Bang also argued that the state had a responsibility to take care of its citizens following mass disasters, just as the reformers of the First Poverty Enlightenment argued that the state had a responsibility to ease the burdens of poverty among its citizens.

The Humanitarian Big Bang's reliance on arguments for natural rights and the common dignity of all persons also created a reason for people to band together into associations devoted to reducing drownings, preventing shipwrecks, improving prison conditions, abolishing slavery, and increasing women's political voice. Christian leaders argued for the universality of human dignity, inspiring evangelizing Christians to attend better to the material and intellectual needs of those whose souls they wished to save. Edmund Burke's argument that Britain's colonialist pursuits could not ignore the natural rights of indigenous peoples planted a seed of reason that eventually grew into national unease about the entire colonialist project. The assertion that soldiers had a right to medical care and humane treatment, and that civilians had a right to protection from the horrors of war, led to the founding of the Red Cross, the ratification of the Geneva Conventions, and the proliferation of efforts to care for the people of a Europe ravaged by World War II.

Following the war, the world's rich democratic countries were swayed by President Harry Truman's argument that assisting developing countries would help them secure two of their most highly valued goals: halting the global creep of communism and preventing a third world war. Later, in the 1970s and 1980s, ethicists and political theorists such as Peter Singer, Onora O'Neill, and Charles Beitz provided reasons for compassion that were all based in one way or another on the assertion that poverty was a natural evil that every person had a moral obligation to take seriously. Everyday people were also charmed by the argument from official agencies and private relief organizations that anyone—even someone watching television at home—could make a difference in the lives of the world's distant poor. And at the beginning of the twenty-first century, empirical evidence that the Millennium Development Goals had been effective at reducing the burden of global poverty began to persuade some development experts that assistance didn't inevitably turn into Dead Aid: Big Pushes can actually work.

During the twenty-first century's Age of Impact, effective altruists and philanthrocapitalists fell under the influence of the utilitarian ethicists, who argued that if it is good to use a dollar to help people in need, then it is even better to squeeze as much help out of that dollar as possible. People also became increasingly open to the power of the social sciences, with their randomized controlled trials and cost-effectiveness analyses, to help us get the biggest possible bang from each philanthropic buck. Finally, the Age of Impact—with its new electronic technologies for transferring information, donating resources, and pooling expertise—held out the possibility that we all could use our money and our minds to become bathrobe humanitarians.

In history's torrent of arguments for helping strangers, we can discern three distinct types of reasons. Reasons of the first type get their force by appealing to naked self-interest. When I realize that helping you today will make you feel obliged to do a favor for me in the future, or bring me glory in the eyes of admiring bystanders, or keep my head out of the guillotine, or win God's approval, I am motivated to help by self-interested goals that I have come to believe are well served by helping. It is tempting to think that the help we provide in search of return favors or admiration is motivated by our evolved instincts for direct and indirect reciprocity, and surely it often is. However, it can also be driven by reason alone. We merely reason our way to the insight that when we help people, the beneficiaries of our generosity will repay us, and witnesses of our generosity will admire us.

Reasons of the second type also appeal to self-interest, but less directly: they focus on the benefits to the collective rather than to the individual. These are reasons that emphasize the desirability of cities that are free of disease and disorder, of nations that are economically competitive, and of a world of comity and prosperity. Arguments based on these reasons should appeal to all people who can appreciate how their own well-being is tied up with the well-being of their countries and communities, and even how, in a more globalized world, the well-being of their own nation is tied up with the well-being of other nations.

Reasons of the third type appeal to self-interest as well, though only in a stretched sense of what "self-interest" means: these reasons appeal to the listener's desire for integrity—that is, for not being a hypocrite, for walking the walk and not just talking the talk, for believing and acting in ways that are consistent with logical, mathematical, and ethical principles. These principles include abstractions such as natural rights and fairness, the economic law of diminishing marginal utility, and the utilitarian proposition that an action's ethical status is determined by its effects on the suffering of sentient beings.

Throughout history, one of the most important and yet underappreciated of these abstractions has been the concept of identity: things that are identical with respect to any particular attribute should be treated the same with respect to that attribute. Despite the orthographic differences between the numeral 1 and the numeric expression 3 - 2—they look different—they are identical in all of the ways that matter to mathematics. That's what the equals sign in the equation 3 - 2 = 1 actually means. If you treat them as different things, you'll get the arithmetic problem wrong.

Similarly, things that are identical in every feature that is relevant to ethics should be treated identically with respect to ethics. The Axial Age's Golden Rule, with its injunction to treat others as one would wish to be treated, is underwritten by the a priori truth that things that are the same in a given domain should be treated as the same within that domain. If you feel entitled to have others respond compassionately to your desires and needs, the only way your claim can have any bite, save through coercion and raw power, is for you to accept that they possess the identical right, and that you possess the identical obligation. If you treat them as different things, you'll get the ethical problem wrong. Kant's proposition that all persons everywhere share a fundamental human dignity in equal amounts also relies on the concept of identity. Because we all possess a common dignity, we are all entitled to be treated as ends withal, and never as means. Consequently, we are all obligated to extend the same treatment to others, not as a matter of charity but as a matter of duty. Emmerich de Vattel's argument that nations are endowed with the same basic slate of natural rights as individuals extends Kant's reasoning to international relations.

Adam Smith and Charles Dickens each extended Kantian reasoning as well, relying on moral identity to propagate the spirit of the First Poverty Enlightenment in economics and fiction, respectively. They did so by attacking the age-old belief that poverty was a punishment for folly and vice. The working poor, they asserted, were no less intelligent, resourceful, diligent, or virtuous than the people of fashion: they just happened to be poor.

People reason best when they reason together instead of muddling through on their own. When people reason in isolation, they are especially vulnerable to the foibles and biases that the social and cognitive psychologists revel in exposing. One of the most pernicious of these foibles, called the "myside bias," manifests itself in the ease with which we find reasons that support our own convictions and in the difficulty with which we find reasons to question our own convictions. When people reason together, in good faith, about issues on which they genuinely disagree, we call it an argument: each party to the argument offers reasons for his or her convictions (some of which will be thoroughly suffused with myside bias), while the other party considers those reasons and tries to come up with objections. As these reasoners climb a staircase of arguments, counterarguments, and revised arguments, they often abandon poorly formed opinions for better opinions.[12] Reasoning together can take us to better beliefs.

When it comes to foreign aid, for example, Americans are some of the most pessimistic people in the world. More than half of Americans think the United States spends too much on foreign aid, in no small measure because they think it sucks up 25 percent of gross domestic income. However, when they're supplied with a handful of facts about the actual costs (which, in fact, amount to much less than 1 percent of gross national income), and about the high effectiveness of the programs we fund to help other nations with food, medicine, and development—particularly when those facts are verified by a nonpartisan group of experts—Americans' support for foreign aid rises.[13] Even more impressive changes in attitude are possible when people reason together. In one study, people were provided with a slate of basic facts about the role of the United States in international aid and development, and then asked either to consider those facts in isolation or to deliberate about them in small groups of their peers. Those who reasoned together became more supportive of America's investments in official development aid than those who contemplated the facts independently.[14] Reasons matter, and when we reason together, they matter even more.

Darwin averred that "the simplest reason" would eventually lead humanity to extend its sympathies to the people of all nations and races. The simplest reason, however—even when paired with our evolved instincts for altruism—can only give us the resolve to help. If we lack tools for rendering effective help, the simplest reason has little to offer beyond crocodile tears. Fortunately, three centuries of progress in technology, science, and trade have given us tools aplenty.

Science and technology have contributed indispensably to humanity's efforts to improve the lives of distant strangers. Many of these contributions have come through improved communication and transportation. Consider, for example, how breakthroughs in communication have shortened the time it takes for news of humanitarian crises to reach us. News of Lisbon's seismic destruction in 1755 didn't reach the rest of Europe for weeks. Within 130 years, improvements in telegraph technology had slashed the speed of information transfer by 99 percent. Thanks to an undersea telegraph cable that was laid on the ocean floor in 1870, news of the magnitude-6 volcanic eruption on the island of Krakatoa in 1883 raced from Java to London at a rate of eight words per minute. London knew about the cataclysm within fourteen hours.[15] By January 12, 2010, digital communications technology had become so fast that news of Haiti's devastating magnitude-7 earthquake on that day reached the rest of the world within minutes. All told, news of humanitarian crises now travels several orders of magnitude faster than it did three centuries ago.

Innovations in mass communication have aided our efforts to respond to slower-moving tragedies as well. Through the eighteenth century, as printing technology became more efficient, it became ever easier to lay one's hands on a newspaper dispatch from a battlefield reporter, or an eyewitness account of a worker demonstration, or Dickens's most recent installment of *Oliver Twist* in *Benton's Miscellany*, or Henry Mayhew's latest peek into the lives of London's street folk in the *Morning Chronicle*, or working-class viewpoints on current events in the *Poor Man's Guardian*. Today, the digital technologies on our desks and in our pockets can bring us details about the lives of the unfortunate with a finger tap, a swipe, or a verbal command to a cheerful and ever-attentive AI program.

Innovations in transportation have also made it easier to address humanity's wants and woes, starting with eighteenth- and nineteenth-century improvements in shipbuilding. As steel hulls replaced wooden ones, as rivets replaced nails, and as steam replaced wind, the transportation of food, medicine, supplies, and technology became faster and cheaper. The trains, planes, and automobiles of the nineteenth and twentieth centuries further slashed the time and money required to get help to where it was needed. And with the advent of the Internet, the speed with which we could donate to worthy causes approached singularity: no longer did we need to throw money into a poor box or an offering plate.

As important as these innovations in communication and transportation have been for expanding the reach of human compassion, smartphones and supply chains are just the beginning of science and technology's contributions. Advances in chemistry and genetics, for instance, paved the way for the high-yield, pesticide-resistant grains that have saved millions of people from starvation and stunting. Advances in chemistry have given us the fertilizers to help them grow. Breakthroughs in microbiology have given us the vaccines and antibiotics that save millions of people from infectious diseases each year, and advances in epidemiology and demography have shown us where to send those resources and whom to give them to.

You know which branch of knowledge doesn't get nearly enough credit for widening humanity's concern for humanity? Mathematics. Progress in math, statistics, and probability theory during the seventeenth, eighteenth, and nineteenth centuries gave us tools for calculating compound interest with precision. With this knowledge in hand, it became possible to confidently estimate how the assets in an insurance or pension fund would grow over time. Complementary advances in the construction of life tables enabled us to estimate how long a person with a specific set of characteristics (for example, age, sex, health, and occupation) could be expected to live or stay healthy. These innovations equipped us to establish social insurance and social welfare programs that could stay solvent without imposing unnecessary burdens on taxpayers.[16]

The not-so-dismal-after-all science of economics has also made invaluable contributions. Economic thought from the late nineteenth and early twentieth centuries gave us a variety of useful concepts, including the idea that poverty reduction does not hinder but instead promotes the production of wealth.[17] Economics gave us other tools for thinking straight about poverty and its alleviation as well, including the poverty line, the poverty trap, the price index, the Gini index, cost-benefit analysis, and purchasing power parity.[18] Without these concepts, it would be difficult to study poverty in theoretically rigorous and empirically tractable ways.

Unlike most other methods for seeking truth, science thrives on criticism and second-guessing. This hospitality to criticism is the very reason why scientific claims are so important for coherent arguments about how to improve people's lives, especially in political environments in which social spending is a contentious issue. Consider the argumentative value of scientific evidence about the cost-effectiveness of various policies and programs. People of many political persuasions are happy to offer a helping hand to people with genuine needs, but some are more cost-conscious than others, and reluctant to waste their money in a world of difficult trade-offs. Science is the human enterprise that is best suited to providing reliable guidance on cost-effectiveness, so it has an important role to play in building nonpartisan consensus. Recent evaluations of whether the Millennium Development Goals improved life in the developing world are important examples of how science can show us what works, but science also has a lot to teach about the effectiveness and ineffectiveness of domestic policies.[19] Scientific knowledge should therefore be front and center in any argument about where our resources should be placed to promote human welfare.

The final human endeavor that deserves credit for expanding humanity's concern for humanity is international trade. Thanks to growth in trade, the world now produces nearly one hundred times as much wealth each year as it did in 1800. With such an explosion in growth, it is hardly surprising that the global rate of extreme poverty has declined from 90 percent to 10 percent over the same time period.[20] Trade has had several felicitous effects on humanity's regard for humanity.

First, trade makes its participants interdependent in a way that few other institutions can. I am better off when I use some of my money to buy from you a good or service that I cannot produce for myself, and you are better off for the money I have paid you for that good or service. As a consequence of this non-zero-sum phenomenon—trade makes both parties better off—buyers and sellers can end up appearing as if they are trying to help each other even when they're doing nothing more than pursuing what's good for themselves. As Adam Smith famously put it in *The Wealth of Nations*, "it is not from the benevolence of the butcher, the brewer, or the baker that we expect our dinner, but from their regard to their own interest."[21]

Second, once two parties become dependent on each other for buying and selling, they encounter additional incentives to treat each other kindly. A hardship that endangers the brewer's business also endangers my access to his beer. Our interdependence gives me an incentive to look after his well-being more directly (for example, buying more of his beer than I otherwise might). The trading powers of Western Europe learned this lesson in the wake of the Great Lisbon Disaster of 1755, when Portugal's derailment from earthquake, fire, and flood threatened to hamper global trade. To prevent this undesirable outcome, Spain, England, France, and Hamburg all offered to reach into their public purses, swollen with the spoils of international commerce, in order to assist their struggling trading partner. Benevolence can arise from self-interest after all, as the American writer Robert Wright memorably illustrated:

> If you ask me why don't I think it's a good idea to bomb the Japanese, I'd say, "For one thing, because they built my minivan." I'm proud to say I have some more high-minded reasons as well, but I do think this basic, concrete interdependence forces people to accord one another at least minimal respect, to think a little about the welfare of people halfway around the world.[22]

Trade's third effect on our regard for others comes from its ability to generate money and free time that people, associations, and nations can devote to good works. In this important sense, the gains of trade are what make large-scale beneficence possible at all. The assertion that trade leads to kindness is frequently derided by liberal politicians and left-wing academics, who sometimes prefer to argue instead that trade inevitably brings dehumanization, exploitation, subjugation, environmental spoliation, and a variety of other social ills that rhyme with agitation. In his 1887 lectures on the history of England's industrial revolution, the British historian Arnold Toynbee could hardly have been more critical in this respect of the political economists who celebrated the virtues of free trade. Toynbee charged David Ricardo, one of the architects of classical economics (and a prominent proponent for reform of the Elizabethan Poor Laws) with the crime of depicting human society as a "world of gold-seeking animals, stripped of every human affection."[23] More recently, the psychoanalyst Adam Phillips and the historian Barbara Taylor wrote in *On Kindness* that "capitalism is no system for the kind-hearted."[24]

The historical and scientific evidence gives lie to this sort of skepticism about the relationship between trade and kindness. From the Axial Age to the Age of Impact, upticks in generosity have followed from upticks in wealth, much of which historically has come through the gains of trade. It is also worthwhile to remember that people give more to charitable causes when the stock market is performing well.[25] Likewise, increases in home equity and large bumps in salaries tend to increase charitable donations.[26] In a monumental review of previous research on this topic, the sociologists Pamela Wiepking and René Bekkers found seventy different studies showing positive correlations between wealth and charitable giving and only a handful showing the opposite. Wealthy people give more.[27]

Speaking of wealth, and as circular as it might sound, the best way to lift people out of poverty is to give them the means to lift themselves out of it. Trade is absolutely essential for getting people out of poverty and keeping them there. Wherever you find countries that are chronically poor, you find countries where trade is constrained by bad geography, bad governance, bad policies, or a bad reputation. Countries without access to the ocean are vulnerable to economic stagnation because they cannot get their products to market without shipping them through neighboring countries, which raises costs and reduces competitiveness. Countries with bad leaders and weak institutions find that the greed and narcissism of even an autocrat can destroy a developing economy. Countries with bad policies—in particular, artificial barriers to free trade, such as government subsidies and tariffs—do as much to undermine their own economic prosperity as bad geography and bad governance can. Finally, countries with histories of these problems develop reputations that scare off global investors who might otherwise take risks in order to take advantage of an emerging market. When investors think of Uganda, they don't think of its scrappy, ambitious entrepreneurs. They think of its madman dictators and messianic warlords.[28]

The evidence that freer trade lifts people out of poverty is what compelled the economists involved in the Copenhagen Consensus Conference (which was devoted to evaluating the cost-effectiveness of the UN's

2015 Sustainable Development Goals) to conclude that removing artificial trade barriers and investing in transportation infrastructure would create trillions of dollars in wealth, half of which would go to the world's poorest countries. Indeed, they identified the liberalization of trade as the single most cost-effective means of reducing global poverty: broad reforms could create $2,100 to $4,700 in poor countries for every dollar spent on putting those reforms into practice.[29] For people who want to do something really significant to promote global development, demolishing barriers to trade ought to be a top priority.[30]

If artificial barriers to trade are so bad, why don't governments work harder to eliminate them? The biggest hindrances to the liberalization of trade, actually, are the politicians themselves. Government subsidies and trade wars appeal to our natural intuition that international trade is a zero-sum game between enemies rather than a win-win game between partners.[31] For this reason, policy barriers to free trade tend to be quite popular in many countries, and politicians can score political points with tough talk about tariffs, trade wars, and job protections. Trade liberalization can also be bad for dictators' net worth if it dams up the rivers of illicit wealth they have learned how to divert into their own offshore accounts.[32]

Few politicians in recent history have been more openly skeptical of free trade than Donald Trump, whom the American economist Larry Summers called a "protectionist demagogue."[33] It was a skepticism that resonated with Trump's electoral base from day one of his quest to become president of the United States, when he announced his candidacy in New York City on June 16, 2015:

> Our country is in serious trouble. We don't have victories anymore. We used to have victories, but we don't have them. When was the last time anybody saw us beating, let's say, China in a trade deal? They kill us. I beat China all the time. All the time. . . . When did we beat Japan at anything? They send their cars over by the millions, and what do we do? When was the last time you saw a Chevrolet in Tokyo? It doesn't exist, folks. They beat us all the time. When do we beat Mexico at the border? They're laughing at us, at our stupidity. And now they are beating us economically. They are not our friend, believe me. But they're killing us economically.[34]

Economic theory, empirical data, and the annals of history all point to the folly of such zero-sum thinking about trade, but our us-versus-them intuitions are hard to shake. As a result, a politician can win a reputation as a "champion of the little guy" by supporting policies that protect domestic industry in the short run at the expense of long-term wealth creation.

The fundamental problem is not that people are so nationalistic that they will gladly impoverish themselves to defend the quality of their country's products, or to make a point about economic sovereignty: instead, it's that most people have never even seriously considered whether free trade brings them benefits. Only about one-third of Americans know about the benefits of free trade, whereas two-thirds know that free trade has sent American jobs offshore. The message that free trade has clear economic benefits is not reaching the masses, while the message that free trade has some economic costs is reaching them just fine.[35] And lest one think that ignorance of the benefits of free trade is a uniquely American problem, it's worth pointing out that knowledge about these benefits is not much better in other countries than it is in the United States.[36]

The good news is that when people become acquainted with the benefits of free trade for aggregate national wealth, they warm up to the idea, especially when they are assured that low-skilled workers (who often do, it must be said, lose their jobs in the transition) will receive additional support from the government during the adjustment period.[37] So why support free trade? Because it strengthens national competitiveness and makes the average person better off (and, when handled with care, without inflicting long-term economic pain on low-skilled workers). Here, as everywhere else when we are puzzling about how to improve others' welfare, good reasons matter.[38]

In a book about the psychology and history of humanity's concern for humankind, I haven't had much to say about the future. The omission was intentional. However difficult it might be to explain the past, it is even more difficult to predict the future. That said, we know enough about the future that we will likely face over the next few decades to make some good guesses about how our concern for others is going to be tested.

First, it will be tested by the persistence of poverty in South Asia, South America, and especially sub-Saharan Africa. There is much to celebrate in how global trade has lifted people out of poverty, particularly in China and India,[39] but many countries have not yet enjoyed those economic miracles. If present trends continue, the World Bank estimates that 87 percent of the half-billion people in the world who face extreme poverty will be located in sub-Saharan Africa. Helping these persistently poor countries find their way out of poverty will require more than free trade. As the development economists Paul Collier and Jeffrey Sachs have argued, the countries that have been left behind are mired in their own sticky blends of geographic barriers, political entropy, and reputational stigma that discourage trade and scare off investors. Each of these distressed countries will require a tailor-made package of assistance along with patient, committed engagement from the developed countries.[40]

Our concern for others will also be tested by climate change. Although the entire world has begun to feel its consequences, the countries where climate change will hit the hardest are the very countries that are least equipped to adapt to it. In fact, nine of the ten most vulnerable countries are in sub-Saharan Africa, where they are already locked in development traps that prevent them from trading their way out of poverty.[41]

Climate change will also test our capacity for compassion because of the long time lag between the emission of greenhouse gases and their effects on the global climate. The climate impacts we are experiencing today are due to greenhouse gases that were emitted years or even decades ago, and most of the people who will bear the consequences of our greenhouse gas emissions have not yet been born. Thinking straight about what we should pay now in order to benefit others in the future requires that we figure out what those future benefits are worth in today's dollars. These are not problems that our minds are very good at solving. Yet solve them we must, because if we underpay, future generations will pay much more dearly. And if we overpay, good money will be unnecessarily diverted away from other challenges we could have tackled instead. Either way, we will have failed to do as much good as we can do.[42]

It is tempting to think that the best way for us to respond to climate change is simply to declare by fiat that the world cease all activities that emit greenhouse gases. The consequences of such a moratorium would be calamitous, however, especially for the developing countries that are depending on fossil fuels to climb out of poverty and stay there (at least until some other fuel technology displaces them). A compassionate response to climate change that is not also a scientifically informed response to climate change would be a disastrous response to climate change.

So what would a compassionate but scientifically informed response to climate change look like? First, it would involve abolishing the subsidies to the fossil fuel industry that reward them for continuing on as if their products aren't making the world hotter. Second, it would involve placing a price on carbon dioxide emissions. Once we as a society agree upon that price, we can either tax carbon emissions directly or create a market mechanism that requires companies to pay for the right to emit carbon in the first place. In such a market scheme, the world's governments would place an annual cap on the amount of carbon dioxide that can be released into the atmosphere, and then they would sell tradable credits that permit the owners of those credits to emit small fractions of the cap. If polluters make changes that reduce their emissions, they can save money on credits or else sell their unused credits to companies that can't or won't follow their lead. Whether we use a tax scheme or a cap-and-trade scheme, companies will be incentivized to reduce their emissions so long as they can do so for less money than they would otherwise pay in taxes or permits.[43]

A compassionate yet scientifically informed response to climate change would have several other components as well. It would include improvements in fuel efficiency, for example, so that we get more productivity out of every ton of coal or natural gas we use. It would also include developing sources of energy that emit little or no carbon in the first place, which include biofuels, along with renewables such as wind, solar, and nuclear. And it would include technologies for capturing and storing the carbon that has already been emitted. Our existing technologies for low-carbon fuels and carbon capture cannot yet be deployed at a large enough scale to be practicable, so additional research and development is going to be crucial. Finally, a compassionate but scientific approach would include assistance to the countries that currently lack the capacity to adapt to the fires, floods, droughts, and storms that are already afflicting them.[44]

Climate change will also create the third challenge we are likely to confront in the coming decades: climate-change migration. By 2050, more than 140 million people in sub-Saharan Africa, South Asia, and

South America are expected to move from one city to another within their own countries, or else abandon their home countries altogether, because the climate has become so inhospitable.[45] (These estimates hold aside the millions of refugees, asylum-seekers, and internally displaced persons who flee violence, armed conflict, or political persecution each year.) Our evolved psychology makes us wary of strangers, and this xenophobia extends to the tired, poor, huddled masses that reach our sea-washed, sunset gates.

One manifestation of the stranger-danger that natural selection has placed within us is a nagging worry that the migrants who reach our shores might steal our jobs.[46] In general, however, this is not how migration affects labor markets. With the right policies in place, immigrants with lots of education can fill critical needs in professions with high barriers to entry (such as medicine, engineering, and research), and lower-skilled immigrants can take up work that creates opportunities for existing citizens to move into jobs that are more lucrative and more stimulating. Migrants also grow the economy each time they buy milk and pay rent. Despite our intuitions to the contrary, migration eventually makes both migrants and their hosts better off economically, which is why the Copenhagen Consensus includes enhanced migration among its Phenomenal Development Targets: it returns at least $45 for every dollar spent.[47]

That said, large-scale migration comes with psychic costs that do not necessarily make it into a standard cost-benefit analysis. For instance, many people in host countries harbor the worry that migrants will unwittingly import the cultural beliefs and norms that contributed to the very problems they are fleeing. They worry as well about the costs of the social services those migrants will consume, their ability to help large numbers of migrants to assimilate, and the cultural and linguistic barriers that can turn social interactions into psychologically draining, time-wasting hassles.[48] These worries, whether founded or not, can form a cogent argument against liberal immigration policy—and a counterargument based on the assertion that the people who hold those views are dumb rednecks or white supremacists won't win them over. Better arguments for more liberal migration policy could conceivably be based, however, on good-faith argumentation about stakeholders' concerns, research that identifies which of those concerns have merit and which do not, and policies that help countries to navigate those concerns effectively.[49]

Evolution did not outfit us to care about distant strangers, much less distant strangers who are still unborn or who are trying to cross our borders. But evolution did outfit us to learn, which is why Richard Dawkins insisted in *The Selfish Gene* that compassion must be taught:

> If you wish, as I do, to build a society in which individuals cooperate generously and unselfishly towards a common good, you can expect little help from biological nature. Let us try to *teach* generosity and altruism, because we are born selfish.[50]

Charles Darwin also thought we would find generosity and altruism not in the genes we spread, but in the lessons we teach. As concern for all of humankind comes to be "honoured and practised by some few men," Darwin conjectured in *The Descent of Man*, "it spreads through instruction and example to the young, and eventually becomes incorporated in public opinion."[51]

But what should we teach if we want to teach generosity and altruism? And to whom should we teach it? And how? Should we teach generosity and altruism to preschoolers with cookies and cooing and with time-outs for bad behavior? Should we teach generosity and altruism to school-aged children with inspiring stories and an insistence that character counts? Should we teach generosity and altruism to high school kids by pressing them into community service as a requirement for graduation? Should we teach generosity and altruism to young adults by paying them for a year of service in an underserved community or in a developing country? Well, sure—Why not? These approaches can be used to teach lessons on all sorts of topics. People learn from approval and censure, they learn from respected role models, they learn by following their incentives, and they learn by doing. So by all means, let's use all of these approaches to teach generosity and altruism.

But above all, let's teach the reasons why generosity and altruism are worth the trouble. The reasons that won arguments for generosity and altruism during the Age of Orphans, the Age of Compassion, the Age of Prevention, the First Poverty Enlightenment, the Humanitarian Big Bang, the Second Poverty Enlightenment, and the Age of Impact are no less compelling today than when they were first derived. Compassion can bring us gratitude and glory, it can protect us from the side effects of poverty and desperation, it can grow the economy rather than shrink it, it can help people take responsibility for their own lives, it can bring deep meaning and fulfillment, and—to anyone who views suffering as the locus of moral concern—it is a duty. If any of these propositions turns out to be false in any particular instance—surely some actions and policies really do promote idle dependency, or cost more than they're worth, or violate an important ethical principle in a way that we haven't yet recognized—it is by offering reasons for and against them that we are most likely to discover our errors.

For this reason, it is important that we maintain spaces in private life and in the public sphere that are hospitable to reasoning and argumentation about matters we care about. "As iron sharpens iron," the Book of Proverbs says, "so one person sharpens another."[52] And without venues in which we can discover and debate our reasons for charity, philanthropy, social spending, international assistance, and guardianship of the planet, thoughtful discourse on these issues degenerates into salvos of slogans, zingers, ad hominem attacks, and focus-grouped bullet points. When their arguments become rusty, liberals will turn to gauzy assurances that we can have a clean, green, nuclear-free, socialist Nirvana if we're only willing to tax the hell out of the rich. And when conservatives' own arguments become flabby, they will resort to tripe about welfare queens, deadbeat dads, big-government bogeymen, and a mythical land in which all successful men and women are self-made. If we forget the reasons why we should care about the stranger, future generations may look back on our Golden Age of Generosity and conclude, as its memory tarnishes and fades, that it was a gilded age instead.

# ACKNOWLEDGMENTS

I could not have written *The Kindness of Strangers* without the kindness of scholars. Many of them. Some were friends and some were strangers; all of them were generous with their time and advice. They include Patrick Barclay, Gojko Barjamovic, Michael Barnett, C. Daniel Batson, Nicolas Baumard, Roel Beetsma, Jeanet Sinding Bentzen, Paul Bloom, Mark Borrello, Alec Brandon, Otavio Bueno, Timothy Clutton-Brock, Bernard Crespi, Lee Cronk, Julia Dallman, Rhodri Davies, Daniel Dennett, C. Nathan Dewall, Thomas Dixon, Brad Duchaine, Nancy Eisenberg, Batya Elbaum, Adam Eyre-Walker, Robert Folger, Daniel Forster, Michael French, Randy Gallistel, Andy Gardner, Harry Gensler, Tom Gibbons, Luke Glowacki, Jozien Goense, Cristina Gomes, Matt Grove, Michael Gurven, Oren Harman, Eugene Harris, Nick Hobson, Dan Hoyer, Mickey Inzlicht, Adrian Jaeggi, Vojtěch Kaše, David Kling, Daniel Krupp, David Lahti, Laurent Lehmann, David Lessman, Josh Levin, Mary Lindemann, John List, Marjorie McIntosh, Daniel Messinger, Meghan Meyer, Charles Michener, Susannah Morris, Daniel Mullins, Daniel Nettle, Rick O'Gorman, Cormac Ó Gráda, Craig Packer, Steven Pinker, Roger Riddell, James Rilling, Max Roser, Steven Sandage, Brooke Scelza, Bill Searcy, Harvey Siegel, Peter Singer, Michael Slote, Deborah Small, Christian Smith, Barbara Snedecor, Alexander Stewart, James Swain, J. Albert Uy, Robert Walker, Robyn Walsh, Felix Warneken, Athula Wikramanayake, Timothy Wilson, Jeffrey Winking, Richard Wrangham, and Daniel Zizzo.

I am also grateful for the kindness of Susan Arellano, Michael Barnett, Max Burton-Chellew, Dexter Callender, Claire El-Mouden, Tom Gibbons, William Green, Liana Hone, Robert Kurzban, Debra

Lieberman, Peter Lindert, William McAuliffe, Thomas McCauley, Eric Pedersen, Martin Ravallion, John Paul Russo, Kiara Timpano, Nathan Timpano, and Stuart West for making important suggestions on my initial proposal for the book or on one or more of its chapters. The book is also better for the reactions and advice of the scores of undergraduate students who read portions of it in my seminars over the past few years.

I owe additional thanks to Juliana Berhane, Erika Boone, Brooke Donner, Abigail Johnson, and Billie Koperwas for their research assistance. I am especially grateful to Jordan Fuchs and Brooke Donner for the beautiful figures that enliven Chapters 5, 6, 10, 12, and 13, and to Brooke for her conscientious work on the endnotes and references.

I am indebted to T. J. Kelleher, my editor at Basic Books, for his wisdom, patience, and enthusiasm, and to Rachel Field and Brynn Warriner for their care in shepherding the book through the publication process. I reserve a special note of thanks for my copyeditor Kathy Streckfus, whose erudition, curiosity, and enjoyment of the English language shine through on every page of this book. I must also thank Katinka Matson and Max Brockman, my literary agents, for their guidance and encouragement throughout the project.

I reserve my deepest gratitude for my family. Thank you, Billie, Joel, and Madeleine, for the many ways both great and small that you supported me over the past seven years.

# NOTES

## CHAPTER 1: A GOLDEN AGE OF COMPASSION

1. Darwin (1871) 1952, 319.
2. Faber et al. 2016; Statistics and Clinical Studies of NHS Blood and Transplant 2018.
3. Yad Vashem 2019.
4. Carnegie Hero Fund Commission 2018.
5. Glynn et al. 2003.
6. Charities Aid Foundation 2018.
7. IUPUI Lilly Family School of Philanthropy 2018; Corporation for National and Community Service 2018.
8. Charities Aid Foundation 2019.
9. Organization for Economic Cooperation and Development (OECD) 2016.
10. "Data," (OECD), https://data.oecd.org/oda/net-oda.htm#indicator-chart.
11. Lecky 1890, 371.
12. Boyer 2018.
13. Dawkins 2006, 253.
14. Mercier and Sperber 2017.
15. Darwin (1871) 1952, 304. Unless otherwise noted, emphasis is reproduced from the original.
16. Darwin (1871) 1952, 317.
17. Darwin (1871) 1952, 319.

## CHAPTER 2: ADAM SMITH'S LITTLE FINGER

1. Luke 10:25–37, New International Version.
2. Gansberg 1964, 1.
3. Rosenthal (1964) 1999, 22, 46.
4. Kassin 2017; Manning et al. 2007.
5. Smith (1759) 1984, 135.
6. Smith (1759) 1984, 136–137.
7. Hyman et al. 2010; Hyman et al. 2014; Simons and Chabris 1999.
8. Banjo et al. 2008; Chabris et al. 2011; Puryear and Reysen 2013.

9. Alexopoulos et al. 2012; Gray et al. 2004; Röer et al. 2013; Tacikowski et al. 2014; Turk et al. 2011.
10. B. A. Anderson 2013; Peich et al. 2010.
11. Hume (1739) 1984, 370.
12. Smith (1759) 1984, 9.
13. Batson 1991, 2011; Eisenberg et al. 2010; Goetz et al. 2010.
14. de Waal 2009, 204.
15. Rifkin 2009, 178.
16. Bloom 2013, 2014.
17. Batson 2010, 2011; de Waal 2008; Goetz et al. 2010.
18. Hume (1739) 1984, 368–369.
19. Hume (1777) 1957, 55.
20. Batson 2011.
21. These studies are reviewed extensively in Batson 2011.
22. Batson 2019, 194.
23. McAuliffe, Carter, et al. 2019; McAuliffe et al. 2018.
24. Williams et al. 2000.
25. Meyer et al. 2013.
26. Cikara et al. 2011; Xu et al. 2009. See also Jackson et al. 2005.
27. Bloom 2013, 118.
28. Cameron and Payne 2011.
29. Andreoni et al. 2017; Trachtman et al. 2015.
30. Singer 2009, 3.
31. Singer 2009, 3, 4.
32. Singer 2009.
33. Smith (1759) 1984, 136–137.
34. Finke 1980; Holmes and Mathews 2005; Holmes et al. 2008; Kosslyn 1995.

## CHAPTER 3: EVOLUTION'S GRAVITY

1. Rees 2000.
2. Stott 2012.
3. de Cruz and de Smedt 2014.
4. Paley 1840, 439.
5. Dennett 1995.
6. Dennett 1995, 25.
7. Cairns-Smith 1985.
8. Dawkins 1996, 15.
9. Dawkins 1976, 21.
10. Dennett 1995, 68.
11. Darwin 1872, 143–144.
12. Lamb et al. 2007.
13. Williams 1996, 16.
14. Gallistel and King 2009; Marr 1982.
15. Allen et al. 1975; Buller 2005; Gould and Lewontin 1979; Laland and Brown 2011; Lickliter and Honeycutt 2013; Krubitzer and Stolzenberg 2014; Richardson 2007.

16. Barkow et al. 1992; Carruthers 2006; Dennett 1995; Hagen 2016; Kurzban 2010; Pinker 1997, 2002.

17. I recommend Barrett 2015 and Hagen 2016.

18. Williams 1996, 4.

## CHAPTER 4: IT'S ALL RELATIVE

1. Gardner et al. 2011.
2. Hamilton 1964.
3. Darwin (1871) 1952, 317, 319.
4. Frank 1998.
5. Charnov 1977; Hamilton 1964.
6. Maynard Smith 1964.
7. Dixon 2005, 2008, 2013.
8. West et al. 2007.
9. Clutton-Brock 1991.
10. Power and Schulkin 2016; Oftedal 2002, 2012.
11. Dawkins 1979. See also Penn and Frommen 2010.
12. Russell and Leng 1998.
13. Griffin and West 2003; Strassmann et al. 2011; Chapais 2010.
14. Lieberman et al. 2007; Tal and Lieberman 2007.
15. Lieberman and Lobel 2012. See also Lieberman 2009.
16. Lieberman and Smith 2012; Lieberman et al. 2007; Sznycer et al. 2016.
17. Tal and Lieberman 2007.
18. Anderson 2006; Neel and Weiss 1975; Scelza 2011.
19. Apicella and Marlowe 2004; Billingsley et al. 2018.
20. Mateo 2015; Penn and Frommen 2010; Porter and Moore 1981; Weisfeld et al. 2003.
21. Roberts et al. 2005.
22. Liu et al. 2012.
23. Franklin and Volk 2018; Lopez et al. 2018.
24. Lieberman and Billingsley 2016.
25. Krupp and Taylor 2013.
26. Nan et al. 2012; Silventoinen et al. 2003.
27. Park and Schaller 2005; Park et al. 2008.
28. Sharp et al. 2005.
29. Oates and Wilson 2001. See also Munz et al. 2018.
30. Bowles and Posel 2005; Burton-Chellew and Dunbar 2014; Gurven et al. 2012; Hooper et al. 2015; Kaplan et al. 2000.
31. Koeneman 2013, 53.
32. Secter and Gaines 1999.
33. Bellow 2003, 237.
34. Escresa and Picci 2017; Treisman 2007.
35. Fisher 1994.
36. Darwin (1871) 1952.
37. McClendon 2016; Rosenfeld and Thomas 2012.
38. Qirko 2011, 2013.

## CHAPTER 5: FOR THE LOVE OF SPOCK

1. Proverbs 6:6–8, New International Version.

2. Marx (1867) 2006.

3. Costa 2002.

4. Wilson 2012, 16–17.

5. Wilson 2012, 138.

6. Boomsma and Gawne 2018; Crespi and Yanega 1995.

7. Wilson 2012, 54.

8. Nowak et al. 2010. This work (particularly, the validity of its conclusions about weaknesses in the inclusive fitness approach) was criticized by many other scholars on a number of grounds, most notably in Abbot et al. 2011 and Liao et al. 2015.

9. Wilson 2012, 181–182.

10. Boyd and Richerson 2010; Henrich 2004; Okasha 2007; West et al. 2011.

11. Sallin and Meyer 1982.

12. Deen et al. 2013.

13. Darwin (1859) 1952, 132.

14. Darwin (1859) 1952, 134.

15. Darwin (1871) 1952, 322–323.

16. Borrello 2010.

17. Borrello 2010.

18. Wynne-Edwards 1962.

19. Wynne-Edwards 1993, 1.

20. Wynne-Edwards 1962, 20.

21. Wynne-Edwards 1993, 4.

22. Borrello 2003, 2010.

23. Lack 1966.

24. Borrello 2010.

25. Borrello 2003.

26. Williams 1996, 4–5.

27. Williams 1996, 108.

28. Borrello 2010, 109.

29. Maynard Smith 1976. See also Leigh 1983; Levin and Kilmer 1974.

30. Gardner and Grafen 2009.

31. Wynne-Edwards 1978, 19.

32. Wynne-Edwards 1993.

33. Haldane 1990; Hamilton 1975. Segerstrale's (2013) biography of Hamilton, and Harman's (2010) biography of George Price, document some of the Price-Hamilton alchemy that led Hamilton to this realization.

34. I am omitting an Expectation term from the simple Price equation to simplify it even further. The consequence of doing so is merely to impose the assumption that all parents "breed true" for the trait (that is, that they transmit the trait to their offspring perfectly).

35. Price 1970, 1972.

36. The Price equation and its usefulness for depicting multilevel selection are covered in many books and articles, but none that I have read is clearer or more useful than Okasha 2007. I use his notation here.

37. D. S. Wilson and E. O. Wilson 2007, 345.
38. Wilson 1975.
39. Hamilton 1975; Wilson 1975.
40. West et al. 2011; West et al. 2007.
41. Bernhard et al. 2006; Bowles 2006; Bowles and Gintis 2011.
42. Choi and Bowles 2007.
43. Darwin (1871) 1952, 321.
44. Bernhard et al. 2006; Bowles 2006, 2009; Bowles and Gintis 2011.
45. Lehmann and Feldman 2008.
46. Lehmann and Feldman 2008.
47. Brown 1991. See also Palmer 1989.
48. Fison and Howitt 1880.
49. Gat 2010, 205.
50. Gottschall 2004, 130.
51. Gottschall 2004.
52. "Violence Against Women: War's Overlooked Victims" 2011.

## CHAPTER 6: THE BIG PAYBACK

1. Brown 1991.
2. Jaeggi and Gurven 2013; Martin and Olson 2015.
3. Darwin (1871) 1952, 322.
4. Darwin (1871) 1952, 309.
5. Trivers 1971.
6. Poundstone 1992.
7. Poundstone 1992; Sally 1995.
8. Rapoport and Chammah 1965.
9. Axelrod 1980a.
10. Axelrod 1980b.
11. Axelrod and Hamilton 1981.
12. I first encountered this joke in Pinker 2012.
13. Axelrod and Dion 1988.
14. Axelrod 1984.
15. Nowak and Sigmund 1992, 252; Nowak and Sigmund 1993.
16. Wu and Axelrod 1995.
17. Frean 1994; Hauert and Schuster 1998; Nowak and Sigmund 1994.
18. Hruschka and Henrich 2006.
19. Nowak and May 1992; Doebeli et al. 2004.
20. For example, see Harper et al. 2017; Knight et al. 2018; Reiter et al. 2018.
21. Axelrod 1984, 126.
22. Trivers 1971.
23. For other early examples see Connor and Norris 1982 and Ligon 1983.
24. Clutton-Brock 2009.
25. Wilkinson 1984.
26. Krams et al. 2007.
27. Jaeggi and Gurven 2013; Schino and Aureli 2007, 2010.
28. Carter 2014.

29. Wood and Marlowe 2013.
30. Kaplan et al. 2000.
31. Hawkes et al. 2001; Wood and Marlowe 2013.
32. Gurven et al. 2012.
33. Gurven and Hill 2009, 54; Jaeggi and Gurven 2013.
34. Smith (1776) 1952, 6.
35. Wood and Marlowe 2013.
36. Howe et al. 2016; Sznycer et al. 2019.
37. Krasnow et al. 2013.
38. Darwin (1871) 1952, 309, 311.
39. Apicella et al. 2012; Eisenbruch et al. 2016.
40. Petersen 2012.
41. Barclay 2013.
42. Darwin (1871) 1952, 322.
43. Darwin (1871) 1952, 322.
44. Alexander 1987.
45. Alexander 1987, 94.
46. Krasnow et al. 2012; Molleman et al. 2013.
47. Darwin (1871) 1952, 310.
48. Frank 2009.
49. Oppenheimer 2013.
50. De Freitas et al. 2019.
51. Becker 1973.
52. Barclay 2013.
53. Lazarsfeld and Merton 1954.
54. Apicella et al. 2012.
55. McPherson et al. 2001.
56. Marlowe 2005.
57. Lebzelter 1934, 37.
58. Gusinde 1937, 918.
59. Lebzelter 1934, 37.
60. Gusinde 1937, 919.
61. Proverbs 22:1–2, English Standard Version.

## CHAPTER 7: THE AGE OF ORPHANS

1. Bar-Yosef 1998. Other agricultural revolutions took place in other regions of the world, of course, but the Levantine agricultural revolution is a good case study because (1) it was the first, and (2) the archaeological record is extensive and well studied.
2. Belfer-Cohen and Hovers 2005; Kuijt 2000; Kuijt and Finlayson 2009.
3. Gowdy and Krall 2016.
4. Belfer-Cohen and Hovers 2005; Rosenberg 2008.
5. Kuijt 2008a; Sterelny 2015.
6. See Sterelny 2015.
7. Kuijt 2008a, 2011; Ringen et al. 2019.
8. Gurven 2004.
9. Bowles and Choi 2013.
10. Hill and Kintigh 2009.

11. Gurven et al. 2014; Sterelny 2015.
12. Boehm 1999; Marlowe 2009.
13. Locke 1764, 109.
14. Barnard and Woodburn 1987, 24; Rusch and Voland 2016.
15. Smith et al. 2010.
16. Eff and Dow 2008.
17. Kuijt 2000.
18. Walker 2014; Walker and Bailey 2014.
19. Kuijt 2008b; Bocquentin et al. 2016.
20. Sterelny 2015, 415.
21. Gurven et al. 2012.
22. Yoffee 2012.
23. Frangipane 2007; Knapp 1988.
24. Borgerhof Mulder et al. 2009; Smith et al. 2010.
25. Richardson 2016; Webber and Wildavsky 1986.
26. Liverani 2006; Webber and Wildavsky 1986.
27. Webber and Wildavsky 1986.
28. Foster 1995, 156.
29. Richardson 2016, 755.
30. Bellah 2011; Fensham 1962.
31. Foster 1995.
32. Kramer 1963, 319.
33. Fensham 1962; Morschauser 1995.
34. Foster 1995; Morschauser 1995, 102.
35. Ferguson 2011; Russo 1999.
36. Bellah 2011.
37. Westbrook 1995.
38. Davidson 2015.
39. Knapp 1988.

## CHAPTER 8: THE AGE OF COMPASSION

1. Armstrong 2006; Cline 2014; Knapp and Manning 2016.
2. Jaspers (1953) 2010, 1.
3. Abtahi 2007; Atran 2016; Wright 2009.
4. Kaše 2018; Mirakhor and Askari 2019.
5. Mullins et al. 2018.
6. Armstrong 2006, xiv, emphasis mine.
7. Loewenberg 1995; Neusner and Avery-Peck 2005.
8. Finlay 2009.
9. Homerin 2005. The quotation is from the Qur'an (2:177). See also Bremner 1994.
10. Lewis 2005, 96–97. The quotation is from the Upasakashila Sutra.
11. Bremner 1994; Fitzgerald 2009; Garnsey 1988; Hands 1968; Ierley 1984; Obocock 2008; Webber and Wildavsky 1986.
12. Jaspers (1953) 2010, 4.
13. Momigliano 1975, 8.
14. Bellah 2011.

15. See, for instance, Ober 2015.
16. Hansen 2006.
17. Baumard et al. 2015.
18. Bakija and Heim 2011.
19. List and Peysakhovich 2011.
20. Do and Paley 2012.
21. Baumard and Chevallier 2015; Baumard et al. 2015.
22. Sanderson 2018, 219.
23. Neusner and Chilton 2008; Wattles 1996.
24. Confucius (c. 500 BCE) 1861, Book 15, Chapter 23.
25. Krishna-Dwaipayana Vyasa 1896, Book XII, Section 259, p. 620.
26. Aristotle (350 BCE) 1924, Book 2, Chapter 4; Berchman 2008, 45.
27. Ælius Lampridius 1924.
28. Blackstone 1965.
29. Appiah 2006, 60.
30. Gensler 2013.
31. Gensler 2013, 2.
32. Latané and Darley 1970, 27.
33. Darwin (1871) 1952, 319.
34. Damon and Colby 2015, 13.
35. Kennedy 1963.
36. Ifill 2016.
37. Mercier 2011.
38. Ostrom 2014, 10.
39. Loewenberg 1994.
40. Loewenberg 1994, 2001.
41. Loewenberg 1995.
42. Jaspers (1953) 2010, quoted in Armstrong 2006, 51.
43. Beaudoin 2007.
44. Bremner 1994.
45. Adrados and van Dijk 1999; Mukherji 2006.
46. Hands 1968.
47. Beaudoin 2007; Neusner and Chilton 2005.
48. G. A. Anderson 2013, 7–8.

## CHAPTER 9: THE AGE OF PREVENTION

1. Matthew 5:3, New International Version.
2. Luke 16:19–31, New International Version.
3. McIntosh 2012.
4. Rushton and Sigle-Rushton 2001; van Bavel and Rijpma 2016.
5. Dyer 2012; Geremek 1994; Richardson 2005; Slack 1988.
6. Beaudoin 2007; Tierney 1959.
7. Allen 2001.
8. Dewitte 2004; Geremek 1994; Slack 1988; Tierney 1959.
9. Sanuto (1897), column 148. Translation from Geremek 1994, 132.
10. Versoris 1885, 24. Translation from Geremek 1994, 126–127.

11. More (1516) 1753, 67.
12. Geremek 1994.
13. Keck 2010; Slack 1988; Tournoy 2004.
14. Bataillon 1952.
15. Vives (1526) 1917, 9; Travill 1987.
16. Vives (1526) 1917, 10.
17. Vives (1526) 1917, 9.
18. Vives (1526) 1917, 11.
19. Tournoy 2004; Travill 1987.
20. Vives (1526) 1917, 11.
21. Vives (1526) 1917, 17.
22. Vives (1526) 1917, 15.
23. Tournoy 2004.
24. Vives (1526) 1917, 31.
25. Dewitte 2004.
26. Dewitte 2004; Geremek 1994; Keck 2010; Slack 1988; Tournoy 2004.
27. Fantazzi 2008.
28. Fantazzi 2008; Geremek 1994.
29. Geremek 1994, 204–205.
30. Slack 1988, 122.
31. Geremek 1994; Pound 1971. For historical European wage data, see Allen 2001, 2013; Álvarez-Nogal and de la Escosura 2013; Broadberry et al. 2015; Malanima 2011; Palma and Reis 2019; and van Zanden 2005.
32. McIntosh 2012.
33. Van Bavel and Rijpma 2016.
34. Richardson 2005.
35. Goose 2013.
36. Bittle and Lane 1976, 1978; Hadwin 1978.
37. Slack 1988.
38. King 2000; Tomkins 2006.
39. King 2000.
40. Greif and Iyigun 2013; van Bavel and Rijpma 2016.
41. Greif and Iyigun 2013; Kelly and Ó Gráda 2011.
42. Douglas, n.d., 2.
43. Van Bavel and Rijpma 2016.
44. Meerkerk and Teeuwen 2014.
45. Goose and Looijesteijn 2012; Meerkerk 2012; Meerkerk and Teeuwen 2014.
46. Lindert 2004a, 68.
47. Fouquet and Broadberry 2015.
48. Lindert 2004a.
49. Beaudoin 2007; Lindert 2004a; Ravallion 2015; Rodgers 1968.
50. Townsend (1786) 1971, 23.
51. Ricardo 1817, 111.
52. Malthus (1803) 1992, 101.
53. Lindert 2004a.

54. See Lindert 2004a, Figure 3.1 (p. 46), and van Bavel and Rijpma 2016, Table 1 (p. 171) and Figure 1 (p. 174).

55. Ravallion 2015.

56. Beaudoin 2007.

57. Lubove 1966, 53.

58. Rodgers (1968) 2006, 38.

59. Rodgers 1968.

60. Van Bavel and Rijpma 2016.

61. Clark and Page 2019.

## CHAPTER 10: THE FIRST POVERTY ENLIGHTENMENT

1. Gilens 1999, 1.

2. Nordheimer 1976; Reagan 1981.

3. Jacobson 2011.

4. Keay 1987.

5. Allen 2001.

6. Allen 2009.

7. Clayton and Rowbotham 2008a, 2008b.

8. Ravallion 2015.

9. Haueter 2013; Richardson 2005.

10. Kuhnle and Sander 2010.

11. Kuhnle and Sander 2010; Webber and Wildavsky 1986.

12. Kuhnle and Sander 2010.

13. Low 1908, 209.

14. Low 1908, 210.

15. Kuhnle and Sander 2010; Beaudoin 2007.

16. Kuhnle and Sander 2010.

17. Beaudoin 2007; Kuhnle and Sander 2010.

18. Garve 1796, quoted in Epstein 1966, 78–79.

19. Sadler 2004, 195.

20. Lee and Lee 2016. On north-south differences in US education and literacy in the first few decades of the nineteenth century, see Isenberg 2016.

21. Lindert 2004b.

22. Lee and Lee 2016.

23. Lee and Lee 2016; Lindert 2004a.

24. Beaudoin 2007; Kuhnle and Sander 2010.

25. Nullmeier and Kaufman 2010.

26. United Nations 1947.

27. Renwick 2017.

28. Lindert 2004a; OECD 2018b, 2019c; van Bavel and Rijpma 2016.

29. Google Books Ngram Viewer 2019; Ravallion 2011. Ravallion also examined changes in the use of the word *pauvreté* in French-language books.

30. For Hobbes's ([1651] 1957) views, see Chapter 30. For Montesquieu's ([1748] 1952) views, see Book 23, Chapter 29.

31. Fleischacker 2004, 4. I am deeply indebted in this chapter to Fleischacker's account of how the modern notion of distributive justice evolved.

32. Rousseau (1754) 1952, 333.
33. Rousseau (1755) 1952, 381–382.
34. Rousseau (1755) 1952, 375.
35. Fleischacker 2004, 56.
36. Buchan 2006.
37. Himmelfarb 1984, 46.
38. Buchan 2006, 6.
39. Smith (1776) 1952, 33.
40. Ross 2010.
41. Ravallion 2015, 1982.
42. Smith (1776) 1952, 7–8.
43. Smith (1776) 1952, 346–347.
44. Laqueur 1989.
45. Smith (1776) 1952, 33–34.
46. Griswold 1999; Himmelfarb 1984.
47. Kant (1785) 1952, 268.
48. Kant (1785) 1952, 276.
49. Kant (1785) 1952, 272.
50. Kant 1930, 192.
51. Fleischacker 2004.
52. Kant 1930, 236.
53. Kant 1930, 236.
54. Kant 1930, 193.
55. Kant (1785) 1952, 258.
56. Kant (1785) 1952, 258.
57. Kant (1785) 1952, 258.
58. Kant (1785) 1952, 269. For interpretation, see Sedgwick 2008.
59. Kant 1930, 236.
60. Kant 1930, 192.
61. Kant (1797) 1965, 93.
62. Fleischacker 2004; LeBar 1999.
63. Paine (1792) 1817, 92.
64. Fichte (1797) 2000.
65. Babeuf 1796, translation in Thomson 1947, 33–34.
66. Fleischacker 2004, 79.
67. Webber and Wildavsky 1986.
68. Himmelfarb 1984, 458.
69. Himmelfarb 1984.
70. Mayhew (1851) 1985.
71. Himmelfarb 1984.
72. Himmelfarb 1991.
73. Baciocchi and Lhuissier 2010.
74. Himmelfarb 1991, 213.
75. Himmelfarb 1991, 237.
76. Himmelfarb 1991.
77. Bulmer et al. 1991a.

78. Marshall 1890.

79. Booth 1889–1903.

80. Ravallion (2015) calculated that 21 shillings per week would provide each person with a daily diet equivalent to 1.5 pounds of good wheat, an estimate he compared to the poverty line of India in the 1990s.

81. Rowntree 1901.

82. Himmelfarb 1991.

83. Watson 1922.

84. Husock 1992.

85. Bulmer et al. 1991b.

86. Riis 1890.

87. Kelley 1895.

88. Woods 1898.

89. Du Bois 1899.

90. Hunter 1904.

91. Hammack 1995.

92. Bremner 1992, 156.

93. Hammack 1995.

94. Lindert 2004a, 188–189; Justman and Gradstein 1999.

95. Acemoglu and Robinson 2000; Lizzeri and Persico 2004.

96. Gradstein and Milanovic 2004.

97. Beaudoin 2007.

98. Webber and Wildavsky 1986.

99. Scheve and Stasavage 2016, 135.

100. George 1918, 79–80.

101. Beetsma et al. 2016; Scheve and Stasavage 2016.

## CHAPTER 11: THE HUMANITARIAN BIG BANG

1. Molesky 2015, calculations on pp. 176–179. Figures from Pereira 2009 yield slightly lower estimates.

2. Neiman 2002, 1.

3. Neiman 2002.

4. Barnett 2011.

5. Molesky 2015.

6. Molesky 2015, 261.

7. Shrady 2008, 47.

8. De Vattel (1758) 2008, 262.

9. De Vattel (1758) 2008, 264–265.

10. Pereira 2009, 487.

11. Gupta 2018.

12. Walker 2015.

13. Molesky 2015; Pereira 2009; Shrady 2008; Zack 2015.

14. Tinniswood 2004.

15. Aristotle, who believed earthquakes were caused by a sudden release of wind that had become trapped underground, was a notable exception. Missiakoulis 2008.

16. Bentzen 2019; Sibley and Bulbulia 2012.

17. Kelemen 2004; Kelemen et al. 2013; Rottman et al. 2017.

18. Hafer and Bègue 2005; Lerner 1980.
19. Weiner 1993, 1995.
20. Shrady 2008.
21. Nichols 2014.
22. Neiman 2002.
23. Malagrida quotation from Kendrick 1955, 138.
24. Molesky 2015.
25. Mendonça et al. 2019.
26. Gupta 2018.
27. Shrady 2008, 146.
28. Amador 2004; Reinhardt and Oldroyd 1983.
29. Translation of I. Kant, *Fortgesetzte Betrachtung der seit einiger Zeit wahrgenommen Erderschütterungen*, vol. 1, from Molesky 2015, 340.
30. Strickland 2019.
31. Davies 2014; Haller 2011.
32. Davies 2014.
33. Davies 2014.
34. Carey (1792) 1891, 63–64.
35. Carey (1792) 1891, 69–70.
36. Barnett 2011.
37. Burke 1847, 670–671.
38. Haller 2011.
39. Davey et al. 2013.
40. Dunant (1862) 1959, 72.
41. Dunant (1862) 1959, 115.
42. Barnett 2011, 80.
43. Weber 2006.
44. Weber 2006.
45. Molesky 2015.
46. Black 2011.
47. Barnett 2011, 80.
48. Shrady 2008.
49. Kelly and Ó Gráda 2019.
50. Clark and Feenstra 2003.
51. Huwart and Verdier 2013.
52. Barnett 2011.
53. Barnett 2011.
54. Forsythe 2005 (quotation p. 34).
55. Davey et al. 2013; Barnett 2011.
56. Barnett 2011.
57. Hitchcock 2008; Lowe 2012.
58. Lowe 2012.
59. Lowe 2012.
60. Crossland 2014
61. Fox 1950.
62. McCleary 2009.
63. Hitchcock 2008; McCleary 2009.

64. Hitchcock 2008.
65. Bundy 2016; Lowe 2012.
66. Bundy 2016.
67. Barnett 2011.
68. Barnett 2011.
69. Barnctt 2011, 120.
70. Morris 1996, 18.
71. Morris 1996, 30.
72. Morris 1996, 18.
73. Morris 1996, 29.
74. McCleary 2009.
75. Morris 1996. To reflect its identity as a confederation of autonomous sister organizations in the United States, Europe, Canada, Japan, and Australia, CARE had still one more name change coming. In 1993, it became CARE International, with the acronym standing for the Cooperative for Assistance and Relief Everywhere.

## CHAPTER 12: THE SECOND POVERTY ENLIGHTENMENT

1. Truman 1949.
2. Truman 1949.
3. Beaudoin 2007.
4. Barnett 2011.
5. Riddell 2007.
6. OECD 2018a.
7. Ravallion 2011.
8. Ravallion 2011.
9. Ravallion 2011.
10. Broad 1940.
11. Singer 1972b, 1986.
12. Singer 1986.
13. Singer 1972a.
14. Singer 1972a, 231–232.
15. O'Neill 1975.
16. Rawls 1971.
17. Rawls 1999.
18. Beitz 1975. See also Beitz 1979.
19. Mandelbaum 1982.
20. Bailey 1976.
21. Brown and Minty 2008; Eisensee and Strömberg 2007; Strömberg 2007.
22. Brown and Minty 2008; Eisensee and Strömberg 2007; Lobb et al. 2012; Simon 1997; Strömberg 2007.
23. Eisensee and Strömberg 2007.
24. Waters 1998.
25. Mittelman and Neilson 2011, 393.
26. Wydick et al. 2013.
27. Jones 2019.
28. From the Church News 2015.

29. Franks 2013; J. Wilson 1986.
30. Franks 2013.
31. Philo 1993.
32. Robins 1985.
33. Franks 2013.
34. Sen 1981.
35. Keating 1986.
36. Franks 2013.
37. Dercon and Porter 2014.
38. Central Intelligence Agency, Office of African and Latin American Analysis, 1985; Franks 2013; Keating 1986.
39. Riddell 2007.
40. World Bank 2018.
41. Riddell 2007.
42. Sachs (2005) lays out his projections on p. 300.
43. Dichter 2003; Easterly 2006; Moyo 2009; Yanguas 2018.
44. Moyo 2009, xix.
45. In most countries, most development aid comes from official sources, with aid from private donors making rather small contributions. In contrast, 50 percent of all American development aid and 40 percent of all Canadian development aid comes from such private sources. Indeed, the United States is responsible for more than three-fourths of all private grant aid globally, giving lie to the view that Americans are not very generous toward the developing world. Once private grants are factored in, the United States actually gives 0.38 percent of its GNI to developing countries—a far cry from the much-hoped-for 0.7 percent, but still slightly more than the average of the world's thirty most active donor nations.
46. United Nations Development Programme 2015.
47. Ahimbisibwe and Ram 2019.
48. Asadullah and Savoia 2018.
49. McArthur and Rasmussen 2018.
50. Ahimbisibwe and Ram 2019; Asadullah and Savoia 2018; McArthur and Rasmussen 2018.

## CHAPTER 13: THE AGE OF IMPACT

1. Charities Aid Foundation 2019; Giving USA 2019.
2. Mill 1863, 9–10.
3. Singer 1972a, 273.
4. Wesley 1872.
5. Hollander 2018, 2019; Karnofsky 2019.
6. Giving What We Can 2019.
7. MacAskill 2015; Singer 2015.
8. MacAskill 2015.
9. MacAskill 2015; Singer 2010.
10. Singer 2015, 41.
11. MacAskill 2015, 25.
12. Caviola et al. 2014.

13. Gabriel 2017; Rubenstein 2016; Snow 2015.
14. De Lazari-Radek and Singer 2017; Smart and Williams 1973.
15. Gabriel 2017, 459.
16. Gabriel 2017, 460.
17. Gabriel 2017.
18. Porter and Kramer 1999.
19. Bishop and Green 2006.
20. Giving Pledge 2019.
21. Findley 2018; Qian 2015.
22. Riddell 2007.
23. Kotsadam et al. 2018.
24. Odokonyero et al. 2018.
25. Martorano et al. 2018.
26. Marty et al. 2017; Wayland 2019.
27. Mbowa et al. 2017.
28. Briggs 2017, 2018a, 2018b.
29. Marty et al. 2017.
30. Ravallion 2019.
31. Banerjee and Duflo 2011.
32. Humphreys and Weinstein 2009.
33. Banerjee and Duflo 2011; Bouguen et al. 2019.
34. Oster and Thornton 2011.
35. Baird et al. 2016; Miguel and Kremer 2004.
36. Baird et al. 2016; Miguel and Kremer 2004.
37. Morgan 2017. For reviews, see Ahuja et al. 2017; Pabalan et al. 2018.
38. "The 169 Commandments," 2015.
39. Lomborg 2018.
40. Wright and Pendergrass 2018.
41. Shirky 2010, 27.
42. Chernobrov 2018.
43. Meier 2015; Whittaker et al. 2015.
44. Chernobrov 2018.
45. Sproull et al. 2005.

## CHAPTER 14: GOOD REASONS

1. Hamilton 1964.
2. Darwin (1871) 1952, 322–323.
3. Darwin (1871) 1952, 322.
4. McAuliffe, Burton-Chellew, et al. 2019; Rand and Nowak 2013.
5. Haidt 2001.
6. Prinz 2006; Slote 2010.
7. Krueger and Funder 2004.
8. Malle 2006.
9. Miller and Sanjurjo 2015, 2018.
10. Wallace 2018.
11. Darwin (1871) 1952, 317.

12. Mercier and Sperber 2017.
13. DiJulio et al. 2015; Hurst et al. 2017.
14. Brady et al. 2003; Luskin et al. 2004.
15. Winchester 2003.
16. Lewin and De Valois 2003.
17. Ravallion 2015.
18. Ravallion 2016.
19. Ahimbisibwe and Ram 2019; McArthur and Rasmussen 2018; Hendren and Sprung-Keyser 2019; Hoynes and Schanzenbach 2018.
20. Pinker 2018.
21. Smith (1776) 1952, 7.
22. Wright and Kaplan 2001.
23. Toynbee 1887, 7.
24. Phillips and Taylor 2010, 106.
25. List 2011.
26. List 2011.
27. Wiepking and Bekkers 2012.
28. Anderson 2018; Collier 2007.
29. Anderson 2018.
30. Collier 2007.
31. Boyer and Petersen 2018; Rubin 2003; Swedberg 2018.
32. Collier 2007; Edwards 2006; Guisinger 2017.
33. Summers 2016.
34. Trump 2015.
35. Rho and Tomz 2017.
36. Medrano and Braun 2012.
37. Hays et al. 2005.
38. Rho and Tomz 2017.
39. United Nations Department of Economic and Social Affairs 2019.
40. Collier 2007; Sachs 2005.
41. Rigaud et al. 2018.
42. Caney 2014.
43. Klenert et al. 2018; Stiglitz et al. 2017.
44. Galiana 2018.
45. Rigaud et al. 2018; United Nations High Commissioner for Refugees 2019.
46. Boyer and Petersen 2018.
47. Kohler and Behrman 2018.
48. Bologna Pavlik et al. 2019; Forrester et al. 2019; Hainmueller and Hopkins 2014; Newman et al. 2012; Papademetriou and Banulescu-Bogdan 2016.
49. Papademetriou and Banulescu-Bogdan 2016.
50. Dawkins 1976, 3.
51. Darwin (1871) 1952, 317.
52. Proverbs 27:17, New International Version.

# REFERENCES

Abbot, P., J. Abe, J. Alcock, S. Alizon, J. A. C. Alpedrinha, M. Andersson, J. B. Andre, et al. 2011. "Inclusive Fitness Theory and Eusociality." *Nature* 471:1057–1062. doi:10.1038/nature09831.

Abtahi, H. 2007. "Reflections on the Ambiguous Universality of Human Rights: Cyrus the Great's Proclamation as a Challenge to the Athenian Democracy's Perceived Monopoly on Human Rights." *Denver Journal of International Law and Policy* 36:55–91.

Acemoglu, D., and J. A. Robinson. 2000. "Why Did the West Extend the Franchise? Democracy, Inequality, and Growth in Historical Perspective." *Quarterly Journal of Economics* 115:1167–1199. doi:10.1162/003355300555042.

Adrados, F. R., and G.-J. van Dijk. 1999. *History of the Graeco-Latin Fable*. Vol. 1, *Introduction and from the Origins to the Hellenistic Age*. Leiden: Brill.

Ælius Lampridius. 1924. *Historia Augusta*. Vol. 2, *Caracalla. Geta. Opellius Macrinus. Diadumenianus. Elagabalus. Severus Alexander. The Two Maximini. The Three Gordians. Maximus and Balbinus*. D. Magie, trans. Loeb Classical Library, vol. 140. Cambridge, MA: Harvard University Press.

Ahimbisibwe, I., and R. Ram. 2019. "The Contribution of Millennium Development Goals Towards Improvement in Major Development Indicators, 1990–2015." *Applied Economics* 51:170–180. doi:10.1080/00036846.2018.1494808.

Ahuja, A., S. Baird, J. H. Hicks, M. Kremer, and E. Miguel. 2017. "Economics of Mass Deworming Programs." In D. A. P. Bundy, N. de Silva, S. Horton, D. T. Jamison, and G. C. Patton, eds., *Child and Adolescent Health and Development*, 3rd ed., 413–422. Washington, DC: International Bank for Reconstruction and Development / The World Bank.

Alexander, R. D. 1987. *The Biology of Moral Systems*. New York: Aldine Transaction.

Alexopoulos, T., D. Muller, F. Ric, and C. Marendaz. 2012. "I, Me, Mine: Automatic Attentional Capture by Self-Related Stimuli." *European Journal of Social Psychology* 42:770–779. doi:10.1002/ejsp.1882.

Allen, E., B. Beckwith, J. Beckwith, S. Chorover, D. Culver, M. Duncan, Steven Gould, et al. 1975. "Against 'Sociobiology.'" *New York Review of Books* 22, no. 18 (November 13).

Allen, R. C. 2001. "The Great Divergence in European Wages and Prices from the Middle Ages to the First World War." *Explorations in Economic History* 38:411–447. doi:10.1006/exeh.2001.077.

———. 2009. "Engels' Pause: Technical Change, Capital Accumulation, and Inequality in the British Industrial Revolution." *Explorations in Economic History* 46:418–435. doi:10.1016/j.eeh.2009.04.004.

———. 2013. *Poverty Lines in History, Theory, and Current International Practice*. Discussion Paper No. 685. Department of Economics, University of Oxford.

Álvarez-Nogal, C., and L. P. de la Escosura. 2013. "The Rise and Fall of Spain (1270–1850)." *Economic History Review* 66:1–37. doi:10.1111/j.1468-0289.2012.00656.x.

Amador, F. 2004. "The Causes of 1755 Lisbon Earthquake on Kant." Paper presented at the Historia de las ciencias y de las técnicas, Rioja, Spain.

Anderson, B. A. 2013. "A Value-Driven Mechanism of Attentional Selection." *Journal of Vision* 13 (3): 1–16. doi:10.1167/13.3.7.

Anderson, G. A. 2013. *Charity: The Place of the Poor in the Biblical Tradition*. New Haven, CT: Yale University Press.

Anderson, K. 2018. "Benefits and Costs of the Trade Targets for the Post-2015 Development Agenda." In B. Lomborg, ed., *Prioritizing Development: A Cost Benefit Analysis of the United Nations' Sustainable Development Goals*, 192–215. New York: Cambridge University Press.

Anderson, K. G. 2006. "How Well Does Paternity Confidence Match Actual Paternity?" *Current Anthropology* 47:513–520. doi:10.1086/504167.

Andreoni, J., J. M. Rao, and H. Trachtman. 2017. "Avoiding the Ask: A Field Experiment on Altruism, Empathy, and Charitable Giving." *Journal of Political Economy* 125:625–653. doi:10.1086/691703.

Apicella, C. L., and F. W. Marlowe. 2004. "Perceived Mate Fidelity and Paternal Resemblance Predict Men's Investment in Children." *Evolution and Human Behavior* 25:371–378. doi:10.1016/j.evolhumbehav.2004.06.003.

Apicella, C. L., F. W. Marlowe, J. H. Fowler, and N. A. Christakis. 2012. "Social Networks and Cooperation in Hunter-Gatherers." *Nature* 481:497–501. doi:10.1038/nature10736.

Appiah, K. A. 2006. *Cosmopolitanism: Ethics in a World of Strangers*. New York: W. W. Norton.

Aristotle. (350 BCE) 1924. *Rhetoric*. W. R. Roberts, trans. Oxford: Clarendon Press.

Armstrong, K. 2006. *The Great Transformation: The Beginning of Our Religious Traditions*. New York: Alfred A. Knopf.

Asadullah, M. N., and A. Savoia. 2018. "Poverty Reduction During 1990–2013: Did Millennium Development Goals Adoption and State Capacity Matter?" *World Development* 105:70–82. doi:10.1016/j.worlddev.2017.12.010.

Atran, S. 2016. "Moralizing Religions: Prosocial or a Privilege of Wealth?" *Behavioral and Brain Sciences* 39, e2. doi:10.1017/S0140525X15000321.

Axelrod, R. 1980a. "Effective Choice in the Prisoner's Dilemma." *Journal of Conflict Resolution* 24:3–25. doi:10.1177/002200278002400101.

———. 1980b. "More Effective Choice in the Prisoner's Dilemma." *Journal of Conflict Resolution* 24:379–403. doi:10.1177/002200278002400301.

———. 1984. *The Evolution of Cooperation*. New York: Basic Books.

Axelrod, R., and D. Dion. 1988. "The Further Evolution of Cooperation." *Science* 242:1385–1390. doi:10.1126/science.242.4884.1385.

Axelrod, R., and W. D. Hamilton. 1981. "The Evolution of Cooperation." *Science* 211:1390–1396. doi:10.1126/science.7466396.

Babeuf, G. 1796. "Analyse de la doctrine de Babeuf, tribun du peuple." Europeana Collections, www.europeana.eu/portal/ca/record/9200365/BibliographicResource_1000055557333.html.

Baciocchi, S., and A. Lhuissier. 2010. *Mapping London Charities: Sources, Enumerations, and Classifications (1820–1920)*. Paris: School of Advanced Studies in the Social Sciences.

Bailey, G. 1976. "Television War: Trends in Network Coverage of Vietnam, 1965–1970." *Journal of Broadcasting and Electronic Media* 20:147–158. doi:10.1080/08838157609386385.

Baird, S., J. H. Hicks, M. Kremer, and E. Miguel. 2016. "Worms at Work: Long-Run Impacts of a Child Health Investment." *Quarterly Journal of Economics* 131:1637–1680. doi:10.1093/qje/qjw022.

Bakija, J., and B. T. Heim. 2011. "How Does Charitable Giving Respond to Incentives and Income? New Estimates from Panel Data." *National Tax Journal* 64:615–650. doi:10.17310/ntj.2011.2S.08.

Banerjee, A. V., and E. Duflo. 2011. *Poor Economics: A Radical Rethinking of the Way to Fight Global Poverty*. New York: Public Affairs.

Banjo, O., Y. Hu, and S. S. Sundar. 2008. "Cell Phone Usage and Social Interaction with Proximate Others: Ringing in a Theoretical Model." *Open Communications Journal* 2:127–135. doi:10.2174/1874916X00802010127.

Barclay, P. 2013. "Strategies for Cooperation in Biological Markets, Especially for Humans." *Evolution and Human Behavior* 34:164–175. doi:10.1016/j.evolhumbehav.2013.02.002.

Barkow, J., L. Cosmides, and J. Tooby. 1992. *The Adapted Mind: Evolutionary Psychology and the Generation of Culture*. New York: Oxford University Press.

Barnard, A., and J. Woodburn. 1987. "Property, Power and Ideology in Hunting-Gathering Societies: An Introduction." In T. Ingold, D. Riches, and J. Woodburn, eds., *Hunters and Gatherers*. Vol. 2, *Property, Power, and Ideology*, 4–31. Oxford: Berg.

Barnett, M. 2011. *Empire of Humanity: A History of Humanitarianism*. Ithaca, NY: Cornell University Press.

Barrett, H. C. 2015. *The Shape of Thought: How Mental Adaptations Evolve*. New York: Oxford University Press.

Bar-Yosef, O. 1998. "The Natufian Culture in the Levant, Threshold to the Origins of Agriculture." *Evolutionary Anthropology* 6:159–177. doi:10.1002/(SICI)1520-6505(1998)6:5<159:AID-EVAN4>3.0.CO;2-7.

Bataillon, M. 1952. "J. L. Vivés: Réformateur de la bienfaisance." *Bibliotheque d'Humanisme et Renaissance* 14:141–158.

Batson, C. D. 1991. *The Altruism Question: Toward a Social-Psychological Answer*. Hillsdale, NJ: Erlbaum.

———. 2010. "The Naked Emperor: Seeking a More Plausible Genetic Basis for Psychological Altruism." *Economics and Philosophy* 26:149–164. doi:10.1017/S0266267110000179.

———. 2011. *Altruism in Humans*. New York: Oxford University Press.

———. 2019. *A Scientific Search for Altruism: Do We Care Only About Ourselves?* New York: Oxford University Press.

Baumard, N., and C. Chevallier. 2015. "The Nature and Dynamics of World Religions: A Life-History Approach." *Proceedings of the Royal Society B: Biological Sciences* 282:20151593. doi:10.1098/rspb.2015.1593.

Baumard, N., A. Hyafil, I. Morris, and P. Boyer. 2015. "Increased Affluence Explains the Emergence of Ascetic Wisdoms and Moralizing Religions." *Current Biology* 25:10–15. doi:10.1016/j.cub.2014.10.063.

Beaudoin, S. M. 2007. *Poverty in World History*. New York: Routledge.

Becker, G. S. 1973. "A Theory of Marriage: Part I." *Journal of Political Economy* 81:813–846. doi:10.1086/260084.

Beetsma, R., A. Cukierman, and M. Giuliodori. 2016. "Political Economy of Redistribution in the United States in the Aftermath of World War II—Evidence and Theory." *American Economic Journal: Economic Policy* 8 (4): 1–40. doi:10.1257/pol.20140193.

Beitz, C. R. 1975. "Justice and International Relations." *Philosophy and Public Affairs* 4:360–389.

———. 1979. *Political Theory and International Relations*. Princeton, NJ: Princeton University Press.

Belfer-Cohen, A., and E. Hovers. 2005. "The Groundstone Assemblages of the Natufian and Neolithic Societies in the Levant: A Brief Review." *Journal of the Israel Prehistoric Society* 35:299–308.

Bellah, R. N. 2011. *Religion in Human Evolution: From the Paleolithic to the Axial Age*. Cambridge, MA: Harvard University Press.

Bellow, A. 2003. *In Praise of Nepotism: A Natural History*. New York: Doubleday.

Bentzen, J. S. 2019. "Acts of God? Religiosity and Natural Disasters Across Subnational World Districts." *Economic Journal* 129:2295–2321. doi:10.1093/ej/uez008.

Berchman, R. M. 2008. "The Golden Rule in Greco-Roman Religion and Philosophy." In J. Neusner and B. Chilton, eds., *The Golden Rule: The Ethics of Reciprocity in World Religions*, 40–54. New York: Continuum.

Bernhard, H., U. Fischbacher, and E. Fehr. 2006. "Parochial Altruism in Humans." *Nature* 442:912–915. doi:10.1038/nature04981.

Billingsley, J., J. Antfolk, P. Santtila, and D. Lieberman. 2018. "Cues to Paternity: Do Partner Fidelity and Offspring Resemblance Predict Daughter-Directed Sexual Aversions?" *Evolution and Human Behavior* 39:290–299. doi:10.1016/j.evolhumbehav.2018.02.001.

Bishop, M., and M. Green. 2006. "The Birth of Philanthrocapitalism." *The Economist*, February 25, 6–9, www.economist.com/special-report/2006/02/25/the-birth-of-philanthrocapitalism.

Bittle, W. G., and R. T. Lane. 1976. "Inflation and Philanthropy in England: A Re-assessment of W. K. Jordan's Data." *Economic History Review* 29:203–210. doi:10.2307/2594310.

———. 1978. "A Re-assessment Reiterated." *Economic History Review* 31:124–128. doi:10.1111/j.1468-0289.1978.tb01136.x.

Black, J. 2011. *The English Press in the Eighteenth Century*. Beckenham, UK: Routledge Revivals.

Blackstone, W. T. 1965. "The Golden Rule: A Defense." *Southern Journal of Philosophy* 3:172–177. doi:10.1111/j.2041-6962.1965.tb01706.x.

Bloom, P. 2013. "The Baby in the Well: The Case Against Empathy." *New Yorker*, May 20, www.newyorker.com/magazine/2013/05/20/the-baby-in-the-well.

———. 2014. "Against Empathy." *Boston Review*, September 10, www.bostonreview.net/forum/paul-bloom-against-empathy.

Bocquentin, F., E. Kodas, and A. Ortiz. 2016. "Headless but Still Eloquent! Acephalous Skeletons as Witnesses of Pre-Pottery Neolithic North-South Levant Connections and Disconnections." *Paléorient* 42 (2): 33–52. doi:10.3406/paleo.2016.5719.

Boehm, C. 1999. *Hierarchy in the Forest: The Evolution of Egalitarian Behavior*. Cambridge, MA: Harvard University Press.

Bologna Pavlik, J., E. Lujan Padilla, and B. Powell. 2019. "Cultural Baggage: Do Immigrants Import Corruption?" *Southern Economic Journal* 85:1243–1261. doi:10.1002/soej.12339.

Boomsma, J. J., and R. Gawne. 2018. "Superorganismality and Caste Differentiation as Points of No Return: How the Major Evolutionary Transitions Were Lost in Translation." *Biological Reviews* 93:28–54. doi:10.1111/brv.12330.

Booth, C. 1889–1903. *Life and Labour of the People in London*. London: Macmillan.

Borgerhof Mulder, M., S. Bowles, T. Hertz, A. Bell, J. Beise, G. Clark, I. Fazzio, et al. 2009. "Intergenerational Wealth Transmission and the Dynamics of Inequality in Small-Scale Societies." *Science* 326:682–688. doi:10.1126/science.1178336.

Borrello, M. 2003. "Synthesis and Selection: Wynne-Edwards' Challenge to David Lack." *Journal of the History of Biology* 36:531–566. doi:10.1023/B:HIST.0000004569.71224.65.

———. 2010. *Evolutionary Restraints: The Contentious History of Group Selection*. Chicago: University of Chicago Press.

Bouguen, A., Y. Huang, M. Kremer, and E. Miguel. 2019. "Using Randomized Controlled Trials to Estimate Long-Run Impacts in Development Economics." *Annual Review of Economics* 11:523–561. doi:10.1146/annurev-economics-080218-030333.

Bowles, S. 2006. "Group Competition, Reproductive Leveling, and the Evolution of Human Altruism." *Science* 314:1569–1572. doi:10.1126/science.1134829.

———. 2009. "Did Warfare Among Ancestral Hunter-Gatherers Affect the Evolution of Human Social Behaviors?" *Science* 324:1293–1298. doi:10.1126/science.1168112.

Bowles, S., and J.-K. Choi. 2013. "Coevolution of Farming and Private Property During the Early Holocene." *Proceedings of the National Academy of Sciences* 110:8830–8835. doi:10.1073/pnas.1212149110.

Bowles, S., and H. Gintis. 2011. *A Cooperative Species: Human Reciprocity and Its Evolution*. Princeton, NJ: Princeton University Press.

Bowles, S., and D. Posel. 2005. "Genetic Relatedness Predicts South African Migrant Workers' Remittances to Their Families." *Nature* 434:380–383. doi:10.1038/nature03420.

Boyd, R., and P. J. Richerson. 2010. "Transmission Coupling Mechanisms: Cultural Group Selection." *Philosophical Transactions of the Royal Society B* 365:3787–3795. doi:10.1098/rstb.2010.0046.

Boyer, P. 2018. *Minds Make Societies: How Cognition Explains the World Humans Create.* New Haven, CT: Yale University Press.

Boyer, P., and M. B. Petersen. 2018. "Folk-Economic Beliefs: An Evolutionary Cognitive Model." *Behavioral and Brain Sciences* 41:e158. doi:10.1017/S0140525X17001960.

Brady, H. E., J. S. Fishkin, and R. C. Luskin. 2003. "Informed Public Opinion About Foreign Policy: The Uses of Deliberative Polling." *Brookings Review* 21 (3): 16–19. doi:10.2307/20081112.

Bremner, R. H. 1992. *The Discovery of Poverty in the United States.* New Brunswick, NJ: Transaction.

———. 1994. *Giving: Charity and Philanthropy in History.* New Brunswick, NJ: Transaction.

Briggs, R. C. 2017. "Does Foreign Aid Target the Poorest?" *International Organization* 71:187–206. doi:10.1017/S0020818316000345.

———. 2018a. "Leaving No One Behind? A New Test of Subnational Aid Targeting." *Journal of International Development* 30:904–910. doi:10.1002/jid.3357.

———. 2018b. "Poor Targeting: A Gridded Spatial Analysis of the Degree to Which Aid Reaches the Poor in Africa." *World Development* 103:133–148. doi:10.1016/j.worlddev.2017.10.020.

Broad, C. D. 1940. "Conscience and Conscientious Action." *Philosophy* 15 (58): 115–130. doi:10.1017/S0031819100035907.

Broadberry, S., B. M. S. Campbell, A. Klein, M. Overton, and B. Van Leeuwen. 2015. *British Economic Growth, 1270–1870.* Cambridge: Cambridge University Press.

Brown, D. E. 1991. *Human Universals.* Boston: McGraw-Hill.

Brown, P. H., and J. H. Minty. 2008. "Media Coverage and Charitable Giving After the 2004 Tsunami." *Southern Economic Journal* 75:9–25. doi:10.2307/20112025.

Buchan, J. 2006. *The Authentic Adam Smith: His Life and Ideas.* New York: W. W. Norton.

Buller, D. J. 2005. *Adapting Minds: Evolutionary Psychology and the Persistent Quest for Human Nature.* Cambridge, MA: MIT Press.

Bulmer, M., K. Bales, and K. K. Sklar. 1991a. "The Social Survey in Historical Perspective." In M. Bulmer, K. Bales, and K. K. Sklar, eds., *The Social Survey in Historical Perspective, 1880–1940*, 1–48. Cambridge: Cambridge University Press.

———, eds. 1991b. *The Social Survey in Historical Perspective, 1880–1940.* Cambridge: Cambridge University Press.

Bundy, C. 2016. "Migrants, Refugees, History and Precedents." *Forced Migration Review* 51:5–6.

Burke, E. 1847. *The Modern Orator: The Speeches of the Right Honorable Edmund Burke.* London: Aylott and Jones.

Burton-Chellew, M. N., and R. I. M. Dunbar. 2014. "Hamilton's Rule Predicts Anticipated Social Support in Humans." *Behavioral Ecology* 26:130–137. doi:10.1093/beheco/aru165.

Cairns-Smith, A. G. 1985. *Seven Clues to the Origin of Life: A Scientific Detective Story.* Cambridge: Cambridge University Press.

Cameron, C. D., and B. K. Payne. 2011. "Escaping Affect: How Motivated Emotion Regulation Creates Insensitivity to Mass Suffering." *Journal of Personality and Social Psychology* 100:1–15. doi:10.1037/a0021643.

Caney, S. 2014. "Climate Change, Intergenerational Equity, and the Social Discount Rate." *Politics, Philosophy, and Economics* 13:320–342. doi:10.1177/1470594X14542566.

Carey, W. (1792) 1891. *An Enquiry into the Obligations of Christians to Use Means for the Conversion of the Heathens*. London: Hodder and Stoughton.

Carnegie Hero Fund Commission. 2018. "History of the Carnegie Hero Fund Commission," www.carnegiehero.org/about-the-fund/history.

Carruthers, P. 2006. *The Architecture of the Mind*. Oxford: Oxford University Press.

Carter, G. 2014. "The Reciprocity Controversy." *Animal Behavior and Cognition* 1:368–386. doi:10.12966/abc.08.11.2014.

Caviola, L., N. Faulmüller, J. A. C. Everett, J. Savulescu, and G. Kahane. 2014. "The Evaluability Bias in Charitable Giving: Saving Administration Costs or Saving Lives?" *Judgement and Decision Making* 9:303–315.

Central Intelligence Agency, Office of African and Latin American Analysis. 1985. *Ethiopia: Political and Security Impact of the Drought: An Intelligence Assessment* (ALA 85-10039), www.cia.gov/library/readingroom/docs/CIA-RDP86T00589R000200160004-5.pdf.

Chabris, C. F., A. Weinberger, M. Fontaine, and D. J. Simons. 2011. "You Do Not Talk About Fight Club if You Do Not Notice Fight Club: Inattentional Blindness for a Simulated Real-World Assault." *i-Perception* 2:150–153. doi:10.1068/i0436.

Chapais, B. 2010. "The Deep Structure of Human Society: Primate Origins and Evolution." In P. M. Kappeler and J. B. Silk, eds., *Mind the Gap: Tracing the Origins of Human Universals*, 19–51. Berlin: Springer-Verlag.

Charities Aid Foundation. 2018. *CAF UK Giving 2018: An Overview of Charitable Giving in the UK*, www.cafonline.org/docs/default-source/about-us-publications/caf uk giving 2018 report.pdf.

———. 2019. *CAF UK Giving 2019: An Overview of Charitable Giving in the UK*, www.cafonline.org/docs/default-source/about-us-publications/caf-uk-giving-2019-report-an-overview-of-charitable-giving-in-the-uk.pdf.

Charnov, E. L. 1977. "An Elementary Treatment of the Genetical Theory of Kin Selection." *Journal of Theoretical Biology* 66:541–550. doi:10.1016/0022-5193(77)90301-0.

Chernobrov, D. 2018. "Digital Volunteer Networks and Humanitarian Crisis Reporting." *Digital Journalism* 6:928–944. doi:10.1080/21670811.2018.1462666.

Choi, J. K., and S. Bowles. 2007. "The Coevolution of Parochial Altruism and War." *Science* 318:636–640. doi:10.1126/science.1144237.

Cikara, M., E. G. Bruneau, and R. R. Saxe. 2011. "Us and Them: Intergroup Failures of Empathy." *Current Directions in Psychological Science* 20:149–153. doi:10.1177/0963721411408713.

Clark, G., and R. C. Feenstra. 2003. "Technology in the Great Divergence." In M. D. Bordo, A. M. Taylor, and J. G. Williamson, eds., *Globalization in Historical Perspective*, 277–322. Chicago: University of Chicago Press.

Clark, G., and M. E. Page. 2019. "Welfare Reform, 1834: Did the New Poor Law in England Produce Significant Economic Gains?" *Cliometrica* 13:221–244. doi:10.1007/s11698-018-0174-4.

Clayton, P., and J. Rowbotham. 2008a. "An Unsuitable and Degraded Diet? Part One: Public Health Lessons from the Mid-Victorian Working Class Diet." *Journal of the Royal Society of Medicine* 101:282–289. doi:10.1258/jrsm.2008.080112.

———. 2008b. "An Unsuitable and Degraded Diet? Part Two: Realities of the Mid-Victorian Diet." *Journal of the Royal Society of Medicine* 101:350–357. doi:10.1258/jrsm.2008.080113.

Cline, E. H. 2014. *1177 B.C.: The Year Civilization Collapsed*. Princeton, NJ: Princeton University Press.

Clutton-Brock, T. H. 1991. *The Evolution of Parental Care*. Princeton, NJ: Princeton University Press.

———. 2009. "Cooperation Between Non-Kin in Animal Societies." *Nature* 462:51–57. doi:10.1038/nature08366.

Collier, P. 2007. *The Bottom Billion: Why the Poorest Countries Are Failing and What Can Be Done About It*. New York: Oxford University Press.

Confucius. (c. 500 BCE) 1861. "The Analects of Confucius." J. Legge, trans. [Great Britain]: Pantianos Classics.

Connor, R. C., and K. S. Norris. 1982. "Are Dolphins Reciprocal Altruists?" *American Naturalist* 119:358–374. doi:10.1086/283915.

Corporation for National and Community Service. 2018. "Trends and Highlights Overview," www.nationalservice.gov/vcla/national.

Costa, J. T. 2002. "Scale Models? What Insect Societies Teach Us About Ourselves." *Proceedings of the American Philosophical Society* 146:170–180.

Crespi, B. J., and D. Yanega. 1995. "The Definition of Eusociality." *Behavioral Ecology* 6:109–115. doi:10.1093/beheco/6.1.109.

Crossland, James. 2014. *Britain and the International Committee of the Red Cross, 1939–1945*. England: Palgrave Macmillan.

Damon, W., and A. Colby. 2015. *The Power of Ideals: The Real Story of Moral Choice*. New York: Oxford University Press.

Darwin, C. (1859) 1952. "The Origin of Species." In R. M. Hutchins, ed., *Great Books of the Western World*. Vol. 39, *Charles Darwin*. Chicago: Encyclopedia Britannica.

———. (1871) 1952. "The Descent of Man, and Selection in Relation to Sex." In R. M. Hutchins, ed., *Great Books of the Western World*. Vol. 39, *Charles Darwin*. Chicago: Encyclopedia Britannica.

———. 1872. *The Origin of Species*, 6th ed. London: John Murray.

Davey, E., J. Borton, and M. Foley. 2013. *A History of the Humanitarian System: Western Origins and Foundations*. Humanitarian Policy Group Working Paper. London: Humanitarian Policy Group, www.odi.org/sites/odi.org.uk/files/odi-assets/publications-opinion-files/8439.pdf.

Davidson, A. 2015. "The V.C.s of B.C." *New York Times Magazine*, August 27, www.nytimes.com/2015/08/30/magazine/the-vcs-of-bc.html.

Davies, T. 2014. *NGOs: A New History of Transnational Civil Society*. New York: Oxford University Press.

Dawkins, R. 1976. *The Selfish Gene*. Oxford: Oxford University Press.

———. 1979. "Twelve Misunderstandings of Kin Selection." *Ethology* 51:184–200. doi:10.1111/j.1439-0310.1979.tb00682.x.

———. 1996. *The Blind Watchmaker: Why the Evidence of Evolution Reveals a Universe Without Design*. New York: W. W. Norton.

———. 2006. *The God Delusion*. New York: Houghton Mifflin.

de Cruz, H., and J. de Smedt. 2014. *A Natural History of Natural Theology: The Cognitive Science of Theology and Philosophy of Religion*. Cambridge, MA: MIT Press.

Deen, D., B. Hollis, and C. Zarpentine. 2013. "Darwin and the Levels of Selection." In M. Ruse, ed., *The Cambridge Encyclopedia of Darwin and Evolutionary Thought*, 202–210. New York: Cambridge University Press.

de Freitas, J., P. DeScioli, K. A. Thomas, and S. Pinker. 2019. "Maimonides' Ladder: States of Mutual Knowledge and the Perception of Charitability." *Journal of Experimental Psychology: General* 148:158–173. doi:10.1037/xge0000507.

de Lazari-Radek, K., and P. Singer. 2017. *Utilitarianism: A Very Short Introduction*. Oxford: Oxford University Press.

Dennett, D. C. 1995. *Darwin's Dangerous Idea: Evolution and the Meanings of Life*. New York: Simon and Schuster.

Dercon, S., and C. Porter. 2014. "Live Aid Revisited: Long-Term Impacts of the 1984 Ethiopian Famine on Children." *Journal of the European Economic Association* 12:927–948. doi:10.1111/jeea.12088.

de Vattel, E. (1758) 2008. *The Law of Nations, or, Principles of the Law of Nature Applied to the Conduct and Affairs of Nations and Sovereigns, with Three Early Essays on the Origin and Nature of Natural Law and on Luxury*. T. Nugent, trans., and B. Kapossy and R. Whatmore, eds. Indianapolis: Liberty Fund.

de Waal, F. B. M. 2008. "Putting the Altruism Back into Altruism: The Evolution of Empathy." *Annual Review of Psychology* 59:279–300. doi:10.1146/annurev.psych.59.103006.093625.

———. 2009. *The Age of Empathy: Nature's Lessons for a Kinder Society*. New York: Harmony Books.

Dewitte, A. 2004. "Poverty and Poverty Control in Bruges Between 1250 and 1590." *City* 8:258–265. doi:10.1080/1360481042000242210.

Dichter, T. W. 2003. *Despite Good Intentions: Why Development Assistance to the Third World Has Failed*. Amherst: University of Massachusetts Press.

DiJulio, B., J. Firth, and M. Brodie. 2015. "Data Note: Americans' Views on the U.S. Role in Global Health." Kaiser Family Foundation (KFF), January 23, www.kff.org/global-health-policy/poll-finding/data-note-americans-views-on-the-u-s-role-in-global-health.

Dixon, T. 2005. "The Invention of Altruism: August Comte's *Positive Polity* and Respectable Unbelief in Victorian Britain." In D. M. Knight and M. D. Eddy, eds., *Science and Beliefs: From Natural Philosophy to Natural Science, 1700–1900*, 195–211. Hampshire, UK: Ashgate.

———. 2008. *The Invention of Altruism: Making Moral Meanings in Victorian Britain*. Oxford: Oxford University Press.

———. 2013. "Altruism: Morals from History." In M. A. Nowak and S. Coakley, eds., *Evolution, Games, and God: The Principle of Cooperation*, 60–81. Cambridge, MA: Harvard University Press.

Do, C., and I. Paley. 2012. "Altruism from the House: The Impact of Home Equity on Charitable Giving." *Review of Economics of the Household* 10:375–393. doi:10.1007/s11150-011-9123-8.

Doebeli, M., C. Hauert, and T. Killingback. 2004. "The Evolutionary Origin of Cooperators and Defectors." *Science* 306:859–862. doi:10.1126/science.1101456.

Douglas, P. n.d. *Dissing the Dutch: All's Fair in Love and War.* Albany, NY: New Netherland Institute.

Du Bois, W. E. B. 1899. *The Philadelphia Negro: A Social Study.* Philadelphia: University of Pennsylvania Press.

Dunant, H. (1862) 1959. *A Memory of Solferino.* Geneva: International Committee of the Red Cross.

Dyer, C. 2012. "The Experience of Being Poor in Late Medieval England." In A. M. Scott, ed., *Experiences of Poverty in Late Medieval and Early Modern England and France*, 19–39. Surrey, UK: Ashgate.

Easterly, W. 2006. *The White Man's Burden: Why the West's Efforts to Aid the Rest Have Done So Much Ill and So Little Good.* New York: Penguin.

Edwards, M. S. 2006. "Public Opinion Regarding Economic and Cultural Globalization: Evidence from a Cross-National Survey." *Review of International Political Economy* 13:587–608. doi:10.1080/09692290600839857.

Eff, E. A., and M. M. Dow. 2008. *Do Markets Promote Prosocial Behavior? Evidence from the Standard Cross-Cultural Sample.* Working Paper No. 200803. Department of Economics and Finance, Middle Tennessee State University, Murfreesboro.

Eisenberg, N., N. D. Eggum, and L. Di Giunta. 2010. "Empathy-Related Responding: Associations with Prosocial Behavior, Aggression, and Intergroup Relations." *Social Issues and Policy Review* 4:143–180. doi:10.1111/j.1751-2409.2010.01020.x.

Eisenbruch, A. B., R. L. Grillot, D. Maestripieri, and J. R. Roney. 2016. "Evidence of Partner Choice Heuristics in a One-Shot Bargaining Game." *Evolution and Human Behavior* 37:429–439. doi:10.1016/j.evolhumbehav.2016.04.002.

Eisensee, T., and D. Strömberg. 2007. "News Droughts, News Floods, and US Disaster Relief." *Quarterly Journal of Economics* 122:693–728. doi:10.1162/qjec.122.2.693.

Epstein, K. 1966. *The Genesis of German Conservatism.* Princeton, NJ: Princeton University Press.

Escresa, L., and L. Picci. 2017. "A New Cross-National Measure of Corruption." *World Bank Economic Review* 31:196–219. doi:10.1093/wber/lhv031.

Faber, D. A., S. Joshi, and G. Ciancio. 2016. "Demographic Characteristics of Non-directed Altruistic Kidney Donors in the United States." *Journal of Kidney* 2 (2): 1–4. doi:10.4172/2472-1220.1000121.

Fantazzi, C. 2008. "Vives and the Emarginati." In C. Fantazzi, ed., *A Companion to Juan Luis Vives*, 65–112. Leiden: Brill.

Fensham, F. C. 1962. "Widow, Orphan, and the Poor in Ancient Near Eastern Legal and Wisdom Literature." *Journal of Near Eastern Studies* 21:129–139. doi:10.2307/543887.

Ferguson, N. 2011. *Civilization: The West and the Rest.* New York: Penguin.

Fichte, J. G. (1797) 2000. *Foundations of Natural Right.* M. Baur, trans., and F. Neuhauser, ed. Cambridge: Cambridge University Press.

Findley, M. G. 2018. "Does Foreign Aid Build Peace?" *Annual Review of Political Science* 21:359–384. doi:10.1146/annurev-polisci-041916-015516.

Finke, R. A. 1980. "Levels of Equivalence in Imagery and Perception." *Psychological Review* 87:113–132. doi:10.1037/0033-295X.87.2.113.

Finlay, T. 2009. "Alms. I: Hebrew Bible / Old Testament." In C.-L. Seow and H. Spieckermann, eds., *Encyclopedia of the Bible and Its Reception*. Vol. 1, *Aaron–Aniconism*, 830. Berlin: Walter de Gruyter.

Fisher, H. 1994. "'Wilson,' They Said, 'You're All Wet!'" *New York Times*, October 16.

Fison, L., and A. W. Howitt. 1880. *Kamilaroi and Kurnai: Group-Marriage and Relationship, and Marriage by Elopement. Also the Kurnai Tribe: Their Customs in Peace and War.* Melbourne: George Robertson.

Fitzgerald, J. T. 2009. "Alms. III: Greco-Roman Antiquity." In C.-L. Seow and H. Spieckermann, eds., *Encyclopedia of the Bible and Its Reception*. Vol. 1, *Aaron–Aniconism*, 835–836. Berlin: Walter de Gruyter.

Fleischacker, S. 2004. *A Short History of Distributive Justice.* Cambridge, MA: Harvard University Press.

Forrester, A. C., B. Powell, A. Nowrasteh, and M. Landgrave. 2019. "Do Immigrants Import Terrorism?" *Journal of Economic Behavior and Organization* 166:529–543. doi:10.1016/j.jebo.2019.07.019.

Forsythe, D. P. 2005. *The Humanitarians: The International Committee of the Red Cross.* Cambridge: Cambridge University Press.

Foster, B. R. 1995. "Social Reform in Ancient Mesopotamia." In K. D. Irani and M. Silver, eds., *Social Justice in the Ancient World*, 165–177. Westport, CT: Greenwood Press.

Fouquet, R., and S. Broadberry. 2015. "Seven Centuries of European Economic Growth and Decline." *Journal of Economic Perspectives* 29:227–244. doi:10.1257/jep.29.4.227.

Fox, G. 1950. "The Origins of UNRRA." *Political Science Quarterly* 65:561–584. doi:10.2307/2145664.

Frangipane, M. 2007. "Different Types of Egalitarian Societies and the Development of Inequality in Early Mesopotamia." *World Archaeology* 39:151–176. doi:10.1080/00438240701249504.

Frank, R. 2009. "Rich Feel Guilty About Giving to Charity." *Wall Street Journal*, May 18, https://blogs.wsj.com/wealth/2009/05/18/rich-feel-guilty-about-giving-to-charity.

Frank, S. A. 1998. *Foundations of Social Evolution.* Princeton, NJ: Princeton University Press.

Franklin, P., and A. A. Volk. 2018. "A Review of Infants' and Children's Facial Cues' Influence on Adults' Perceptions and Behaviors." *Evolutionary Behavioral Sciences* 12:296–321. doi:10.1037/ebs0000108.

Franks, S. 2013. *Reporting Disasters: Famine, Aid, Politics and the Media.* London: Hurst.

Frean, M. R. 1994. "The Prisoner's Dilemma Without Synchrony." *Proceedings of the Royal Society B: Biological Sciences* 257:75–79. doi:10.1098/rspb.1994.0096.

From the Church News. 2015. "Viewpoint: 1985 Fast Marked Beginning of LDS Charities." Church of Jesus Christ of Latter-day Saints, January 25, www.church

ofjesuschrist.org/church/news/viewpoint-1985-fast-marked-beginning-of-lds-charities?lang=eng.

Gabriel, I. 2017. "Effective Altruism and Its Critics." *Journal of Applied Philosophy* 34:457–473. doi:10.1111/japp.12176.

Galiana, I. 2018. "Benefits and Costs of the Climate Change Targets for the Post-2015 Development Agenda." In B. Lomborg, ed., *Prioritizing Development: A Cost Benefit Analysis of the United Nations' Sustainable Development Goals*, 54–63. New York: Cambridge University Press.

Gallistel, C. R., and A. P. King. 2009. *Memory and the Computational Brain: Why Cognitive Science Will Transform Neuroscience*. Chichester, UK: Wiley-Blackwell.

Gansberg, M. 1964. "37 [*sic*] Who Saw Murder Didn't Call the Police." *New York Times*, March 27.

Gardner, A., and A. Grafen. 2009. "Capturing the Superorganism: A Formal Theory of Group Adaptation." *Journal of Evolutionary Biology* 22:659–671. doi:10.1111/j.1420-9101.2008.01681.x.

Gardner, A., S. A. West, and G. Wild. 2011. "The Genetical Theory of Kin Selection." *Journal of Evolutionary Biology* 24:1020–1043. doi:10.1111/j.1420-9101.2011.02236.x.

Garnsey, P. 1988. *Famine and Food Supply in the Graeco-Roman World: Responses to Risk and Crisis*. Cambridge: Cambridge University Press.

Garve, C. 1796. "Über den Charakter der Bauern und ihr Verhältniss gegen die Gutsherrn und gegen die Regierung," vol. 1. Wrocław, Poland: Wilhelm Gottlieb Korn.

Gat, A. 2010. "Why War? Motivations for Fighting in the Human State of Nature." In P. M. Kappeler and J. B. Silk, eds., *Mind the Gap: Tracing the Origins of Human Universals*, 197–220. New York: Springer.

Gensler, H. J. 2013. *Ethics and the Golden Rule*. New York: Routledge.

George, D. L. 1918. *The Great Crusade: Extracts from Speeches Delivered During the War*. F. L. Stevenson, ed. New York: George H. Doran.

Geremek, B. 1994. *Poverty: A History*. Oxford: Blackwell.

Gilens, M. 1999. *Why Americans Hate Welfare: Race, Media, and the Politics of Antipoverty Policy*. Chicago: University of Chicago Press.

Giving Pledge. 2019. "History of the Pledge," https://givingpledge.org/About.aspx.

Giving USA. 2019. "Giving USA 2019: Americans Gave $427.71 Billion to Charity in 2018 amid Complex Year for Charitable Giving," https://givingusa.org/giving-usa-2019-americans-gave-427-71-billion-to-charity-in-2018-amid-complex-year-for-charitable-giving.

Giving What We Can. 2019. "A Growing Movement," www.givingwhatwecan.org/about-us/#a-growing-movement.

Glynn, S. A., M. P. Busch, G. B. Schreiber, E. L. Murphy, D. J. Wright, Y. Tu, and S. H. Kleinman. 2003. "Effect of a National Disaster on Blood Supply and Safety: The September 11 Experience." *JAMA* 289:2246–2253. doi:10.1001/jama.289.17.2246.

Goetz, J. L., D. Keltner, and E. Simon-Thomas. 2010. "Compassion: A Theoretical Analysis and Empirical Review." *Psychological Bulletin* 136:351–374. doi:10.1037/a0018807.

Google Books Ngram Viewer. 2019. https://books.google.com/ngrams.

Goose, N. 2013. "Review of the Book *Poor Relief in England, 1350–1600*, by M. McIntosh." *Reviews in History*, https://reviews.history.ac.uk/review/1404.

Goose, N., and H. Looijesteijn. 2012. "Almshouses in England and the Dutch Republic Circa 1350–1800: A Comparative Perspective." *Journal of Social History* 45:1049–1073. doi:10.1093/jsh/shr146.

Gottschall, J. 2004. "Explaining Wartime Rape." *Journal of Sex Research* 41:129–136. doi:10.2307/3813647.

Gould, S. J., and R. C. Lewontin. 1979. "The Spandrels of San Marco and the Panglossian Paradigm: A Critique of the Adaptationist Programme." *Proceedings of the Royal Society B* 205:581–598. doi:10.1098/rspb.1979.0086.

Gowdy, J., and L. Krall. 2016. "The Economic Origins of Ultrasociality." *Behavioral and Brain Sciences* 39:E92. doi:10.1017/S0140525X1500059X.

Gradstein, M., and B. Milanovic. 2004. "Does Liberté = Égalité? A Survey of the Empirical Links Between Democracy and Inequality with Some Evidence on the Transition Economies." *Journal of Economic Surveys* 18:515–537. doi:10.1111/j.0950-0804.2004.00229.x.

Gray, H. M., N. Ambady, W. T. Lowenthal, and P. Deldin. 2004. "P300 as an Index of Attention to Self-Relevant Stimuli." *Journal of Experimental Social Psychology* 40:216–224. doi:10.1016/S0022-1031(03)00092-1.

Greif, A., and M. Iyigun. 2013. *What Did the Old Poor Law Really Accomplish? A Redux*. Discussion Paper No 7398. Bonn: Institute for the Study of Labor.

Griffin, A. S., and S. A. West. 2003. "Kin Discrimination and the Benefit of Helping in Cooperatively Breeding Vertebrates." *Science* 302:634–636. doi:10.1126/science.1089402.

Griswold, C. L. 1999. *Adam Smith and the Virtues of Enlightenment*. Cambridge: Cambridge University Press.

Guisinger, A. 2017. *American Opinion on Trade: Preferences Without Politics*. doi:10.1093/acprof:oso/9780190651824.003.0003.

Gupta, H. K. 2018. "1755 Lisbon Tsunami and the Birth of Seismology." *Journal of the Geological Society of India* 92:255–258. doi:10.1007/s12594-018-1000-0.

Gurven, M. 2004. "To Give and to Give Not: An Evolutionary Ecology of Human Food Transfers." *Behavioral and Brain Sciences* 27:543–583. doi:10.1017/S0140525X04000123.

Gurven, M., and K. Hill. 2009. "Why Do Men Hunt?" *Current Anthropology* 50:51–74. doi:10.1086/595620.

Gurven, M., J. Stieglitz, P. L. Hooper, C. Gomes, and H. Kaplan. 2012. "From the Womb to the Tomb: The Role of Transfers in Shaping the Evolved Human Life History." *Experimental Gerontology* 47:807–813. doi:10.1016/j.exger.2012.05.006.

Gurven, M., C. Von Rueden, J. Stieglitz, H. Kaplan, and D. E. Rodriguez. 2014. "The Evolutionary Fitness of Personality Traits in a Small-Scale Subsistence Society." *Evolution and Human Behavior* 35:17–25. doi:10.1016/j.evolhumbehav.2013.09.002.

Gusinde, M. 1937. "Yahgan: The Life and Thought of the Water Nomads of Cape Horn." F. Schütze, trans. In *Die Feuerland-Indianer*. Vol. 2, *Die Yamana*. Mödling, Austria: Anthropos-Bibliotek.

Hadwin, J. F. 1978. "Deflating Philanthropy." *Economic History Review* 31:105–117. doi:10.1111/j.1468-0289.1978.tb01133.x.

Hafer, C. L., and L. Bègue. 2005. "Experimental Research on Just-World Theory: Problems, Developments, and Future Challenges." *Psychological Bulletin* 131:128–167. doi:10.1037/0033-2909.131.1.128.

Hagen, E. H. 2016. "Evolutionary Psychology and Its Critics." In D. M. Buss. ed., *The Handbook of Evolutionary Psychology*. Vol. 1, *Foundations*, 2nd ed., 136–160. Hoboken, NJ: Wiley.

Haidt, J. 2001. "The Emotional Dog and Its Rational Tail: A Social Intuitionist Approach to Moral Judgment." *Psychological Review* 108:814–834. doi:10.1037/0033-295x.108.4.814.

Hainmueller, J., and D. J. Hopkins. 2014. "Public Attitudes Toward Immigration." *Annual Review of Political Science* 17:225–249. doi:10.1146/annurev-polisci-102512-194818.

Haldane, J. B. S. 1990. *The Causes of Evolution*. Princeton, NJ: Princeton University Press.

Haller, J. S. 2011. *Battlefield Medicine: A History of the Military Ambulance from the Napoleonic Wars Through World War I*. Carbondale: Southern Illinois University Press.

Hamilton, W. D. 1964. "The Genetical Evolution of Social Behaviour. I, II." *Journal of Theoretical Biology* 7:1–52. doi:10.1016/0022-5193(64)90038-4.

———. 1975. "Innate Social Aptitudes of Man: An Approach from Evolutionary Genetics." In R. Fox, ed., *Biosocial Anthropology*, 133–155. New York: John Wiley and Sons.

Hammack, D. C. 1995. "A Center of Intelligence for the Charity Organization Movement: The Foundation's Early Years." In D. C. Hammack and S. Wheeler, eds., *Social Science in the Making: Essays on the Russell Sage Foundation, 1907–1972*, 1–33. New York: Russell Sage Foundation.

Hands, A. R. 1968. *Charities and Social Aid in Greece and Rome*. Ithaca, NY: Cornell University Press.

Hansen, M. H. 2006. *The Shotgun Method: The Demography of the Ancient Greek City-State Culture*. Columbia: University of Missouri Press.

Harman, O. 2010. *The Price of Altruism: George Price and the Search for the Origins of Kindness*. New York: W. W. Norton.

Harper, M., V. Knight, M. Jones, G. Koutsovoulos, N. E. Glynatsi, and O. Campbell. 2017. "Reinforcement Learning Produces Dominant Strategies for the Iterated Prisoner's Dilemma." *PLoS One* 12 (12). doi:10.1371/journal.pone.0188046.

Hauert, C., and H. G. Schuster. 1998. "Extending the Iterated Prisoner's Dilemma Without Synchrony." *Journal of Theoretical Biology* 192:155–166. doi:10.1006/jtbi.1997.0590.

Haueter, N. V. 2013. *Swiss Re: A History of Insurance*. Zurich: Swiss Re Group, www.swissre.com/dam/jcr:638f00a0-71b9-4d8e-a960-dddaf9ba57cb/150_history_of_insurance.pdf.

Hawkes, K., J. F. O'Connell, and N. G. Blurton Jones. 2001. "Hadza Meat Sharing." *Evolution and Human Behavior* 22:113–142. doi:10.1016/S1090-5138(00)00066-0.

Hays, J. C., S. D. Ehrlich, and C. Peinhardt. 2005. "Government Spending and Public Support for Trade in the OECD: An Empirical Test of the Embedded Liberalism

Thesis." *International Organization* 59:473–494. doi:10.1017/S002081830505 0150.

Hendren, N., and B. D. Sprung-Keyser. 2019. *A Unified Welfare Analysis of Government Policies*. Working Paper No. 26144. National Bureau of Economic Research.

Henrich, J. 2004. "Cultural Group Selection, Coevolutionary Processes and Large-Scale Cooperation." *Journal of Economic Behavior and Organization* 53:3–35. doi:10.1016/S0167-2681(03)00094-5.

Hill, K., and K. Kintigh. 2009. "Can Anthropologists Distinguish Good and Poor Hunters? Implications for Hunting Hypotheses, Sharing Conventions, and Cultural Transmission." *Current Anthropology* 50:369–378. doi:10.1086/597981.

Himmelfarb, G. 1984. *The Idea of Poverty: England in the Early Industrial Age*. New York: Alfred A. Knopf.

———. 1991. *Poverty and Compassion: The Moral Imagination of the Late Victorians*. New York: Alfred A. Knopf.

Hitchcock, W. I. 2008. *The Bitter Road to Freedom: A New History of the Liberation of Europe*. New York: Free Press.

Hobbes, T. (1651) 1957. *Leviathan*. New York: Oxford University Press.

Hollander, C. 2018. "Our Recommendation to Good Ventures." Give Well, November 27, https://blog.givewell.org/2018/11/26/our-recommendation-to-good-ventures.

———. 2019. "Review of GiveWell's Work in 2018." GiveWell, May 15, https://blog.givewell.org/2019/05/15/review-of-givewells-work-in-2018.

Holmes, E. A., and A. Mathews. 2005. "Mental Imagery and Emotion: A Special Relationship?" *Emotion* 5:489–497. doi:10.1037/1528-3542.5.4.489.

Holmes, E. A., A. Mathews, B. Mackintosh, and T. Dalgleish. 2008. "The Causal Effect of Mental Imagery on Emotion Assessed Using Picture-Word Cues." *Emotion* 8:395–409. doi:10.1037/1528-3542.8.3.395.

Homerin, T. E. 2005. "Altruism in Islam." In J. Neusner and B. Chilton, eds., *Altruism in World Religions*, 67–87. Washington, DC: Georgetown University Press.

Hooper, P. L., M. Gurven, J. Winking, and H. S. Kaplan. 2015. "Inclusive Fitness and Differential Productivity Across the Life Course Determine Intergenerational Transfers in a Small-Scale Human Society." *Proceedings of the Royal Society B: Biological Sciences* 282 (1803): 1–9. doi:10.1098/rspb.2014.2808.

Howe, E. L., J. J. Murphy, D. Gerkey, and C. T. West. 2016. "Indirect Reciprocity, Resource Sharing, and Environmental Risk: Evidence from Field Experiments in Siberia." *PLoS One* 11 (7). doi:10.1371/journal.pone.0158940.

Hoynes, H. W., and D. W. Schanzenbach. 2018. *Safety Net Investments in Children*. Working Paper No. 24594. National Bureau of Economic Research.

Hruschka, D. J., and J. Henrich. 2006. "Friendship, Cliquishness, and the Emergence of Cooperation." *Journal of Theoretical Biology* 239:1–15. doi:10.1016/j.jtbi.2005.07.006.

Hume, D. (1739) 1984. *A Treatise of Human Nature*. New York: Penguin.

———. (1777) 1957. *An Enquiry Concerning the Principles of Morals*, vol. 4. New York: Liberal Arts Press.

Humphreys, M., and J. M. Weinstein. 2009. "Field Experiments and the Political Economy of Development." *Annual Review of Political Science* 12:367–378. doi:10.1146/annurev.polisci.12.060107.155922.

Hunter, R. 1904. *Poverty*. New York: Macmillan.

Hurst, R., T. Tidwell, and D. Hawkins. 2017. "Down the Rathole? Public Support for US Foreign Aid." *International Studies Quarterly* 61:442–454. doi:10.1093/isq/sqx019.

Husock, H. 1992. "Bringing Back the Settlement House." *The Public Interest* 109:53–72.

Huwart, J. Y., and L. Verdier. 2013. *Economic Globalisation: Origins and Consequences*. Paris: Organization for Economic Cooperation and Development, www.oecd-ilibrary.org/economics/economic-globalisation_9789264111905-en.

Hyman, I. E., S. M. Boss, B. M. Wise, K. McKenzie, and J. M. Caggiano. 2010. "Did You See the Unicycling Clown? Inattentional Blindness While Walking and Talking on a Cell Phone." *Applied Cognitive Psychology* 24:597–607. doi:10.1002/acp.1638.

Hyman, I. E., B. A. Sarb, and B. M. Wise-Swanson. 2014. "Failure to See Money on a Tree: Inattentional Blindness for Objects That Guided Behavior." *Frontiers in Psychology* 5:356. doi:10.3389/fpsyg.2014.00356.

Ierley, M. 1984. *With Charity for All: Welfare and Society, Ancient Times to the Present*. New York: Praeger.

Ifill, G. 2016. "Questions for President Obama: A Town Hall Special." *PBS NewsHour*. NewsHour Productions.

Isenberg, N. 2016. *White Trash: The 400-Year Untold History of Class in America*. New York: Viking.

IUPUI Lilly Family School of Philanthropy. 2018. *Highlights: An Overview of Giving in 2017*. Giving USA, https://store.givingusa.org/products/giving-usa-2018-report-highlights?variant=12366640775247.

Jackson, P. L., A. N. Meltzoff, and J. Decety. 2005. "How Do We Perceive the Pain of Others: A Window into the Neural Processes Involved in Empathy." *NeuroImage* 24:771–779. doi:10.1016/j.neuroimage.2004.09.00.

Jacobson, L. 2011. "Newt Gingrich Says You Can Use Food Stamps to Get to Hawaii." *PolitiFact*, December 1, www.politifact.com/truth-o-meter/statements/2011/dec/01/newt-gingrich/Gingrich-says-use-food-stamps-Hawaii.

Jaeggi, A. V., and M. Gurven. 2013. "Reciprocity Explains Food Sharing in Humans and Other Primates Independent of Kin Selection and Tolerated Scrounging: A Phylogenetic Meta-Analysis." *Proceedings of the Royal Society B: Biological Sciences* 280 (1768): 20131615. doi:10.1098/rspb.2013.1615.

Jaspers, K. (1953) 2010. *The Origin and Goal of History*. M. Bullock, trans. London: Routledge.

Jones, A. 2019. "The Unknown Famine: Television and the Politics of British Humanitarianism." In M. Lawrence and R. Tavernor, eds., *Global Humanitarianism and Media Culture*, 122–142. Manchester: Manchester University Press.

Justman, M., and M. Gradstein. 1999. "The Industrial Revolution, Political Transition, and the Subsequent Decline in Inequality in 19th-Century Britain." *Explorations in Economic History* 36:109–127. doi:10.1006/exeh.1999.0713.

Kant, I. (1785) 1952. "General Introduction to the Metaphysic of Morals." W. Hastie, trans. In R. Maynard, ed., *Great Books of the Western World*. Vol. 42, *Kant*, 383–394. Chicago: Encyclopedia Britannica.

———. (1797) 1965. *The Metaphysical Elements of Justice: Part I of the Metaphysics of Morals*. J. Ladd, trans. Indianapolis: Bobbs-Merrill.

———. 1930. *Lectures on Ethics*. L. Infield, trans. London: Methuen.

Kaplan, H., K. Hill, J. Lancaster, and A. M. Hurtado. 2000. "A Theory of Human Life History Evolution: Diet, Intelligence, and Longevity." *Evolutionary Anthropology* 9:156–184. doi:10.1002/1520-6505(2000)9:4<156:AID-EVAN5>3.0.CO;2-7.

Karnofsky, H. 2019. "Our Progress in 2018 and Plans for 2019." Good Ventures, April 15, www.goodventures.org/blog/our-progress-in-2018-and-plans-for-2019.

Kaše, V. 2018. *The Emergence of Big Gods in the Ancient Mediterranean*. [Manuscript]. Retrieved from osf.io/preprints/socarxiv/3tjb7/.

Kassin, S. 2017. "The Killing of Kitty Genovese: What Else Does This Case Tell Us?" *Perspectives on Psychological Science* 12:374–381. doi:10.1177/1745691616679465.

Keating, R. 1986. "Live Aid: The Terrible Truth." *Spin* 2 (4): 74–80, www.spin.com/featured/live-aid-the-terrible-truth-ethiopia-bob-geldof-feature.

Keay, D. 1987. "AIDS, Education, and the Year 2000!" *Woman's Own*, October 31, 8–10, www.margaretthatcher.org/document/106689.

Keck, A. 2010. "The Change of Philosophical Motives of Care from Thomas Aquinas' Notion of Alms to Juan Luis Vives' 'De Subventione Pauperum.'" *European Journal of Social Work* 13:127–130. doi:10.1080/13691451003621184.

Kelemen, D. 2004. "Are Children 'Intuitive Theists'?: Reasoning About Purpose and Design in Nature." *Psychological Science* 15:295–301. doi:10.1111/j.0956-7976.2004.00672.x.

Kelemen, D., J. Rottman, and R. Seston. 2013. "Professional Physical Scientists Display Tenacious Teleological Tendencies: Purpose-Based Reasoning as a Cognitive Default." *Journal of Experimental Psychology: General* 142:1074–1083. doi:10.1037/a0030399.

Kelley, F. 1895. *Hull-House Maps and Papers*. New York: Thomas Y. Crowell.

Kelly, M., and C. Ó Gráda. 2011. "The Poor Law of Old England: Institutional Innovation and Demographic Regimes." *Journal of Interdisciplinary History* 41:339–366. doi:10.1162/JINH_a_00105.

———. 2019. "Speed Under Sail During the Early Industrial Revolution (c. 1750–1830)." *Economic History Review* 72:459–480. doi:10.1111/ehr.12696.

Kendrick, T. D. 1955. *The Lisbon Earthquake*. Philadelphia: J. B. Lippincott.

Kennedy, J. F. 1963. "Excerpt, Report to the American People on Civil Rights." June 11, 1963. Washington, DC: Columbia Broadcasting System. Archived at John F. Kennedy Presidential Library and Museum, www.jfklibrary.org/asset-viewer/archives/TNC/TNC-262-EX/TNC-262-EX.

King, S. 2000. *Poverty and Welfare in England, 1700–1850*. Manchester: Manchester University Press.

Klenert, D., L. Mattauch, E. Combet, O. Edenhofer, C. Hepburn, R. Rafaty, and N. Stern. 2018. "Making Carbon Pricing Work for Citizens." *Nature Climate Change* 8:669–677. doi:10.1038/s41558-018-0201-2.

Knapp, A. B. 1988. *The History and Culture of Ancient Western Asia and Egypt*. Chicago: Dorsey Press.

Knapp, A. B., and S. W. Manning. 2016. "Crisis in Context: The End of the Late Bronze Age in the Eastern Mediterranean." *American Journal of Archaeology* 120:99–149. doi:10.3764/aja.120.1.0099.

Knight, V., M. Harper, N. E. Glynatsi, and O. Campbell. 2018. "Evolution Reinforces Cooperation with the Emergence of Self-Recognition Mechanisms: An Empirical

Study of Strategies in the Moran Process for the Iterated Prisoner's Dilemma." *PLoS One* 13 (10). doi:10.1371/journal.pone.0204981.

Koeneman, K. 2013. *First Son: The Biography of Richard M. Daley*. Chicago: University of Chicago Press.

Kohler, H.-P., and J. R. Behrman. 2018. "Benefits and Costs of the Population and Demography Targets for the Post-2015 Development Agenda." In B. Lomborg, ed., *Prioritizing Development: A Cost Benefit Analysis of the United Nations' Sustainable Development Goals*, 375–394. New York: Cambridge University Press.

Kosslyn, S. M. 1995. "Visual Cognition." In D. N. Osherson and S. M. Kosslyn, eds., *An Invitation to Cognitive Science*. Vol. 2, *Visual Cognition*, 267–296. Cambridge, MA: MIT Press.

Kotsadam, A., G. Østby, S. A. Rustad, A. F. Tollefsen, and H. Urdal. 2018. "Development Aid and Infant Mortality: Micro-Level Evidence from Nigeria." *World Development* 105:59–69. doi:10.1016/j.worlddev.2017.12.022.

Kramer, S. H. 1963. *The Sumerians: Their History, Culture, and Character*. Chicago: University of Chicago Press.

Krams, I., T. Krama, K. Igaune, and R. Mänd. 2007. "Experimental Evidence of Reciprocal Altruism in the Pied Flycatcher." *Behavioral Ecology and Sociobiology* 62:599–605. doi:10.1007/s00265-007-0484-1.

Krasnow, M. M., L. Cosmides, E. J. Pedersen, and J. Tooby. 2012. "What Are Punishment and Reputation For?" *PLoS One* 7 (9). doi:10.1371/journal.pone.0045662.

Krasnow, M. M., A. W. Delton, J. Tooby, and L. Cosmides. 2013. "Meeting Now Suggests We Will Meet Again: Implications for Debates on the Evolution of Cooperation." *Scientific Reports* 3:1747. doi:10.1038/srep01747.

Krishna-Dwaipayana Vyasa. 1896. "Mahābhārata." K. M. Ganguli, trans. Kolkata, India: Pratap Chandra Roy.

Krubitzer, L., and D. S. Stolzenberg. 2014. "The Evolutionary Masquerade: Genetic and Epigenetic Contributions to the Neocortex." *Current Opinion in Neurobiology* 24:157–165. doi:10.1016/j.conb.2013.11.010.

Krueger, J. I., and D. C. Funder. 2004. "Towards a Balanced Social Psychology: Causes, Consequences, and Cures for the Problem-Seeking Approach to Social Behavior and Cognition." *Behavioral and Brain Sciences* 27:313–327. doi:10.1017/S0140525X04000081.

Krupp, D. B., and P. D. Taylor. 2013. "Enhanced Kin Recognition Through Population Estimation." *American Naturalist*, 181:707–714. doi:10.1086/670029.

Kuhnle, S., and A. Sander. 2010. "The Emergence of the Western Welfare State." In F. G. Castles, S. Leibfried, J. Lewis, H. Obinger, and C. Pierson, eds., *The Oxford Handbook of the Welfare State*, 61–80. New York: Oxford University Press.

Kuijt, I. 2000. "People and Space in Early Agricultural Villages: Exploring Daily Lives, Community Size, and Architecture in the Late Pre-Pottery Neolithic." *Journal of Anthropological Archaeology* 19:75–102. doi:10.1006/jaar.1999.0352.

———. 2008a. "Demography and Storage Systems During the Southern Levantine Neolithic Demographic Transition." In J. P. Bocquet-Appel and O. Bar-Yosef, eds., *The Neolithic Demographic Transition and Its Consequences*, 287–313. New York: Springer.

———. 2008b. "The Regeneration of Life: Neolithic Structures of Symbolic Remembering and Forgetting." *Current Anthropology* 49:171–197. doi:10.1086/526097.

———. 2011. "Home Is Where We Keep Our Food: The Origins of Agriculture and Late Pre-Pottery Neolithic Food Storage." *Paléorient* 37:137–152. doi:10.3406/paleo.2011.5444.

Kuijt, I., and B. Finlayson. 2009. "Evidence for Food Storage and Predomestication Granaries 11,000 Years Ago in the Jordan Valley." *Proceedings of the National Academy of Sciences* 106:10966–10970. doi:10.1073/pnas.0812764106.

Kurzban, R. 2010. *Why Everyone (Else) Is a Hypocrite: Evolution and the Modular Mind*. Princeton, NJ: Princeton University Press.

Lack, D. 1966. *Population Studies of Birds*. Oxford: Oxford University Press.

Laland, K. N., and G. R. Brown. 2011. *Sense and Nonsense: Evolutionary Perspectives on Human Behaviour*, 2nd ed. Oxford: Oxford University Press.

Lamb, T. D., C. P. Collin, and E. N. Pugh, Jr. 2007. "Evolution of the Vertebrate Eye: Opsins, Photoreceptors, Retina and Eye Cup." *Nature Reviews Neuroscience* 8:960–976. doi:10.1038/nrn2283.

Landry, C. E., A. Lange, J. A. List, M. K. Price, and N. G. Rupp. 2006. "Toward an Understanding of the Economics of Charity: Evidence from a Field Experiment." *Quarterly Journal of Economics* 121 (2): 747–782.

Laqueur, T. 1989. "Bodies, Details, and the Humanitarian Narrative." In L. Hunt, ed., *The New Cultural History*, 176–204. Berkeley: University of California Press.

Latané, B., and J. M. Darley. 1970. *The Unresponsive Bystander: Why Doesn't He Help?* New York: Appleton-Century-Crofts.

Lazarsfeld, P. F., and R. K. Merton. 1954. "Friendship as a Social Process: A Substantive and Methodological Analysis." In M. Berger, T. Abel, and C. H. Page, eds., *Freedom and Control in Modern Society*, 18–66. New York: Van Nostrand.

LeBar, M. 1999. "Kant on Welfare." *Canadian Journal of Philosophy* 29:225–250. doi:10.1080/00455091.1999.10717512.

Lebzelter, V. 1934. *Native Cultures in Southwest and South Africa*, vol. 2. R. Neuse, trans. Leipzig: Karl W. Hiersemann.

Lecky, W. E. H. 1890. *History of European Morals from Augustus to Charlemagne in Two Volumes*, vol. 1, 9th ed. London: Longmans, Green.

Lee, J.-W., and H. Lee. 2016. "Human Capital in the Long Run." *Journal of Development Economics* 122:147–169. doi:10.1016/j.jdeveco.2016.05.006.

Lehmann, L., and M. W. Feldman. 2008. "War and the Evolution of Belligerence and Bravery." *Proceedings of the Royal Society B: Biological Sciences* 275:2877–2885. doi:10.1098/rspb.2008.0842.

Leigh, E. G. 1983. "When Does the Good of the Group Override the Advantage of the Individual?" *Proceedings of the National Academy of Sciences* 80:2985–2989. doi:10.1073/pnas.80.10.2985.

Lerner, M. J. 1980. *Belief in a Just World: A Fundamental Delusion*. New York: Plenum Press.

Levin, B. R., and W. L. Kilmer. 1974. "Interdemic Selection and the Evolution of Altruism: A Computer Simulation Study." *Evolution* 28:527–545. doi:10.1111/j.1558-5646.1974.tb00787.x.

Lewin, C., and M. De Valois. 2003. "History of Actuarial Tables." In M. Campbell-Kelly, M. Croarken, R. Flood, and E. Robson, eds., *The History of Mathematical Tables: From Sumer to Spreadsheets*. Oxford: Oxford University Press, 79–104.

Lewis, T. 2005. "Altruism in Classical Buddhism." In J. Neusner and B. Chilton, eds., *Altruism in World Religions*, 88–114. Washington, DC: Georgetown University Press.

Liao, X., S. Rong, and D. C. Queller. 2015. "Relatedness, Conflict, and the Evolution of Eusociality." *PLoS Biology* 13:e1002098. doi:10.1371/journal.pbio.1002098.

Lickliter, R., and H. Honeycutt. 2013. "A Developmental Evolutionary Framework for Psychology." *Review of General Psychology* 17:184–189. doi:10.1037/a0032932.

Lieberman, D. 2009. "Rethinking the Taiwanese Minor Marriage Data: Evidence the Mind Uses Multiple Kinship Cues to Regulate Inbreeding Avoidance." *Evolution and Human Behavior* 30:153–160. doi:10.1016/j.evolhumbehav.2008.11.003.

Lieberman, D., and J. Billingsley. 2016. "Current Issues in Sibling Detection." *Current Opinion in Psychology* 7:57–60. doi:10.1016/j.copsyc.2015.07.014.

Lieberman, D., and T. Lobel. 2012. "Kinship on the Kibbutz: Coresidence Duration Predicts Altruism, Personal Sexual Aversions and Moral Attitudes Among Communally Reared Peers." *Evolution and Human Behavior* 33:26–34. doi:10.1016/j.evolhumbehav.2011.05.002.

Lieberman, D., and A. Smith. 2012. "It's All Relative: Sexual Aversions and Moral Judgments Regarding Sex Among Siblings." *Current Directions in Psychological Science* 21:243–247. doi:10.1177/0963721412447620.

Lieberman, D., J. Tooby, and L. Cosmides. 2007. "The Architecture of Human Kin Detection." *Nature* 445:727–731. doi:10.1038/nature05510.

Ligon, J. D. 1983. "Cooperation and Reciprocity in Avian Social Systems." *American Naturalist* 121:366–384. doi:10.1086/284066.

Lindert, P. H. 2004a. *Growing Public: Social Spending and Economic Growth Since the Eighteenth Century*. Vol. 1, *The Story*. Cambridge: Cambridge University Press.

———. 2004b. *Growing Public: Social Spending and Economic Growth Since the Eighteenth Century*. Vol. 2, *Further Evidence*. Cambridge: Cambridge University Press.

List, J. A. 2011. "The Market for Charitable Giving." *Journal of Economic Perspectives* 25:157–180. doi:10.1257/jep.25.2.157.

List, J. A., and Y. Peysakhovich. 2011. "Charitable Donations Are More Responsive to Stock Market Booms Than Busts." *Economics Letters* 110:166–169. doi:10.1016/j.econlet.2010.10.016.

Liu, F., F. van der Lijn, C. Schurmann, G. Zhu, M. M. Chakravarty, P. G. Hysi, A. Wollstein, et al. 2012. "A Genome-Wide Association Study Identifies Five Loci Influencing Facial Morphology in Europeans." *PLoS Genetics* 8 (9): e1002932. doi:10.1371/journal.pgen.1002932.

Liverani, M. 2006. *Uruk: The First City*. London: Equinox.

Lizzeri, A., and N. Persico. 2004. "Why Did the Elites Extend the Suffrage? Democracy and the Scope of Government, with an Application to Britain's 'Age of Reform.'" *Quarterly Journal of Economics* 119:707–765. doi:10.1162/0033553041382175.

Lobb, A., N. Mock, and P. L. Hutchinson. 2012. "Traditional and Social Media Coverage and Charitable Giving Following the 2010 Earthquake in Haiti." *Prehospital and Disaster Medicine* 27:319–324. doi:10.1017/S1049023X12000908.

Locke, J. 1764. *Two Treatises of Government*. London: A. Millar.

Loewenberg, F. M. 1994. "On the Development of Philanthropic Institutions in Ancient Judaism: Provisions for Poor Travelers." *Nonprofit and Voluntary Sector Quarterly* 23:193–207. doi:10.1177/089976409402300302.

———. 1995. "Financing Philanthropic Institutions in Biblical and Talmudic Times." *Nonprofit and Voluntary Sector Quarterly* 24:307–320. doi:10.1177/0899764 09502400404.

———. 2001. *From Charity to Social Justice: The Emergence of Communal Institutions for the Support of the Poor in Ancient Judaism*. New Brunswick, NJ: Transaction.

Lomborg, B. 2018. "Introduction." In B. Lomborg, ed., *Prioritizing Development: A Cost Benefit Analysis of the United Nations' Sustainable Development Goals*, 1–12. New York: Cambridge University Press.

Lopez, M. B., A. Hadid, E. Boutellaa, J. Goncalves, V. Kostakos, and S. Hosio. 2018. "Kinship Verification from Facial Images and Videos: Human Versus Machine." *Machine Vision and Applications* 29:873–890. doi:10.1007/s00138-018-0943-x.

Low, A. M. 1908. "Foreign Affairs: A Century of Constitutions." *The Forum* 40:205–212.

Lowe, K. 2012. *Savage Continent: Europe in the Aftermath of World War II*. New York: St. Martin's Press.

Lubove, R., ed. 1966. *Social Welfare in Transition: Selected English Documents, 1834–1909*. Pittsburgh: University of Pittsburgh Press.

Luskin, R. C., J. S. Fishkin, and S. Iyengar. 2004. *Considered Opinions on US Foreign Policy: Face-to-Face Versus Online Deliberative Polling*. Paper presented at the International Communication Association, New Orleans.

MacAskill, W. 2015. *Doing Good Better: Effective Altruism and How You Can Make a Difference*. New York: Penguin.

Malanima, P. 2011. "The Long Decline of a Leading Economy: GDP in Central and Northern Italy, 1300–1913." *European Review of Economic History* 15:169–219. doi:10.1017/S136149161000016X.

Malle, B. F. 2006. "The Actor-Observer Asymmetry in Attribution: A (Surprising) Meta-Analysis." *Psychological Bulletin* 132:895–919S. doi:10.1037/0033-2909 .132.6.895.

Malthus, T. (1803) 1992. *An Essay on the Principle of Population*, 2nd ed. D. Winch, ed. London: Ward, Lock.

Mandelbaum, M. 1982. "Vietnam: The Television War." *Daedalus* 111:157–169.

Manning, R., M. Levine, and A. Collins. 2007. "The Kitty Genovese Murder and the Social Psychology of Helping: The Parable of the 38 Witnesses." *American Psychologist* 62:555–562. doi:10.1037/0003-066X.62.6.555.

Marlowe, F. W. 2005. "Hunter-Gatherers and Human Evolution." *Evolutionary Anthropology* 14:54–67. doi:10.1002/evan.20046.

———. 2009. "Hadza Cooperation: Second-Party Punishment, Yes; Third-Party Punishment, No." *Human Nature* 20:417–430. doi:10.1007/s12110-009-9072-6.

Marr, D. 1982. *Vision*. San Francisco: W. H. Freeman.

Marshall, A. 1890. *Principles of Economics*. London: Macmillan.

Martin, A., and K. R. Olson. 2015. "Beyond Good and Evil: What Motivations Underlie Children's Prosocial Behavior?" *Perspectives on Psychological Science* 10:159–175. doi:10.1177/1745691615568998.

Martorano, B., L. Metzger, and M. Sanfilippo. 2018. *Chinese Development Assistance and Household Welfare in Sub-Saharan Africa*. AidData Working Paper No. 50. AidData at William and Mary. Williamsburg, VA.

Marty, R., C. B. Dolan, M. Leu, and D. Runfola. 2017. "Taking the Health Aid Debate to the Subnational Level: The Impact and Allocation of Foreign Health Aid in Malawi." *BMJ Global Health* 2 (1). doi:10.1136/bmjgh-2016-000129.

Marx, K. (1867) 2006. *Capital: A Critique of Political Economy*. London: Penguin.

Mateo, J. M. 2015. "Perspectives: Hamilton's Legacy: Mechanisms of Kin Recognition in Humans." *Ethology* 121:419–427. doi:10.1111/eth.12358.

Mayhew, H. (1851) 1985. *London Labour and the London Poor*. London: Penguin.

Maynard Smith, J. 1964. "Group Selection and Kin Selection." *Nature* 201:1145–1147. doi:10.1038/2011145a0.

———. 1976. "Group Selection." *Quarterly Review of Biology* 51:277–283. doi:doi.org/10.1086/409311.

Mbowa, S., T. Odokonyero, T. Muhumuza, and E. Munyambonera. 2017. "Does Coffee Production Reduce Poverty? Evidence from Uganda." *Journal of Agribusiness in Developing and Emerging Economies* 7:260–274. doi:10.1108/JADEE-01-2016-0004.

McArthur, J. W., and K. Rasmussen. 2018. "Change of Pace: Accelerations and Advances During the Millennium Development Goal Era." *World Development* 105:132–143. doi:10.1016/j.worlddev.2017.12.030.

McAuliffe, W. H. B., M. N. Burton-Chellew, and M. E. McCullough. 2019. "Cooperation and Learning in Unfamiliar Situations." *Current Directions in Psychological Science* 28:436–440. doi:10.1177/0963721419848673.

McAuliffe, W. H. B., E. C. Carter, J. Berhane, A. C. Snihur, and M. E. McCullough. 2019. "Is Empathy the Default Response to Suffering? A Meta-Analytic Evaluation of Perspective-Taking Instructions' Effects on Empathic Concern." *Personality and Social Psychology Review*. https://doi.org/10.1177/1088868319887599.

McAuliffe, W. H. B., D. E. Forster, J. Philippe, and M. E. McCullough. 2018. "Digital Altruists: Resolving Key Questions About the Empathy-Altruism Hypothesis in an Internet Sample." *Emotion* 18:493–506. doi:10.1037/emo0000375.

McCleary, R. M. 2009. *Global Compassion: Private Voluntary Organizations and U.S. Foreign Policy Since 1939*. New York: Oxford University Press.

McClendon, D. 2016. "Religion, Marriage Markets, and Assortative Mating in the United States." *Journal of Marriage and Family* 78:1399–1421. doi:10.1111/jomf.12353.

McIntosh, M. K. 2012. *Poor Relief in England, 1350–1600*. Cambridge: Cambridge University Press.

McPherson, M., L. Smith-Lovin, and J. M. Cook. 2001. "Birds of a Feather: Homophily in Social Networks." *Annual Review of Sociology* 27:415–444. doi:10.1146/annurev.soc.27.1.415.

Medrano, J. D., and M. Braun. 2012. "Uninformed Citizens and Support for Free Trade." *Review of International Political Economy* 19:448–476. doi:10.1080/09692290.2011.561127.

Meerkerk, E. V. N. 2012. "The Will to Give: Charitable Bequests, Inter Vivos Gifts and Community Building in the Dutch Republic, c. 1600–1800." *Continuity and Change* 27:241–270. doi:10.1017/S0268416012000124.

Meerkerk, E. V. N., and D. Teeuwen. 2014. "The Stability of Voluntarism: Financing Social Care in Early Modern Dutch Towns Compared with the English Poor Law, c. 1600–1800." *European Review of Economic History* 18:82–105. doi:10.1093/ereh/het014.

Meier, P. 2015. *Digital Humanitarians: How BIG DATA Is Changing the Face of Humanitarian Response*. Boca Raton, FL: CRC Press.

Mendonça, D., I. Amorim, and M. Kagohara. 2019. "An Historical Perspective on Community Resilience: The Case of the 1755 Lisbon Earthquake." *International Journal of Disaster Risk Reduction* 34:363–374. doi:10.1016/j.ijdrr.2018.12.006.

Mercier, H. 2011. "What Good Is Moral Reasoning?" *Mind and Society* 10:131–148. doi:10.1007/s11299-011-0085-6.

Mercier, H., and D. Sperber. 2017. *The Enigma of Reason*. Cambridge, MA: Harvard University Press.

Meyer, M. L., C. L. Masten, Y. Ma, C. Wang, Z. Shi, N. I. Eisenberger, and S. Han. 2013. "Empathy for the Social Suffering of Friends and Strangers Recruits Distinct Patterns of Brain Activation." *Social Cognitive and Affective Neuroscience* 8:446–454. doi:10.1093/scan/nss019.

Michel, J.-B., Y. K. Shen, A. P. Aiden, A. Veres, M. K. Gray, J. P. Pickett, Google Books Team, et al. 2011. "Quantitative Analysis of Culture Using Millions of Digitized Books." *Science* 331:176–182. doi:10.1126/science.1199644.

Miguel, E., and M. Kremer. 2004. "Worms: Identifying Impacts on Education and Health in the Presence of Externalities." *Econometrica* 72:159–217. doi:10.1111/j.1468-0262.2004.00481.x.

Mill, J. S. 1863. *Utilitarianism*. London: Parker, Son and Bourn.

Miller, J. B., and A. Sanjurjo. 2015. *Is It a Fallacy to Believe in the Hot Hand in the NBA Three-Point Contest?* Working Paper No. 548. Innocenzo Gasparini Institute for Economic Research.

———. 2018. "Surprised by the Gambler's and Hot Hand Fallacies? A Truth in the Law of Small Numbers." *Econometrica* 86:2019–2047. doi:10.2139/ssrn.2627354.

Mirakhor, A., and H. Askari. 2019. *Conceptions of Justice from Earliest History to Islam*. New York: Palgrave Macmillan.

Missiakoulis, S. 2008. "Aristotle and Earthquake Data: A Historical Note." *International Statistical Review* 76:130–133. doi:10.111 1/j.1751-5823.2007.00040.

Mittelman, R., and L. C. Neilson. 2011. "Development Porn? Child Sponsorship Advertisements in the 1970s." *Journal of Historical Research in Marketing* 3:370–401. doi:10.1108/17557501111157788.

Molesky, M. 2015. *This Gulf of Fire: The Destruction of Lisbon, or Apocalypse in the Age of Science and Reason*. New York: Alfred A. Knopf.

Molleman, L., E. van den Broek, and M. Egas. 2013. "Personal Experience and Reputation Interact in Human Decisions to Help Reciprocally." *Proceedings of the Royal Society B: Biological Sciences* 280 (1757). doi:10.1098/rspb.2012.3044.

Momigliano, A. 1975. *Alien Wisdom: The Limits of Hellenization*. Cambridge: Cambridge University Press.

Montesquieu, C. (1748) 1952. "The Spirit of Laws." T. Nugent and J. V. Prichard, trans. In R. M. Hutchins, ed., *Great Books of the Western World*. Vol. 38, *Montesquieu/Rousseau*, 1–315. Chicago: Encyclopedia Britannica.

More, Sir Thomas. (1516) 1753. *Utopia: Containing an Impartial History of the Manners, Customs, Polity, Government, etc. of That Island*. G. Barnet, trans. London: T. Carnan.

Morgan, G. 2017. "2016 Costs of School-Based Deworming: A Best-Buy Development Intervention." Evidence Action, www.evidenceaction.org/blog-full/deworm-cost-per-child.

Morris, D. 1996. *A Gift from America: The First 50 Years of CARE*. Atlanta: Longstreet Press.

Morschauser, S. N. 1995. "The Ideological Basis for Social Justice / Responsibility in Ancient Egypt." In K. D. Irani and M. Silver, eds., *Social Justice in the Ancient World*, 101–113. Westport, CT: Greenwood Press.

Moyo, D. 2009. *Dead Aid: Why Aid Is Not Working and How There Is a Better Way for Africa*. New York: Farrar, Straus and Giroux.

Mukherji, P. 2006. "The Indian Influence on Chinese Literature: A Folk Literary Perspective." In P. A. George, ed., *East Asian Literatures (Japanese, Chinese and Korean): An Interface with India*, 183–190. New Delhi: Northern Book Centre.

Mullins, D., D. Hoyer, C. Collins, T. Currie, K. Feeney, P. François, P. Savage, H. Whitehouse, and P. Turchin. 2018. "A Systematic Assessment of 'Axial Age' Proposals Using Global Comparative Historical Evidence." *American Sociological Review* 83:596–626. doi:10.1177/0003122418772567.

Munz, K., M. Jung, and A. Alter. 2018. *Name Similarity Encourages Generosity: A Field Experiment in Email Personalization*. Manuscript. Department of Marketing, Leonard N. Stern School of Business, New York University.

Nan, C., B. Guo, C. Warner, T. Fowler, T. Barrett, D. Boomsma, T. Nelson, et al. 2012. "Heritability of Body Mass Index in Pre-adolescence, Young Adulthood and Late Adulthood." *European Journal of Epidemiology* 27:247–253. doi:10.1007/s10654-012-9678-6.

Neel, J. V., and K. M. Weiss. 1975. "The Genetic Structure of a Tribal Population, the Yanomama Indians: XII. Biodemographic Studies." *American Journal of Physical Anthropology* 42:25–52. doi:10.1002/ajpa.1330420105.

Neiman, S. 2002. *Evil in Modern Thought: An Alternative History of Philosophy*. Princeton, NJ: Princeton University Press.

Neusner, J., and A. J. Avery-Peck. 2005. "Altruism in Classical Judaism." In J. Neusner and B. Chilton, eds., *Altruism in World Religions*, 31–52. Washington, DC: Georgetown University Press.

Neusner, J., and B. Chilton, eds. 2005. *Altruism in World Religions*. Washington, DC: Georgetown University Press.

———, eds. 2008. *The Golden Rule: The Ethics of Reciprocity in World Religions*. New York: Continuum.

Newman, B. J., T. K. Hartman, and C. S. Taber. 2012. "Foreign Language Exposure, Cultural Threat, and Opposition to Immigration." *Political Psychology* 33:635–657. doi:10.1111/j.1467-9221.2012.00904.x.

Nichols, R. 2014. "Re-evaluating the Effects of the 1755 Lisbon Earthquake on Eighteenth-Century Minds: How Cognitive Science of Religion Improves Intellectual History with Hypothesis Testing Methods." *Journal of the American Academy of Religion* 82:970–1009. doi:10.1093/jaarel/lfu033.

Nordheimer, J. 1976. "Reagan Is Picking His Florida Spots: His Campaign Aides Aim for New G.O.P. Voters in Strategic Areas." *New York Times*, February 5.

Nowak, M. A., and R. M. May. 1992. "Evolutionary Games and Spatial Chaos." *Nature* 359:826–829. doi:10.1038/359826a0.

Nowak, M. A., and K. Sigmund. 1992. "Tit for Tat in Heterogeneous Populations." *Nature* 355:250–252. doi:10.1038/355250a0.

———. 1993. "A Strategy of Win-Stay, Lose-Shift That Outperforms Tit-for-Tat in the Prisoner's Dilemma Game." *Nature* 364:56–58. doi:10.1038/364056a0.

———. 1994. "The Alternating Prisoner's Dilemma." *Journal of Theoretical Biology* 168:219–226. doi:10.1006/jtbi.1994.1101.

Nowak, M. A., C. E. Tarnita, and E. O. Wilson. 2010. "The Evolution of Eusociality." *Nature* 466:1057–1062. doi:10.1038/nature09205.

Nullmeier, F., and F. Kaufman. 2010. "Post-War Welfare State Development." In F. G. Castles, S. Liebfried, J. Lewis, H. Obinger, and C. Pierson, eds., *The Oxford Handbook of the Welfare State*, 81–101. Oxford: Oxford University Press.

Oates, K., and M. Wilson. 2001. "Nominal Kinship Cues Facilitate Altruism." *Proceedings of the Royal Society B: Biological Sciences* 269:105–109. doi:10.1098/rspb.2001.1875.

Ober, J. 2015. *The Rise and Fall of Classical Greece.* Princeton, NJ: Princeton University Press.

Obocock, P. 2008. "Introduction: Vagrancy and Homelessness in Global and Historical Perspective." In A. L. Beier and P. Obocock, eds., *Cast Out: Vagrancy and Homelessness in Global and Historical Perspective*, 1–34. Athens: Ohio University Press.

Odokonyero, T., R. Marty, T. Muhumuza, A. T. Ijjo, and G. Owot Moses. 2018. "The Impact of Aid on Health Outcomes in Uganda." *Health Economics* 27:733–745. doi:10.1002/hec.3632.

Oftedal, O. T. 2002. "The Mammary Gland and Its Origin During Synapsid Evolution." *Journal of Mammary Gland Biology and Neoplasia* 7:225–252. doi:1083-3021/02/0700-0225/0.

———. 2012. "The Evolution of Milk Secretion and Its Ancient Origins." *Animal* 6:355–368. doi:10.1017/S1751731111001935.

Okasha, S. 2007. *Evolution and the Levels of Selection.* New York: Oxford University Press.

"The 169 Commandments: The Proposed Sustainable Development Goals Would Be Worse Than Useless." 2015. *The Economist*, March 28, www.economist.com/leaders/2015/03/26/the-169-commandments.

O'Neill, O. 1975. "Lifeboat Earth." *Philosophy and Public Affairs* 4:273–292.

Oppenheimer, M. 2013. In Big-Dollar Philanthropy, (Your Name Here) vs. Anonymity. *New York Times*, May 11, A20.

Organization for Economic Cooperation and Development (OECD). 1985. "Social Expenditure, 1960–1990: Problems of Growth and Control." *Journal of Public Policy* 5:133–168. doi:10.1017/S0143814X00003007.

———. 2016. *Social Expenditure Update 2016: Social Spending Stays at Historically High Levels in Many OECD Countries*, www.oecd.org/els/soc/OECD2016-Social-Expenditure-Update.pdf.

———. 2018a. "The 0.7% ODA/GNI Target—A History," www.oecd.org/dac/stats/the07odagnitarget-ahistory.htm.

———. 2018b. *Education at a Glance 2018: OECD Indicators*, www.oecd-ilibrary.org/content/publication/eag-2018-en.
———. 2019a. "Grants by Private Agencies and NGOs (Indicator)," https://data.oecd.org/drf/grants-by-private-agencies-and-ngos.htm#indicator-chart.
———. 2019b. "Net ODA (Indicator)," https://data.oecd.org/oda/net-oda.htm.
———. 2019c. *Social Expenditure Update 2019: Public Social Spending Is High in Many OECD Countries*, www.oecd.org/els/soc/OECD2019-Social-Expenditure-Update.pdf.
———. 2019d. "Social Expenditure: Aggregated Data" (Publication no. 10.1787/data-00166-en). OECD Social and Welfare Statistics, www.oecd-ilibrary.org/content/data/data-00166-en.

Ortiz-Ospina, E., and M. Roser. 2019. "Government Spending," https://ourworldindata.org/government-spending.

Oster, E., and R. Thornton. 2011. "Menstruation, Sanitary Products, and School Attendance: Evidence from a Randomized Evaluation." *American Economic Journal: Applied Economics* 3:91–100. doi:10.1257/app.3.1.91.

Ostrom, E. 2014. "Do Institutions for Collective Action Evolve?" *Journal of Bioeconomics* 16:3–30. doi:10.1007/s10818-013-9154-8.

Pabalan, N., E. Singian, L. Tabangay, H. Jarjanazi, M. J. Boivin, and A. E. Ezeamama. 2018. "Soil-Transmitted Helminth Infection, Loss of Education and Cognitive Impairment in School-Aged Children: A Systematic Review and Meta-Analysis." *PLoS Neglected Tropical Diseases* 12 (1). doi:10.1371/journal.pntd.0005523.

Paine, T. (1792) 1817. *The Rights of Man*, vol. 2. London: W. T. Sherwin.

Paley, W. 1840. *The Works of William Paley, D.D.: Archdeacon of Carlisle*. Edinburgh: Thomas Nelson.

Palma, N., and J. Reis. 2019. "From Convergence to Divergence: Portuguese Economic Growth, 1527–1850." *Journal of Economic History* 79:477–506. doi:10.1017/S0022050719000056.

Palmer, C. 1989. "Is Rape a Cultural Universal? A Re-examination of the Ethnographic Data." *Ethnology* 28:1–16. doi:10.2307/3773639.

Papademetriou, D. G., and N. Banulescu-Bogdan. 2016. *Understanding and Addressing Public Anxiety About Immigration*. Washington, DC: Migration Policy Institute, www.immigrationresearch.org/system/files/TCM_Trust_CouncilStatement-FINAL.pdf.

Park, J. H., and M. Schaller. 2005. "Does Attitude Similarity Serve as a Heuristic Cue for Kinship? Evidence of an Implicit Cognitive Association." *Evolution and Human Behavior* 26:158–170. doi:10.1016/j.evolhumbehav.2004.08.013.

Park, J. H., M. Schaller, and M. Van Vugt. 2008. "Psychology of Human Kin Recognition: Heuristic Cues, Erroneous Inferences, and Their Implications." *Review of General Psychology* 12:215–235. doi:10.1037/1089-2680.12.3.215.

Peich, R. M., M. T. Pastorino, and D. H. Zald. 2010. "All I Saw Was the Cake. Hunger Effects on Attentional Capture by Visual Food Cues." *Appetite* 54:579–582. doi:10.1016/j.appet.2009.11.003.

Penn, D. J., and J. G. Frommen. 2010. "Kin Recognition: An Overview of Conceptual Issues, Mechanisms and Evolutionary Theory." In P. Kappeler, ed., *Animal Behaviour: Evolution and Mechanisms*, 55–85. Heidelberg: Springer.

Pereira, A. S. 2009. "The Opportunity of a Disaster: The Economic Impact of the 1755 Lisbon Earthquake." *Journal of Economic History* 69:466–499. doi:10.1017/S0022050709000850.

Petersen, M. B. 2012. "Social Welfare as Small-Scale Help: Evolutionary Psychology and the Deservingness Heuristic." *American Journal of Political Science* 56:1–16. doi:10.1111/j.1540-5907.2011.00545.x.

Phillips, A., and B. Taylor. 2010. *On Kindness*. New York: Penguin.

Philo, G. 1993. "From Buerk to Band Aid: The Media and the 1984 Ethiopian Famine." In J. Eldridge, ed., *Getting the Message: News, Truth and Power*, 104–125. London: Routledge.

Pinker, S. 1997. *How the Mind Works*. New York: W. W. Norton.

———. 2002. *The Blank Slate: The Modern Denial of Human Nature*. New York: Penguin.

———. 2012. "The False Allure of Group Selection." *Edge*, June 18, www.edge.org/conversation/steven_pinker-the-false-allure-of-group-selection.

———. 2018. *Enlightenment Now: The Case for Reason, Science, Humanism, and Progress*. New York: Viking.

Porter, M. E., and M. R. Kramer. 1999. "Philanthropy's New Agenda: Creating Value." *Harvard Business Review* 77:121–131.

Porter, R. H., and J. D. Moore. 1981. "Human Kin Recognition by Olfactory Cues." *Physiology and Behavior* 27:493–495. doi:10.1016/0031-9384(81)90337-1.

Pound, J. 1971. *Poverty and Vagrancy in Tudor England*. London: Longman.

Poundstone, W. 1992. *Prisoner's Dilemma: John Von Neumann, Game Theory, and the Puzzle of the Bomb*. New York: Doubleday.

Power, M. L., and J. Schulkin. 2016. *Milk: The Biology of Lactation*. Baltimore: Johns Hopkins University Press.

Price, G. R. 1970. "Selection and Covariance." *Nature* 227:520–521. doi:10.1038/227520a0.

———. 1972. "Extension of Covariance Selection Mathematics." *Annals of Human Genetics* 36:485–490. doi:10.1111/j.1469-1809.1957.tb01874.x.

Prinz, J. 2006. "The Emotional Basis of Moral Judgments." *Philosophical Explorations* 9:29–43. doi:10.1080/13869790500492466.

Puryear, C., and S. Reysen. 2013. "A Preliminary Examination of Cell Phone Use and Helping Behavior." *Psychological Reports* 113:1001–1003. doi:10.2466/17.21.PR0.113x31z4.

Qian, N. 2015. "Making Progress on Foreign Aid." *Annual Review of Economics* 7:277–308. doi:10.1146/annurev-economics-080614-115553.

Qirko, H. N. 2011. "Fictive Kinship and Induced Altruism." In C. A. Salmon and T. K. Shackelford, eds., *The Oxford Handbook of Evolutionary Family Psychology*, 310–328. New York: Oxford University Press.

———. 2013. "Induced Altruism in Religious, Military, and Terrorist Organizations." *Cross-Cultural Research* 47:131–161. doi:10.1177/1069397112471804.

Rand, D. G., and M. A. Nowak. 2013. "Human Cooperation." *TRENDS in Cognitive Sciences* 17:413–425. doi:10.1016/j.tics.2013.06.003.

Rapoport, A., and A. M. Chammah. 1965. *Prisoner's Dilemma: A Study in Conflict and Cooperation*. Ann Arbor: University of Michigan Press.

Ravallion, M. 2011. "The Two Poverty Enlightenments: Historical Insights from

Digitized Books Spanning Three Centuries." *Poverty and Public Policy* 3:1–46. doi:10.2202/1944-2858.1173.

———. 2015. "The Idea of Antipoverty Policy." In A. B. Atkinson and F. Bourguignon, eds., *Handbook of Income Distribution*, vol. 2B, 1967–2061. Oxford: Elsevier.

———. 2016. *The Economics of Poverty*. New York: Oxford University Press.

———. 2019. *Should the Randomistas (Continue to) Rule?* Washington, DC: Center for Global Development.

Rawls, J. 1971. *A Theory of Justice*. Cambridge, MA: Harvard University Press.

———. 1999. *The Law of Peoples, with the Idea of Public Reason Revisited*. Cambridge, MA: Harvard University Press.

Reagan, R. 1981. "Remarks in an Interview with Managing Editors on Domestic Issues," December 3, Ronald Reagan Presidential Library and Museum. www.reaganlibrary.gov/research/speeches/120381e.

Rees, M. 2000. *Just Six Numbers: The Deep Forces That Shape the Universe*. New York: Basic Books.

Reinhardt, O., and D. R. Oldroyd. 1983. "Kant's Theory of Earthquakes and Volcanic Action." *Annals of Science* 40:247–272. doi:10.1080/00033798300200221.

Reiter, J. G., C. Hilbe, D. G. Rand, K. Chatterjee, and M. A. Nowak. 2018. "Crosstalk in Concurrent Repeated Games Impedes Direct Reciprocity and Requires Stronger Levels of Forgiveness." *Nature Communications* 9. doi:10.1038/s41467-017-02721-8.

Renwick, C. 2017. *Bread for All: The Origins of the Welfare State*. London: Penguin.

Rho, S., and M. Tomz. 2017. "Why Don't Trade Preferences Reflect Economic Self-Interest?" *International Organization* 71:S85–S108. doi:10.1017/S0020818316000394.

Ricardo, D. 1817. *Principles of Political Economy and Taxation*. London: John Murray.

Richardson, G. 2005. "The Prudent Village: Risk Pooling Institutions in Medieval English Agriculture." *Journal of Economic History* 65:316–413. doi:10.1017/S0022050705000136.

Richardson, R. C. 2007. *Evolutionary Psychology as Maladapted Psychology*. Cambridge, MA: MIT Press.

Richardson, S. 2016. "Obedient Bellies: Hunger and Food Security in Ancient Mesopotamia." *Journal of the Economic and Social History of the Orient* 59:750–792. doi:10.1163/15685209-12341413.

Riddell, R. C. 2007. *Does Foreign Aid Really Work?* New York: Oxford University Press.

Rifkin, J. 2009. *The Empathic Civilization: The Race to Global Consciousness in a World in Crisis*. New York: Jeremy P. Tarcher / Penguin.

Rigaud, K. K., A. de Sherbinin, B. Jones, J. Bergmann, V. Clement, K. Ober, J. Schewe, et al. 2018. *Groundswell: Preparing for Internal Climate Migration*. Washington, DC: World Bank, available at https://openknowledge.worldbank.org/handle/10986/29461.

Riis, J. A. 1890. *How the Other Half Lives: Studies Among the Tenements of New York*. New York: Charles Scribner's Sons.

Ringen, E. J., P. Duda, and A. V. Jaeggi. 2019. "The Evolution of Daily Food Sharing: A Bayesian Phylogenetic Analysis." *Evolution and Human Behavior* 40:375–384. doi:10.1016/j.evolhumbehav.2019.04.003.

Roberts, S. C., L. M. Gosling, T. D. Spector, P. Miller, D. J. Penn, and M. Petrie. 2005. "Body Odor Similarity in Noncohabiting Twins." *Chemical Senses* 30:651–656. doi:10.1093/chemse/bji058.

Robins, W. 1985. "Voices Behind an Anti-Hunger Blitz." *Boston Globe*, April 28, B27.

Rodgers, B. (1968) 2006. *The Battle Against Poverty: From Pauperism to Human Rights*. Oxford: Routledge.

Röer, J. P., R. Bell, and A. Buchner. 2013. "Self-Relevance Increases the Irrelevant Sound Effect: Attentional Disruption by One's Own Name." *Journal of Cognitive Psychology* 25:925–931. doi:10.1080/20445911.2013.828063.

Rosenberg, D. 2008. "Serving Meals Making a Home: The PPNA Limestone Vessel Industry of the Southern Levant and Its Importance to the Neolithic Revolution." *Paléorient* 34:23–32. doi:10.3406/paleo.2008.5231.

Rosenfeld, M. J., and R. J. Thomas. 2012. "Searching for a Mate: The Rise of the Internet as a Social Intermediary." *American Sociological Review* 77:523–547. doi:10.1177/0003122412448050.

Rosenthal, A. M. (1964) 1999. *Thirty-Eight Witnesses: The Kitty Genovese Case*. New York: Melville House.

Roser, M., and E. Ortiz-Ospina. 2019. "Financing Education." Our World in Data, https://ourworldindata.org/financing-education.

Ross, I. S. 2010. *The Life of Adam Smith*, 2nd ed. Oxford: Oxford University Press.

Rottman, J., L. Zhu, W. Wang, R. Seston Schillaci, K. J. Clark, and D. Kelemen. 2017. "Cultural Influences on the Teleological Stance: Evidence from China." *Religion, Brain and Behavior* 7:17–26. doi:10.1080/2153599X.2015.1118402.

Rousseau, J.-J. (1754) 1952. "A Discourse on the Origin of Inequality." G. D. H. Cole, trans. In R. M. Hutchins, ed., *Great Books of the Western World*. Vol. 38, *Montesquieu/Rousseau*, 323–366. Chicago: Encyclopedia Britannica.

———. (1755) 1952. "A Discourse on Political Economy." G. D. H. Cole, trans. In R. M. Hutchins, ed., *Great Books of the Western World*. vol. 38, *Montesquieu/Rousseau*, 367–385. Chicago: Encyclopedia Britannica.

Rowntree, B. S. 1901. *Poverty: A Study of Town Life*. London: Macmillan.

Rubenstein, J. C. 2016. "The Lessons of Effective Altruism." *Ethics and International Affairs* 30:511–526. doi:10.1017/S0892679416000484.

Rubin, P. 2003. "Folk Economics." *Southern Economic Journal* 70:157–171. doi:10.2307/1061637.

Rusch, H., and E. Voland. 2016. "Human Agricultural Economy Is, and Likely Always Was, Largely Based on Kinship—Why?" *Behavioral and Brain Sciences* 39:e112. doi:10.1017/S0140525X15001168.

Rushton, N. S., and W. Sigle-Rushton. 2001. "Monastic Poor Relief in Sixteenth-Century England." *Journal of Interdisciplinary History* 32:193–216. doi:10.1162/002219501750442378.

Russell, J. A., and G. Leng. 1998. "Sex, Parturition, and Motherhood Without Oxytocin?" *Journal of Endocrinology* 157:343–359. doi:10.1677/joe.0.1570343.

Russo, J. P. 1999. "The Sicilian Latifundia." *Italian Americana* 17:40–57.

Sachs, J. D. 2005. *The End of Poverty: Economic Possibilities for Our Time*. New York: Penguin.

Sadler, M. 2004. *Representative Sadleriana: Sir Michael Sadler (1861–1943) on English, French, German and American School and Society*. J. Sislian, ed. New York: Nova Science Publishers.

Sallin, R. (producer), and N. Meyer (director). 1982. *Star Trek II: The Wrath of Khan*. Paramount Pictures.

Sally, D. 1995. "Conversation and Cooperation in Social Dilemmas: A Meta-Analysis of Experiments from 1958 to 1992." *Rationality and Society* 7:58–92. doi:10.1177/1043463195007001004.

Sanderson, S. K. 2018. *Religious Evolution and the Axial Age: From Shamans to Priests to Prophets*. London: Bloomsbury Academic.

Sanuto, M. 1897. *I diarii di Marino Sanuto*, vol. 47. Venice: Visentini Cav Federico.

Scelza, B. A. 2011. "Female Choice and Extra-Pair Paternity in a Traditional Human Population." *Biology Letters* 7:889–891. doi:10.1098/rsbl.2011.0478.

Scheve, K., and D. Stasavage. 2016. *Taxing the Rich: A History of Fiscal Fairness in the United States and Europe*. Princeton, NJ: Princeton University Press.

Schino, G., and F. Aureli. 2007. "Grooming Reciprocation Among Female Primates: A Meta-Analysis." *Biology Letters* 4:9–11. doi:10.1098/rsbl.2007.0506.

———. 2010. "The Relative Roles of Kinship and Reciprocity in Explaining Primate Altruism." *Ecology Letters* 13:45–50. doi:10.1111/j.1461-0248.2009.01396.x.

Secter, B., and W. Gaines. 1999. "Daley Inc.: Patronage Blooms on the Family Tree." *Chicago Tribune*, June 13.

Sedgwick, S. 2008. *Kant's Groundwork of the Metaphysics of Morals: An Introduction*. Cambridge: Cambridge University Press.

Segerstrale, U. 2013. *Nature's Oracle: The Life and Work of W. D. Hamilton*. Oxford: Oxford University Press.

Sen, A. 1981. *Poverty and Famines: An Essay on Entitlement and Deprivation*. New York: Oxford University Press.

Sharp, S. P., A. McGowan, M. J. Wood, and B. J. Hatchwell. 2005. "Learned Kin Recognition Cues in a Social Bird." *Nature* 434:1127–1130. doi:10.1038/nature03522.

Shirky, C. 2010. *Cognitive Surplus: Creativity and Generosity in a Connected Age*. New York: Penguin.

Shrady, N. 2008. *The Last Day: Wrath, Ruin, and Reason in the Great Lisbon Earthquake of 1755*. New York: Viking.

Sibley, C. G., and J. Bulbulia. 2012. "Faith After an Earthquake: A Longitudinal Study of Religion and Perceived Health Before and After the 2011 Christchurch New Zealand Earthquake." *PLoS One* 7 (12). doi:10.1371/journal.pone.0049648.

Silventoinen, K., S. Sammalisto, M. Perola, D. I. Boomsma, B. K. Cornes, C. Davis, L. Dunka, et al. 2003. "Heritability of Adult Body Height: A Comparative Study of Twin Cohorts in Eight Countries." *Twin Research and Human Genetics* 6:399–408. doi:10.1375/twin.6.5.399.

Simon, A. F. 1997. "Television News and International Earthquake Relief." *Journal of Communication* 47:82–93. doi:10.1111/j.1460-2466.1997.tb02718.x.

Simons, D. J., and C. F. Chabris. 1999. "Gorillas in Our Midst: Sustained Inattentional Blindness for Dynamic Events." *Perception* 28:1059–1074. doi:10.1068/p281059.

Singer, P. 1972a. "Famine, Affluence, and Morality." *Philosophy and Public Affairs* 1:229–243.

———. 1972b. "Moral Experts." *Analysis* 32:115–117. doi:10.2307/3327906.

———. ed. 1986. *Applied Ethics*. New York: Oxford University Press.

———. 2009. *The Life You Can Save: Acting Now to End World Poverty*. New York: Random House.

———. 2010. *The Life You Can Save: How to Do Your Part to End World Poverty*. New York: Random House.

———. 2015. *The Most Good You Can Do: How Effective Altruism Is Changing Ideas About Living Ethically*. New Haven, CT: Yale University Press.

Slack, P. 1988. *Poverty and Policy in Tudor and Stuart England*. London: Longman.

Slote, M. 2010. *Moral Sentimentalism*. New York: Oxford University Press.

Smart, J. J. C., and B. Williams. 1973. *Utilitarianism: For and Against*. Cambridge: Cambridge University Press.

Smith, A. (1759) 1984. *The Theory of Moral Sentiments*, 6th ed. Indianapolis: Liberty Fund.

———. (1776) 1952. "An Inquiry into the Nature and Causes of the Wealth of Nations." In R. M. Hutchins, ed., *Great Books of the Western World*. Vol. 39, *Adam Smith*, 1–468. Chicago: Encyclopedia Britannica.

Smith, E. A., M. B. Mulder, S. Bowles, M. Gurven, T. Hertz, and M. Shenk. 2010. "Production Systems, Inheritance, and Inequality in Premodern Societies." *Current Anthropology* 51:85–94. doi:10.1086/649029.

Snow, M. 2015. "Against Charity." *Jacobin*, August 25, www.jacobinmag.com/2015/08/peter-singer-charity-effective-altruism.

Sproull, L., C. Conley, and J. Y. Moon. 2005. "Prosocial Behavior on the Net." In Y. Amichai-Hamburger, ed., *The Social Net: Understanding Human Behavior in Cyberspace*, 139–161. Oxford: Oxford University Press.

Statistics and Clinical Studies of NHS Blood and Transplant. 2018. *Organ Donation and Transplantation: Activity Report 2017/2018*, National Health Service, UK, https://nhsbtdbe.blob.core.windows.net/umbraco-assets-corp/12300/transplant-activity-report-2017-2018.pdf.

Sterelny, K. 2015. "Optimizing Engines: Rational Choice in the Neolithic?" *Philosophy of Science* 82:402–423. doi:10.1086/681602.

Stiglitz, J. E., N. Stern, M. Duan, O. Edenhofer, G. Giraud, G. M. Heal, E. Lèbre la Rovere, et al. 2017. *Report of the High-Level Commission on Carbon Prices*. Washington, DC: Carbon Pricing Leadership Coalition, www.carbonpricingleadership.org/report-of-the-highlevel-commission-on-carbon-prices.

Stott, R. 2012. *Darwin's Ghosts: The Secret History of Evolution*. New York: Spiegel and Grau.

Strassmann, J. E., O. M. Gilbert, and D. C. Queller. 2011. "Kin Discrimination and Cooperation in Microbes." *Annual Review of Microbiology* 65:349–367. doi:10.1146/annurev.micro.112408.134109.

Strickland, L. 2019. "Staying Optimistic: The Trials and Tribulations of Leibnizian Optimism." *Journal of Modern Philosophy* 1:1–21. doi:10.32881/jomp.3.

Strömberg, D. 2007. "Natural Disasters, Economic Development, and Humanitarian Aid." *Journal of Economic Perspectives* 21:199–222. doi:10.1257/jep.21.3.199.

Summers, L. 2016. "Donald Trump Is a Serious Threat to American Democracy." *Washington Post*, March 1.

Swedberg, R. 2018. "Folk Economics and Its Role in Trump's Presidential Campaign: An Exploratory Study." *Theory and Society* 47:1–36. doi:10.1007/s11186-018-9308-8.

Sznycer, D., A. W. Delton, T. E. Robertson, L. Cosmides, and J. Tooby. 2019. "The Ecological Rationality of Helping Others: Potential Helpers Integrate Cues of Recipients' Need and Willingness to Sacrifice." *Evolution and Human Behavior* 40:34–45. doi:10.1016/j.evolhumbehav.2018.07.005.

Sznycer, D., D. De Smet, J. Billingsley, and D. Lieberman. 2016. "Coresidence Duration and Cues of Maternal Investment Regulate Sibling Altruism Across Cultures." *Journal of Personality and Social Psychology* 111:159–177. doi:10.1037/pspi0000057.

Tacikowski, P., H. B. Cygan, and A. Nowicka. 2014. "Neural Correlates of Own and Close-Other's Name Recognition: ERP Evidence." *Frontiers in Human Neuroscience* 8. doi:10.3389/fnhum.2014.00194.

Tal, I., and D. Lieberman. 2007. "Kin Detection and the Development of Sexual Aversions: Toward an Integration of Theories of Family Sexual Abuse." In C. A. Salmon and T. K. Shackelford, eds., *Family Relationships: An Evolutionary Perspective*, 205–229. New York: Oxford University Press.

Tanzi, V., and L. Schuknecht. 2000. *Public Spending in the 20th Century: A Global Perspective*. New York: Cambridge University Press.

Thomson, D. 1947. *The Babeuf Plot: The Making of a Republican Legend*. London: Kegan Paul, Trench, Trubner.

Tierney, B. 1959. *Medieval Poor Law: A Sketch of Canonical Theory and Its Application in England*. Berkeley: University of California Press.

Tinniswood, A. 2004. *By Permission of Heaven: The True Story of the Great Fire of London*. New York: Riverhead Books.

Tomkins, A. 2006. *The Experience of Urban Poverty, 1723–1782: Parish, Charity and Credit*. Manchester: Manchester University Press.

Tournoy, G. 2004. "Towards the Roots of Social Welfare." *City* 8:266–273. doi:10.1080/1360481042000242229.

Townsend, J. (1786) 1971. *A Dissertation on the Poor Laws, by a Well-Wisher to Mankind*. Berkeley: University of California Press.

Toynbee, A. 1887. *Lectures on the Industrial Revolution of the 18th Century in England*, 2nd ed. London: Covington.

Trachtman, H., A. Steinkruger, M. Wood, A. Wooster, J. Andreoni, J. J. Murphy, and J. M. Rao. 2015. "Fair Weather Avoidance: Unpacking the Costs and Benefits of 'Avoiding the Ask.'" *Journal of the Economic Science Association* 1:8–14. doi:10.1007/s40881-015-0006-2.

Travill, A. A. 1987. "Juan Luis Vives: The De Subventione Pauperum." *Canadian Bulletin of Medical History* 4:165–181. doi:10.3138/cbmh.4.2.165.

Treisman, D. 2007. "What Have We Learned About the Causes of Corruption from Ten Years of Cross-National Empirical Research?" *Annual Review of Political Science* 10:211–244. doi:10.1146/annurev.polisci.10.081205.095418.

Trivers, R. L. 1971. "The Evolution of Reciprocal Altruism." *Quarterly Review of Biology* 46:35–57. doi:10.1086/406755.

Truman, H. S. 1949. "Inaugural Address." Harry S. Truman Library and Museum, www.trumanlibrary.gov/library/public-papers/19/inaugural-address.

Trump, D. 2015. "Remarks Announcing Candidacy for President in New York City." American Presidency Project, University of California Santa Barbara, www

.presidency.ucsb.edu/documents/remarks-announcing-candidacy-for-president-new-york-city.

Turk, D. J., K. van Bussel, J. L. Brebner, A. S. Toma, O. Krigolson, and T. C. Handy. 2011. "When 'It' Becomes 'Mine': Attentional Biases Triggered by Object Ownership." *Journal of Cognitive Neuroscience* 23:3725–3733. doi:10.1162/jocn_a_00101.

United Nations. 1947. *Yearbook of the United Nations, 1946–1947.* Lake Success, NY: United Nations Publications.

United Nations Department of Economic and Social Affairs. 2019. *The Sustainable Development Goals Report, 2019.* New York: United Nations, https://unstats.un.org/sdgs/report/2019/The-Sustainable-Development-Goals-Report-2019.pdf.

United Nations Development Programme. 2015. *The Millennium Development Goals Report 2015: Summary.* New York: United Nations, www.undp.org/content/undp/en/home/librarypage/mdg/the-millennium-development-goals-report-2015.html.

United Nations High Commissioner for Refugees. 2019. *Global Trends: Forced Displacement in 2018.* Geneva: UNHCR, www.unhcr.org/5d08d7ee7.pdf.

van Bavel, B., and A. Rijpma. 2016. "How Important Were Formalized Charity and Social Spending Before the Rise of the Welfare State? A Long-Run Analysis of Selected Western European Cases, 1400–1850." *Economic History Review* 69:159–187. doi:10.1111/ehr.12111.

van Zanden, J. L. 2005. "What Happened to the Standard of Living Before the Industrial Revolution? New Evidence from the Western Part of the Netherlands." In R. C. Allen, T. Bengtsson, and M. Dribe, eds., *Living Standards in the Past: New Perspectives on Well-Being in Asia and Europe*, 173–194. Oxford: Oxford University Press.

Versoris, N. 1885. *Livre de raison de maître Nicolas Versoris, avocat au parlement de Paris, 1519–1530*, vol. 12. Paris: G. Fagniez.

"Violence Against Women: War's Overlooked Victims." 2011. *The Economist*, January 13, www.economist.com/node/17900482.

Vives, J. L. (1526) 1917. *Concerning the Relief of the Poor*, vol. 11. M. M. Sherwood, trans. New York: New York School of Philanthropy.

Walker, R. S. 2014. "Amazonian Horticulturalists Live in Larger, More Related Groups Than Hunter-Gatherers." *Evolution and Human Behavior* 35:384–388. doi:10.1016/j.evolhumbehav.2014.05.003.

Walker, R. S., and D. H. Bailey. 2014. "Marrying Kin in Small-Scale Societies." *American Journal of Human Biology* 26:384–388. doi:10.1002/ajhb.22527.

Walker, T. D. 2015. "Enlightened Absolutism and the Lisbon Earthquake: Asserting State Dominance over Religious Sites and the Church in Eighteenth-Century Portugal." *Eighteenth-Century Studies* 48:307–328. doi:10.1353/ecs.2015.0016.

Wallace, J. 2018. "Practical Reason." In E. N. Zalta, ed., *The Stanford Encyclopedia of Philosophy* (Spring ed.).

Waters, K. 1998. "How World Vision Rose from Obscurity to Prominence: Television Fundraising, 1972–1982." *American Journalism* 15:69–93. doi:10.1080/08821127.1998.10739142.

Watson, F. D. 1922. *The Charity Organization Movement in the United States: A Study in American Philanthropy.* New York: Macmillan.

Wattles, J. 1996. *The Golden Rule*. New York: Oxford University Press.

Wayland, J. 2019. "Constraints on Aid Effectiveness in the Water, Sanitation, and Hygiene (WASH) Sector: Evidence from Malawi." *African Geographical Review* 38:140–156. doi:10.1080/19376812.2017.1340169.

Webber, C., and A. Wildavsky. 1986. *A History of Taxation and Expenditure in the Western World*. New York: Simon and Schuster.

Weber, J. 2006. "Strassburg, 1605: The Origins of the Newspaper in Europe." *German History* 24:387–412. doi:10.1191/0266355406gh380oa.

Weiner, B. 1993. "On Sin Versus Sickness: A Theory of Perceived Responsibility and Social Motivation." *American Psychologist* 48:957–965. doi:10.1037/0003-066X.48.9.957.

———. 1995. *Judgments of Responsibility: Foundations for a Theory of Social Conduct*. New York: Guilford.

Weisfeld, G. E., T. Czilli, K. A. Phillips, J. A. Gall, and C. M. Lichtman. 2003. "Possible Olfaction-Based Mechanisms in Human Kin Recognition and Inbreeding Avoidance." *Journal of Experimental Child Psychology* 85:279–295. doi:10.1016/S0022-0965(03)00061-4.

Wesley, J. 1872. "Sermon 50: The Use of Money." *The Sermons of John Wesley*, Wesley Center Online, http://wesley.nnu.edu/john-wesley/the-sermons-of-john-wesley-1872-edition/sermon-50-the-use-of-money.

West, S. A., C. El Mouden, and A. Gardner. 2011. "Sixteen Common Misconceptions About the Evolution of Cooperation in Humans." *Evolution and Human Behavior* 32:231–262. doi:10.1016/j.evolhumbehav.2010.08.001.

West, S. A., A. S. Griffin, and A. Gardner. 2007. "Social Semantics: Altruism, Cooperation, Mutualism, Strong Reciprocity and Group Selection." *Journal of Evolutionary Biology* 20:415–432. doi:10.1111/j.1420-9101.2006.01258.x.

Westbrook, R. 1995. "Social Justice in the Ancient Near East." In K. D. Irani and M. Silver, eds., *Social Justice in the Ancient World*, 149–163. Westport, CT: Greenwood Press.

Whittaker, J., B. McLennan, and J. Handmer. 2015. "A Review of Informal Volunteerism in Emergencies and Disasters: Definition, Opportunities and Challenges." *International Journal of Disaster Risk Reduction* 13:358–368. doi:10.1016/j.ijdrr.2015.07.010.

Wiepking, P., and R. Bekkers. 2012. "Who Gives? A Literature Review of Predictors of Charitable Giving. Part Two: Gender, Family Composition and Income." *Voluntary Sector Review* 3:217–245. doi:10.1332/204080512X649379.

Wilkinson, G. S. 1984. "Reciprocal Food Sharing in the Vampire Bat." *Nature* 308:181–184. doi:10.1038/308181a0.

Williams, G. C. 1996. *Adaptation and Natural Selection: A Critique of Some Current Evolutionary Thought*. Princeton, NJ: Princeton University Press.

Williams, K. D., C. K. T. Cheung, and W. Choi. 2000. "Cyberostracism: Effects of Being Ignored over the Internet." *Journal of Personality and Social Psychology* 79:748–762. doi:10.1037//O022-3514.79.5.748.

Wilson, D. S. 1975. "A Theory of Group Selection." *Proceedings of the National Academy of Sciences* 72:143–146. doi:10.1073/pnas.72.1.143.

Wilson, D. S., and E. O. Wilson. 2007. "Rethinking the Theoretical Foundation of Sociobiology." *Quarterly Review of Biology* 82:327–348. doi:10.1086/522809.

Wilson, E. O. 2012. *The Social Conquest of Earth*. New York: Liveright.

Wilson, J. 1986. "'We Are the World' Passes Goal; States Getting 'Hands' Money." *Gainesville Sun*, October 9, 7A.

Winchester, S. 2003. *Krakatoa: The Day the World Exploded*. New York: HarperCollins.

Wood, B. M., and F. W. Marlowe. 2013. "Household and Kin Provisioning by Hadza Men." *Human Nature* 24:280–317. doi:10.1007/s12110-013-9173-0.

Woods, R. A., ed. 1898. *The City Wilderness: A Settlement Study*. Boston: Houghton Mifflin.

World Bank. 2018. "Heavily Indebted Poor Country (HIPC) Initiative." World Bank, January 11, www.worldbank.org/en/topic/debt/brief/hipc.

Wright, M. F., and W. S. Pendergrass. 2018. "Online Prosocial Behaviors." In M. Khosrow-Pour. ed., *Encyclopedia of Information Science and Technology*, 4th ed., 7077–7087. Hershey, PA: IGI Global.

Wright, R. 2009. *The Evolution of God*. New York: Little, Brown.

Wright, R., and R. Kaplan. 2001. "Mr. Order Meets Mr. Chaos." *Foreign Policy* 124 (May/June): 50–60.

Wu, J., and R. Axelrod. 1995. "How to Cope with Noise in the Iterated Prisoner's Dilemma." *Journal of Conflict Resolution*, 39:183–189. doi:10.1177/002200279503900 1008.

Wydick, B., P. Glewwe, and L. Rutledge. 2013. "Does International Child Sponsorship Work? A Six-Country Study of Impacts on Adult Life Outcomes." *Journal of Political Economy* 121:393–436. doi:10.1086/670138.

Wynne-Edwards, V. C. 1962. *Animal Dispersion in Relation to Social Behavior*. London: Oliver and Boyd.

———. 1978. "Intrinsic Population Control: An Introduction." In F. J. Ebling and D. M. Stoddart. eds., *Population Control by Social Behaviour*, 1–22. London: Institute of Biology.

———. 1993. "A Rationale for Group Selection." *Journal of Theoretical Biology* 162:1–22. doi:10.1006/jtbi.1993.1073.

Xu, X., X. Zuo, X. Wang, and S. Han. 2009. "Do You Feel My Pain? Racial Group Membership Modulates Empathic Neural Responses." *Journal of Neuroscience* 29:8525–8529. doi:10.1523/JNEUROSCI.2418-09.2009.

Yad Vashem. 2019. "Names of Righteous by Country." Yad Vashem, www.yadvashem .org/righteous/statistics.html.

Yanguas, P. 2018. *Why We Lie About Aid: Development and the Messy Politics of Change*. London: Zed Books.

Yoffee, N. 2012. "Deep Pasts: Interconnections and Comparative History in the Ancient World." In D. Northrop, ed., *A Companion to World History*. Oxford: Wiley Blackwell, 156–170.

Zack, M. 2015. "Rebuilding Mathematically: A Study of the Rebuilding of Lisbon and London." *Nexus Network Journal* 17:571–586. doi:10.1007/s00004-015-0248-6.

# INDEX

*Alain Martinez*

**Michael McCullough** is a professor of psychology at the University of California, San Diego, where he directs the Evolution and Human Behavior Laboratory. The winner of numerous distinctions for his research and writing, he is a fellow of the American Psychological Association and the Society for Personality and Social Psychology. He lives in La Jolla, California.